Learning for Adaptive and Reactive Robot Control

Intelligent Robotics and Autonomous Agents
Edited by Ronald C. Arkin

A complete list of the books in the Intelligent Robotics and Autonomous Agents series appears at the back of this book

Learning for Adaptive and Reactive Robot Control

A Dynamical Systems Approach

Aude Billard, Sina Mirrazavi, and Nadia Figueroa

The MIT Press
Cambridge, Massachusetts
London, England

The MIT Press would like to thank the anonymous peer reviewers who provided comments on drafts of this book. The generous work of academic experts is essential for establishing the authority and quality of our publications. We acknowledge with gratitude the contributions of these otherwise uncredited readers.

This book was set in Times New Roman by Westchester Publishing Services. Printed and bound in the United States of America.

Library of Congress Cataloging-in-Publication Data

Names: Billard, Aude, author. | Mirrazavi, Sina, author. | Figueroa, Nadia, author.
Title: Learning for adaptive and reactive robot control : a dynamical systems
   approach / Aude Billard, Sina Mirrazavi, Nadia Figueroa.
Description: Cambridge, Massachusetts : The MIT Press, [2021] | Series:
   Intelligent robotics and autonomous agents series | Includes bibliographical
   references and index.
Identifiers: LCCN 2021005086 | ISBN 9780262046169 (hardcover)
Subjects: LCSH: Robots--Control systems--Mathematical models. | Machine learning. |
   Autonomous robots.
Classification: LCC TJ211.35 .B55 2021 | DDC 629.8/92631--dc23
LC record available at https://lccn.loc.gov/2021005086

10   9   8   7   6   5   4   3   2   1

To all our colleagues worldwide, who believed in our work and us

Contents

Preface

For a world where robots will be agile, smart, and safe.

To appreciate the present, we look at the past.

Humans have never ceased to build tools and machines, whether to make doing daily chores easier, speed up productivity, or just to enjoy the sheer pleasure of creating. The 20th century has been particularly prolific in this respect, and had led to the creation of a plethora of robots with all sorts of shapes and functions. Many have been entranced by the multitude of tasks that robots could soon achieve, but the truth is that robots remain largely machines that can perform only routine tasks, in a repetitive manner.

Today, the autonomy and decision power of robots rarely amounts to more than moving from one point to another. While this was sufficient when robots were confined to industrial settings, it falls short to meet the expectations of the twenty-first century. Many would like to deploy robots everywhere: in the streets, as cars, wheelchairs, and other mobility devices; in our homes, to cook, clean, and entertain us; on the body, to replace a lost limb or to augment its capabilities. For these robots to become reality, they need to depart from their ancestors in one dramatic way: They must escape from the comfortable, secluded, and largely predictable industrial world. To cope with constant, often unexpected changes in their environment, robots will need to **adapt** their paths *rapidly* and *appropriately*, without endangering humans.

Consider a robot wheelchair tasked with the tedious challenge to negotiate its path through a crowd without hitting any pedestrians. As it modifies its path to keep away from these nervy pedestrians, it should avoid braking too abruptly, which could throw its user out of it. Similarly, an arm prosthesis tasked to carry a tray as its owner walks through a cafeteria would need to adapt its posture as it avoids other nearby consumers, while making sure that the tray remains horizontal. It would further need to adapt its forces as more dishes are piled onto the tray. Avoiding obstacles in crowds or catching a dish on the verge of falling leaves little time for much thinking; both the wheelchair and the prosthesis would need to react within milliseconds.

This book presents control approaches that can re-plan at run time to adapt to new environmental constraints. It seeks to endow robots with the necessary reactivity to adapt their path *at time-critical situations*.

Determining what matters and what does not when acting in the world is the result of years of expertise. For instance, we know that trays are to be held horizontally because we have learned that holding them differently would cause dishes to fall to the floor. This learning process took place under the patient guidance of caretakers who were kind enough to wipe the floor and hold the tray for us as long as required. It is unlikely that robots' owners would be as willing to support robots through similar years of practice. It is, hence, crucial that **robots learn from those who know**, and that they can learn *from very few examples*. This book presents methods by which robots can learn control laws from only a handful of examples. We provide theoretical guarantees that ensure that robots can generalize their acquired knowledge beyond the provided examples.

Our approach to controlling robots: To generate movements, robots traditionally rely on planning techniques to compute feasible paths [89]. In the early days, planning was slow, taking hours to determine a path. Recent advances, however, have allowed robots to generate complex plans in only a few seconds [68]. However, the feasibility of these plans depends on having accurate models of the environment. Robotics today acknowledges that no real-world model will ever be either sufficiently accurate or valid, for a long period of time. The environment and its dynamics both change.

Objects may change texture and mass as a result of being manipulated, and in ways that cannot be easily modeled mathematically. The robot's own dynamics, while known when the robot comes out of the factory, is bound to change as the mechanics depreciates. To address these issues, the field started incorporating models of uncertainty in both planning and control [110]. The robustness of these approaches, however, depends on having good models of uncertainty. As both the model of the world and the models of uncertainty may change, robots must be able to learn and update their models based on their experience. Moreover, they must be able to generate motor plans that can cope with uncertainty at run time, without a need for replanning.

Redundancy to tackle uncertainty in path planning; feasibility preferred to optimality: Most tasks entail redundancy in the ways they can be achieved. When acquiring skills, humans do not learn a single approach to performing a task, but rather a variety of ways in which the task is fulfilled. For instance, while we grab a cup of hot tea by keeping the cup vertical, we allow for variability in the way we move the cup in the horizontal plane. This can help us avoid unexpected disturbances, such as obstacles in the way, and to take a comfortable posture of the arm. Redundancy is advantageous, in that it offers multiple ways to solve the task, a flexibility required to adapt to perturbations. This book presents robot controllers that overcome uncertainty in the environment by exploiting redundancy in the way that one can achieve the task. This redundancy of solutions can be embedded in a single control law that offers a diverse set of paths to achieve the same goal. We, hence, take the view that feasibility is preferred over optimality in order to adapt to real-world uncertainty.

From planning to acting: This book offers a new set of control laws to enable robots to perform complex actions with seamless control and at very high speed. This online reactivity results from the use of time-invariant dynamical systems (DSs). The use of DSs to solve motion-planning problems in robotics has become a staple, thanks to their ability to generate online motion plans inherently robust to uncertainties and changes in dynamic environments [123, 125, 58, 8, 78]. DSs are also key to modeling human movements [140, 130, 129, 63, 132]. Their use for robot control, hence, simplifies the control of robots when moving in synch with humans. It also eases the transfer of the motion model from human to robot.

Online reactivity is not just a matter of ensuring that there is a good-enough central processing unit (CPU) on board the robot. It requires inherently robust control laws that can provide multiple solutions. Control based on DSs offers a closed-form solution, with no need for further optimization. DSs offer theoretical guarantees, such as convergence to a goal, nonpenetration of obstacles, and passivity. Moreover, they allow one to easily synchronize multiple robotic systems, which can ensure coordinated control toward a target. They can also be used to switch rapidly across modes of control.

Compliance to tackle uncertainty in force control: One major challenge faced by robots today is to cope with sudden unexpected forces that, if not properly handled, may raise serious risks to people nearby. These unexpected contact forces may arise from external unexpected impacts; for instance, someone places an object in the robot's way or a human bumps into the robot. They may also arise from unwanted breakage of a piece as the robot manipulates it. To mitigate risks of damage or injury arising from unexpected forces, one approach is to make the robot compliant. Compliance allows the robot to absorb part of the impact. One powerful approach to control the robot's compliance is *impedance control* [55], in which the extent to which the robot motion absorbs or resists external forces is controlled through the stiffness of the impedance law. For a long time, compliance of a controller was fixed. Recent studies, however, have explored how one could vary this compliance, allowing the robot to be stiff at times and compliant at others [25, 148]. When precision is required, such as when moving into a narrow passage, one may request that the robot displays high stiffness. Once in free space, the robot can become compliant again. One can also make the robot directionally compliant. This is useful, for instance, if some directions of control must remain stiff (such as to balance a load), while other directions can be used to absorb the disturbance. This book presents different methods by which one can learn task-relative compliance and its variation during task completion, as well as how these can be embedded in DS-base control to benefit from the ability to replan a trajectory on the fly.

Learning requires data: In this book, we assume that data is made available to the robot either through a human providing demonstrations of the task or through other means, such as by solving the problem through optimal control to generate sets of feasible control laws.

To generate data from human demonstration, there are several techniques, which we review briefly in chapter 2; see [16, 120] for more complete reviews. When the robot must be trained by a human, it is crucial that the number of demonstrations be small for the task to be bearable. Hence, the algorithm must learn from sparse data sets. In addition to being sparse, data are often suboptimal due to noise or the lack of a skilled demonstrator. In this book, learning exploits different machine learning techniques that can learn from small data sets. An introduction to these techniques is provided in appendix B.

Another issue when learning from demonstration is that the user must provide explicit labels to tell the robot which demonstration corresponds to which task. We show that we can achieve this through algorithms that can automatically decompose a long flow of demonstrations into subsegments and associate a DS control law to each segment. We also offer examples of techniques to automatically discover the number of underlying dynamics.

Using the book as a textbook: The book is aimed as a support for a graduate course and therefore is constructed in a didactic manner. The technical content of this book is divided into three parts, with a total of 11 chapters; see figure 0.1 for a schematic of the book's structure.

We start with an introductory section composed of two chapters. Chapter 1 gives an overview of all the techniques presented in this book. It introduces the motivation for these techniques using real-world examples of robotic application. Chapter 2 presents techniques to gather data for robots to learn the systems presented in the rest of the book. It focuses primarily on learning from demonstration and optimization. It also briefly touches upon the concept of reinforcement learning.

Part II of the book presents the core techniques to learn control laws with DSs. We present methods to learn a control law composed on a single-attractor DS (chapter 3). We distinguish between methods to learn first- and second-order DSs. Next, we present methods to learn control laws composed of different dynamics with different attractors in chapter 4. In chapter 5, we show how we can sequence attractors and associated dynamics.

Part II presents extensions to part I for trajectory planning with DSs. Chapter 6 develops the concept of coupling for one or more controllers based on DSs, so as to control multiple agents in synch. We show in chapter 7 how this can be used to catch flying objects with one and two robotic arms. Chapter 8 presents methods to modify a known DS by learning locally new dynamics, while chapter 9 shows one application of such a modulation to perform obstacle avoidance.

Part IV introduces methods for compliant and force control using DSs. Chapter 10 starts with a short introduction to compliant control through impedance control. It then presents two methods to learn variable impedance controllers and explains how to combine learning impedance with DSs. Chapter 11 explains how DSs can be used in conjunction with force control to allow flexible control when in contact with moving objects.

Each chapter of the book includes examples of applications of these techniques to control a variety of robots, from arm manipulators, anthropomorphic hands, and whole-body control of humanoid robots. We also present applications for dexterous manipulation of objects with multiple robotic arms to be used for avoiding complex and moving obstacles in real time. We exemplify the advantage of DSs for immediate and fast re-planning in applications where robots catch objects in flight and avoid humans moving toward them. Videos of these applications are available on the book's website, see https://www.epfl.ch/labs/lasa /mit-press-book-learning/.

It is expected that the book will be read in order. Chapters 1 to 3 form the core of the book and should be covered first. Chapters 4 to 11 offer extensions of these core concepts. It is expected that lecturers may select only a subset of these chapters to present in a course. To accompany each chapter, we provide handwritten and programming exercises. Solutions to the exercises and slides are available on request through the pendant webpage of the book.

Prerequisites: The book assumes that the reader has taken a basic course in robot control and is familiar with the following concepts: PD control, Inverse Kinematics (IK) and inverse dynamics, trajectory planning, and optimal control. The book also assumes that the reader is familiar with machine learning, statistics and optimization, and DSs.

The book's appendices offer complementary information on the machine learning techniques used in the book. We also offer a brief introduction to the main theoretical concepts

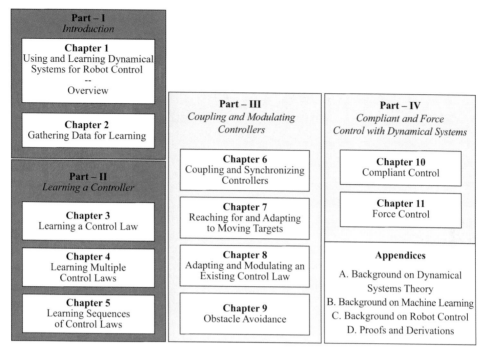

Figure 0.1
Structure of the chapters in this book

of DSs and to robot control in the main chapters. The appendices provide a brief summary of the main definitions in DSs and in robot control relevant for this book. However, the appendices do not serve as a comprehensive textbook on these domains. For a good introduction to the topic of robotics, the reader is encouraged to consult the *Handbook of Robotics* [136], and its main relevant chapters ("Kinematics," "Motion Planning," "Motion Control" and "Force Control"). A good introduction to DSs for robot control is to be found in the book *Applied Non-Linear Control* [137] (in particular, the chapters "Phase Plane Analysis" and "Fundamental of Lyapunov's Theory."

Lecturers are encouraged to introduce key concepts in robot control, machine learning, and DSs, which are required for a good understanding of the book. This can be done either in introductory lectures or as supplementary material, as suited for the audience's background.

Acknowledgments

We are deeply indebted to the researchers who were instrumental to the developments presented in this book. We list them here in alphabetical order: Walid Amanhoud, Lukas Huber, Farshad Khadivar, Mohammad Khansari Zadeh, Mahdi Khoramshahi, Klas Krondander, Ilaria Lauzana, Ashwini Shukla, Nicolas Sommer, and Rui Wu. We also thank Jean-Jacques Slotine and Jose Santos-Victor for their invaluable collaboration. In addition, the authors thank the European Commission, the European Research Council, the Swiss National Science Foundation, and the Swiss State Funding for financial support.

Notation

Next, we list the notation used throughout this book. The text in parentheses denotes the chapters in which these variables appear.

$\mathbb{R}$	Domain of unidimensional, real-valued numbers (All)
$\mathbb{R}^N$	Domain of N-dimensional, real-valued vectors (All)
$N \in \mathbb{N}$	Robot Cartesian task-space (end-effector) dimensionality (All)
$D \in \mathbb{N}$	Robot joint-space dimensionality (i.e. degrees-of-freedom) (All)
$M \in \mathbb{N}$	Number of samples in a data set or reference trajectory (All)
$x \in \mathbb{R}^N$	State of the robot's end-effector (i.e. Cartesian position). (All)
$\dot{x} \in \mathbb{R}^N$	Cartesian velocity of the robot's end-effector (All)
$\ddot{x} \in \mathbb{R}^N$	Cartesian acceleration of the robot's end-effector (chapters 3, 4, 8, 9)
$f(\cdot) : \mathbb{R}^N \to \mathbb{R}^N$	Vector-valued function representing a state-dependent DS (All)
$q \in \mathbb{R}^D$	Robot joint angles (positions) (All)
$\dot{q} \in \mathbb{R}^D$	Robot joint velocities (All)
$\ddot{q} \in \mathbb{R}^D$	Robot joint accelerations (All)
$x^* \in \mathbb{R}^N$	Attractor or target for a single-attractor DS (All)
$A, A^k \in \mathbb{R}^{N \times N}$	Single or kth affine linear system matrix (All)
$b, b^k \in \mathbb{R}^N$	Bias for a linear or kth linear system (chapters 3, 5).
$\gamma_k(\cdot) : \mathbb{R}^N \to \mathbb{R}$	Activation function that selects the kth linear system (chapters 3, 5, 6).
$\boldsymbol{\gamma}(\cdot) : \mathbb{R}^N \to \mathbb{R}^K$	Vector of K activation functions $\boldsymbol{\gamma}(\cdot) = [\gamma_1(\cdot), \dots, \gamma_K(\cdot)]$ (All)
$\mathbf{A}(\boldsymbol{\gamma}) \in \mathbb{R}^{N \times N}$	Mixture of linear system matrices [i.e., $\sum_{k=1}^{K} \gamma_k(\cdot) A^k$] (chapters 3, 6).
$p(\cdot \mid \theta) : \mathbb{R}^N \to \mathbb{R}$	Probability distribution with parameters θ (All)
$\mathcal{N}(\cdot \mid \theta) : \mathbb{R}^N \to \mathbb{R}$	Normal (Gaussian) probability distribution, with $\theta = \{\mu, \Sigma\}$ (All)
$\mathrm{IW}(\cdot \mid \cdot) : \mathbb{R}^N \to \mathbb{R}$	Inverse Wishart probability distribution (chapters 3, 5).
$\mathcal{N}\mathrm{IW}(\cdot \mid \cdot) : \mathbb{R}^N \to \mathbb{R}$	Normal inverse Wishart (NIW) probability distribution (chapters 3,5)
$\mu^k \in \mathbb{R}^N$	Mean for a kth Gaussian distribution (All)
$\Sigma^k \in \mathbb{S}_N^+ \subset \mathbb{R}^{N \times N}$	Covariance matrix for a kth Gaussian distribution (All)

$P, P^k \in \mathbb{S}_N^+ \subset \mathbb{R}^{N \times N}$ Single or kth symmetric positive definite matrices (All)

$\mathbb{S}_N^+ \subset \mathbb{R}^{N \times N}$ Space of symmetric positive definite (SPD) matrices (All)

$\mathbb{I}_M$ M-dimensional identity matrix (All)

$\chi \subset \mathbb{R}^N$ A compact set that is a subset of $\mathbb{R}^N$ (chapters 5, 8)

$\rho \in [0, 1]$ Softness or coordination parameters (chapter 7)

$F_e \in \mathbb{R}^N$ External force in task space (chapter 9)

$F_c \in \mathbb{R}^N$ Control force in task space (chapter 9)

Regarding subscripts and superscripts:

- *Vectors:* A subscript of a vector $x \in \mathbb{R}^N$ (i.e., $x_n \in \mathbb{R}$) refers to the nth dimension of the vector. A superscript (i.e. $x^i \in \mathbb{R}^N$) refers to the index of an ith instance of a vector in $\mathbb{R}^N$.

- *Matrices:* A subscript of a matrix $A \in \mathbb{R}^{N \times N}$ (i.e., $A_{ii} \in \mathbb{R}$), refers to the i, ith index of a matrix. The subscript can also be used to represent a matrix subblock of A. A superscript (i.e. $A^i \in \mathbb{R}^{N \times N}$) refers to the index of an ith instance of a matrix in $\mathbb{R}^{N \times N}$.

I INTRODUCTION

1 Using and Learning Dynamical Systems for Robot Control—Overview

This chapter introduces in a lighthearted manner the core concepts that will be developed at full length in the rest of the book. We start by introducing the problem of planning a trajectory in free space and explain the interest in doing so through the use of dynamical systems (DSs). DSs are ordinary differential equations that control for the temporal evolution of the robot's path in space. Such an analytical (closed-form) solution to robot planning is advantageous, as it offers a ready means to generate new trajectories immediately, without replanning. This is particularly suited to responding rapidly to disturbances encountered when the robot travels along the way. We illustrate this point through various schematic examples and contrast the use of DSs to alternative approaches through handwritten and MATLAB exercises. We further present how DSs make it easy to control multiple robots in a coordinated manner by coupling the robots' control laws. DSs can be modulated through an external multiplicative function. We illustrate how such a modulation can be used to change the original dynamics (e.g., to avoid obstacles or to control for the force applied on objects) and explain how control through DSs can enable closed-loop torque control when combined with impedance control.

DSs can be shaped in a number of ways, either at the design stage or through usage. Shaping is accomplished by modifying the parameters and form of the ordinary differential equation that describes the DS. This is done via learning. As we will see in this book, one can use off-the-shelf machine learning techniques to learn the DS in some cases. Often, however, the machine learning technique must be modified to preserve the mathematical properties of the DS, such as stability and convergence. Each chapter of this book is constructed in a way that illustrates this balance between choosing the DS with appropriate mathematical properties and learning its parameters. In each chapter, we first introduce the type of control that can be generated by different DSs. We then present ways in which the parameters of the controller can be learned.

1.1 Prerequisites and Additional Material

This chapter assumes that readers are acquainted with standard notions of robot control, such as inverse kinematic, inverse dynamics, force, and impedance control. Readers and teachers, using this textbook as a guide, are invited to refresh their memory or that of their students by reading or presenting selected chapters from the *Handbook of Robotics* [136], in particular the chapters on kinematics and motion planning. For completeness, we offer an introduction to DSs and review each machine learning technique used in the book in the appendixes. Readers who are unfamiliar with DSs theory and machine learning are referred to the relevant chapters in annexes. A thorough review of DSs for control can be found in [137].

1.2 Trajectory Planning under Uncertainty

Trajectory planning, also known as *path planning*, is likely one of the oldest problems faced by robotics, and there is a plethora of approaches to plan trajectories [89]. It is concerned with the problem of computing a free trajectory from one point to another. This is also known as *point-to-point planning*. Path planning may also aim at offering optimal coverage of the space, such as when controlling the paths of a vacuum cleaner robot [30]. This book is concerned with point-to-point planning.

When both the environment (obstacles and target position) and the robot (kinematic and dynamics) are known, point-to-point planning can be formulated as a constraint-based optimization. For a robot arm traveling in free space, the problem can be solved in closed form, as we will see in exercise 1.1. Full optimization is necessary, however, when control is applied to a complex kinematic chain, such as when controlling humanoid robots [21].

Issues arise when the environment is only partially known or can change before the robot has completed the planned trajectory. Consider a robot arm extending to grasp an object that is not in the expected location. If the plan is not adapted in time, the arm may hit the object, making it fall. As robots get to be deployed in human-inhabited environments, there is an urgent need to provide them with the ability to immediately recover from erroneous planning to avoid detrimental side effects to humans. To address these issues, trajectory planning has evolved and follows two courses of action: first, it now embeds uncertainty about the environment at onset (i.e., when planning the trajectory). The robot computes a path known to be robust to *expected* uncertainties. For instance, if some sensors are known to provide unreliable measurements, the plan will be carried out in such a way that it will depend solely on sensor measurements that are known to be reliable [110]. This assumes that the robot has at its disposal noiseless measurements. While no sensor is free of noise, the signal-to-noise ratio may be negligible in some applications. For instance, an imprecision of less than a degree in the positioning of a robot arm may be perfectly acceptable and not cause any failure when the robot is meant to swipe a floor. Conversely, the same error may lead to failure when it comes to grasping a small object. Trajectory planning can also update the plan as uncertainties are discovered—for instance, by updating the distribution of feasible paths on the go [24, 92]. Consider planning the gaits and foot placements of a humanoid robot to prevent it from falling. To do so, one may either relax the constraints and allow the planner to produce a path that reaches the vicinity of the target [144, 114], or rely on the availability of a set of alternative plans generated offline and switch to the most appropriate one at run time [65, 124].

All of these can work only *if the planner can plan faster than changes arise. Planning fast is therefore crucial.* This book offers techniques that enable one to replan a path immediately by embedding the plan into a closed-form control law through *time-invariant DSs*.

Time-invariant DSs, also known as autonomous DSs, depends solely on the state of the system and not time explicit dependency. As we show in this book, such time-implicit encoding is ideally suited to enabling robots to react immediately to disturbances. Whereas fast-reactive path planning techniques usually come at the cost of losing convergence guarantees, the approaches presented in this book enable robots to react to new obstacles and to changes in the position of the target, while ensuring that the path will converge correctly to reach the desired target.

The methods presented in this book are based on two assumptions that the robot's kinematic and dynamics (i.e., torques and acceleration limits) are known and sufficient to follow the plan. The kinematic is known for all rigid robots. The dynamics is usually well described, and the intertial and gravitational components of the dynamics are already compensated for by the low-level controller in most commercial robots sold nowadays. However, the precision with which the dynamics is known may not suffice for high-precision tasks. When this issue arises, alternative mechanisms must be sought to better estimate the robot's dynamics (e.g., through learning the inverse dynamics [109]), and to adapt the DS's parameters to suit the robot's hardware limits [45], or to compensate for poorly modeled interaction forces (see chapter 11).

1.2.1   Planning a Path to Grasp an Object

Consider the problem of controlling a robot arm to grasp a static ball (see figure 1.1). Assume that the ball position is given. First, we must choose a trajectory for the arm to travel in space from its initial configuration to the ball's position. There is an infinite number of

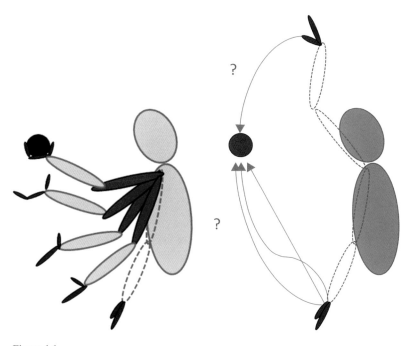

Figure 1.1
A robot is tasked to grasp a static ball (*left*). There is an infinite number of paths, which change depending on the initial configuration (some of which are shown at *right*).

paths that can bring the arm to the ball's position. To decide which path is best, one needs a criterion. In robotics, this problem is often solved through optimization. The simplest approach is to optimize for the shortest path in Cartesian space.

Consider that we are controlling solely for the robot's end-effector and ignoring the orientation. The state of the robot's end-effector is $x \in \mathbb{R}^3$. At initialization, the robot is in position $x(0) = x^0 \in \mathbb{R}^3$ and must travel to a target position $x^*(T) \in \mathbb{R}^3$. If no constraints apply, the shortest path from x^0 to x^* is a straight line. To prove this, let us consider a function $f(x)$, which describes the path traveled from x^0 to x^*. At each time step δt, the robot moves by an increment $\delta x(t)$, the optimization of which can be formulated as follows:

$$\min_{\delta x(2)...\delta x(T-1)} \sum_{i=1}^{T-1} \| \delta x(i)) \|$$

subject to   (1.1)

$$x(0) = x^0,$$

$$x(T) = x^*,$$

where T corresponds to the time taken to reach the target.

By the triangle inequality, we can set that $\sum_{i=1}^{T-1} \| \delta x(i)) \| \geq \| \sum_{i=1}^{T-1} \delta x(i)) \| = \| x^T - x^0 \|$. The solution is the line between the two points. Time T determines the speed at which the arm travels. To set the system to move as fast as possible while avoiding an overshoot, one can use the equation of a *critically damped system*. Such a system determines the acceleration along the path as the composition of two terms following a proportional-derivative (PD) control: a spring term that brings the motion to the target proportionally to the distance, and a damping term that slows it down as the velocity increases:

$$\ddot{x} = Kx - D\dot{x},$$ (1.2)

with K and D as the spring and damping coefficients. A relationship between K and D can ensure that the target value is reached and that no overshoot occurs. As robots cannot move arbitrarily fast, the parameters K and D must be set in such a way to ensure that the path is *dynamically* feasible. When the robot moves along a straight line, bounds on the parameters can be set relatively easily [66]. However, there are many reasons why the straight line may be neither desirable nor feasible—for instance, when the robot must avoid obstacles, or when it must grasp an object with a particular orientation (e.g., grasping the handle of a cup). When the path is nonlinear and the robot needs to accelerate and decelerate along the way, computing the optimal path becomes more challenging. One approach is to decompose the path through a series of straight lines, called *splines*, or a series of polynomials. A large variety of approaches have been offered to perform this decomposition (see [89]). However, spline decomposition works well, so long as the environment is static.

1.2.2   Updating the Plan Online

A plan is valid so long as what has been taken into account remains valid for the entire duration of the execution. If an obstacle gets in the way or the target object moves as the robot follows the planned path, the robot will need to recompute the path. The new path may be shorter or longer than the previous path. If one uses a traditional technique, such as

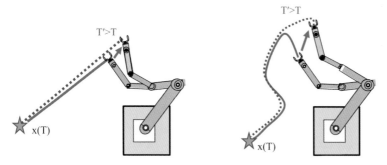

Figure 1.2
When the robot is subjected to a disturbance that sends it away from its planned trajectory, a new path must be computed. When the path is described through an explicit set of points in time, the time T to the target must also be recomputed along with the path, to prevent the robot from exceeding its speed limit. When the path is linear (*left*), computing the new time T' is immediate. When the path is nonlinear (*right*), it is not easy to determine both the new path and the new time, as this depends on many factors, and new constraints may need to be taken into account in the optimization.

spline decomposition, a new decomposition will need to be computed. If one must adapt very rapidly, there may not be enough time to perform this decomposition.

Issues also arise when the robot's plan is computed as an explicit function over time. Assume that the robot's path is defined as a set of desired positions and velocities at each time step $\{\dot{x}(t), x(t)\}_{t=1}^{T}$. If the robot's path length and shape is modified because of an external disturbance, using the previous time T may exceed the speed limit of the robot, as illustrated in figure 1.2. One will need to recompute both the path and the speed profile along the path.

When computing a path for a robot, one must also make sure that this path is *kinematically* feasible; that is, one must verify that a series of contiguous joint configuration for N robot joints exists that allows the end-effector to follow the desired Cartesian path. One should also ensure that the speed at which the joints move is feasible. This can be set through the constraints $\dot{x}(t) = J(q)\dot{q}(t) \; \forall \, t = 0, ..T$, and $\dot{q}(t) \leq \bar{q}$, with $q \in \mathbb{R}^{D}$ being the robot's joints, $J(q)$ the Jacobian of the robot's kinematic function, and $\bar{q}$ an upper bound on the velocity. This is can be solved through quadratic programming (QP) [136].

This optimization may still lead to a path that is suboptimal from an energy viewpoint. To include energy efficiency as a criterion in the optimization, one must estimate the effort generated to produce the movement. A first approximation to estimate and minimize effort is to request that the robot takes the shortest path in joint space, as opposed to the shortest path in Cartesian space. Note that following the shortest path in Cartesian space or joint space may lead to the same or different paths, depending on whether we have a redundant arm or not (see exercises 1.1 and 1.2). To estimate the true effort generated by the robot, one must compute the torque produced at each joint during the entire path. To minimize effort, hence, would lead to finding a trajectory that requires the least amount of torque.

Imagine that you control a four-joint robot, as illustrated in figure 1.3, subjected to gravity. A trajectory that is required to move the joint at the base of the robot may lead to more effort than a trajectory that is required to move only the second joint because this first joint supports the entire arm. Minimizing the total torque produced during the trajectory may

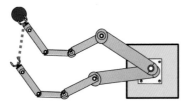 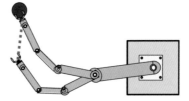

Figure 1.3
Different solutions to move a four-joint robot arm, both leading to a straight path in Cartesian space. When optimizing for minimal effort (minimal torque), the solution on the right, which moves the second joint, may be preferred over the solution on the left, which moves the first joint, as the second joint carries a lower mass than the first joint.

lead to a movement that privileges displacement of the distal joints (i.e., the drawing on the right side of figure 1.3).

Exercise 1.1   *Given a two-joint arm, write the optimization problem to move from an initial posture* x, *given by the forward kinematics* $x = \cos(q_1(0)) + \cos(q_2(0))$ *to a desired position* x^*, *when:*

(a) Minimizing the time taken to reach the target

(b) Following the straightest path in Cartesian space

(c) Following the straightest path in joint space

Programming Exercise 1.1   *The aim of this exercise is to familiarize the reader with basic optimization methods to the motion of a manipulator. Open ch1_ex1.m for exercise 1.1. The code generates a drawing with a four-joint manipulator. Edit the code to do the following:*

1. Compute a closed-form time-dependent trajectory generator based on a third-order polynomial.

2. Write the optimization problems of exercise 1.1 for this redundant manipulator moving in three dimensions, hence with a third-dimensional (3D) attractor. Assume that the joints are attached serially. Initialize the end-effector at different locations in space and compare the trajectory solutions for the three optimizations. Can you find a configuration when the solutions to (b) and (c) are identical?

One issue when expressing path planning as an optimization problem is that the optimization must be solved for each new initial configuration. While the problem may sometimes be convex, as in the case of a QP, that is often not the case, such as when considering constraints such as self-collisions with the robot's joints. To find a feasible and optimal solution, then, may require the solver to be started with new initial conditions several times. The optimization also implies an explicit time dependency (see equation 1.1). This time dependency is an issue both for the optimization and for the control. Too short a time window may not allow for finding a feasible solution that respects the robot's speed and torque constraints. Too large a time window may unnecessarily slow down the movement (see programming exercise 1.2).

Controlling for a specific number of steps in time is also problematic when incurring a disturbance along the way. Imagine a robot stopped in its path due to an unexpected but transient mechanical break. If the controller keeps clicking, time will continue progressing during the break. Once the mechanical break disappears, the control system expects the robot to be at a different location than the real one and generates the wrong motor command. To avoid this, one can take care of resetting the clock, but this requires a separate additional control loop to keep track of real and desired time. The problem is more complicated when the disturbance sends the robot to a different path. For instance, if a human bumps inadvertently into the robot, this may send the robot off track, letting it land in a new location x'. From this new location, a new path must be generated, which may require the optimization to start again. Such time-rescaling or reoptimization slows the robot's reactivity. If we remove time and make the control loop solely *state-dependent*, these issues disappear. However, this comes at a cost—namely, that one must guarantee that the system will stabilize at the required target, as we will see in the next section.

Programming Exercise 1.2   *The aim of this programming exercise is to determine how to recompute a path under disturbances with the optimization techniques seen in programming exercise 1.1. Load the code you created in programming exercise 1.1.*

1. Initialize the end-effector at one of the locations chosen in the previous programming exercise and generate a path following optimization (c).

2. Generate a disturbance midpath by suddenly displacing one of the joints of the robot by 10 degrees.

3. Redo the optimization to generate a new path to the target using the residual time to the target.

In this book, we take an approach in which the trajectory is given by a closed-form mathematical expression, hence eliminating the need for optimization at run time. Furthermore, the trajectory is generated through a state-dependent system to avoid having to restart the clock in the face of a disturbance.

1.3   Computing Paths with DSs

We set that the control law is described by a *deterministic time-invariant DS*. In the rest of this book, we refer to this type of encoding as *DS*. A DS describes the temporal evolution of the robot's path. Let $x \in \mathbb{R}^N$ be the state of the robot. The dynamics of the path in space is given by the state time derivative $\dot{x} \in \mathbb{R}^N$:

$$\dot{x} = f(x), \tag{1.3}$$

where $f : \mathbb{R}^N \to \mathbb{R}^N$ is a smooth continuous function.

Note that the use of a first-order derivative for the control is not constraining. As we will see in chapter 7, the same principle can be applied to second-order control laws for controlling the robot in acceleration.

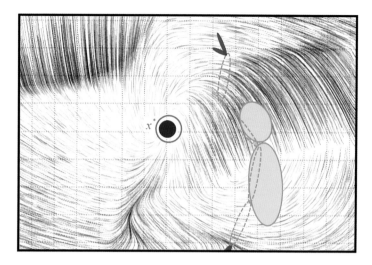

Figure 1.4
Representing the multiplicity of paths to the ball through a time-invariant DS that is asymptotically stable at the target, denoted with x^*. Each line represents the temporal evolution of the DS from an initial state.

The DS in equation (1.3) is said to be *autonomous* or *time-invariant*, as its evolution does not depend explicitly on time. Time is implicit and appears in the time-derivative of the state variable (here, $\dot{x}$). f is deterministic. The temporal evolution of the robot's state, hence, depends solely on the current state, location of the robot, x. $f(x)$ generates the robot's path. Transitions between locations in space during the execution of a task depend solely on the current state of the robot and the environment. Such a control law is said to be *state dependent*. The removal of explicit time dependency in the control law is advantageous, as it frees us from having to recompute the time to target when updating the path. While this leads to the problem of not being able to determine when the robot will reach its target, we will show (in section 1.7 and chapter 7), that this can be circumvented by coupling the system to the dynamics of the target and enforcing that both move in synch.

The robot's path can be obtained by integrating forward $f(x)$ from an initial state $x(0)$. This can be visualized by plotting path integrals through an integration of the vector field of motion when starting from a series of initial states. A visualization of our illustrative two-dimensional (2D) example of how to reach a ball is shown in figure 1.4. Each line denotes a path and is obtained by computing the series of states $\{x(1), x(2), \ldots\}$ that traveled over time from an initial point $x(0)$. We stop the computation once the flow reaches the target.

1.3.1 Stabilizing the System

A time-invariant DS may be unstable if one does not control explicitly for stability. The system may diverge away from the goal if it is initialized outside the basin of attraction of the equilibrium point. This aspect will be described in detail in chapter 3.

To control a robot arm meant to reach a ball with the time-invariant system as described here, we must request that the controller stops at the ball's location, x^*. Furthermore, to make sure that it stops only at the ball, we must request that the location of the ball is the unique stable point of the DS. Formally, this is equivalent to requesting that the system has one single fixed point at x^* [i.e., $f(x) \neq 0$, $\forall x \neq x^*$, and $f(x^*) = 0$], and that

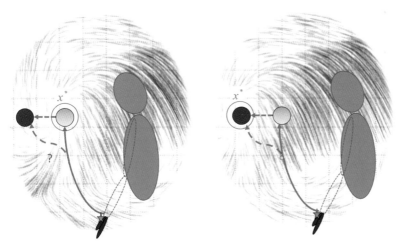

Figure 1.5
(*Left*) The robot is set to follow the dark gray path toward the ball located at x^*, but the ball is moved before the robot reaches the ball. If the new location of the ball is the unique stable point of the DS, and if the DS is asymptotically stable at the ball's location, there is a unique path that will lead the robot to the ball. This path is generated by following the vector field of the DS, as illustrated on the right side.

all paths be asymptotically stable toward this point. If we request that the paths be asymptotically stable to x^*, the center of the ball, we can ensure that all paths will lead to the ball. Hence, we set that:

$$f{:}\mathbb{R}^N \rightarrow \mathbb{R}^N,$$

$$\dot{x} = f(x)$$

$$f(x) \neq 0, \forall x \neq x^*,$$ \hfill (1.4)

$$\dot{x}^* = f(x^*) = 0,$$

$$\lim_{t \to \infty} x = x^*.$$

If the system satisfies equation (1.4), there is a unique path from any point x to the target, x^*. The asymptotic stability and time invariance of the DS offer a natural robustness to perturbations. Imagine that, while on its way to the ball, the robot is pushed away from its original trajectory to a new location, x'. As all trajectories from any point in space eventually converge to the ball, there is no need to reestimate the trajectory. One only needs to follow the flow generated by f (see figure 1.5). To simplify computation, one places the origin of the system on the ball, and hence a displacement of the target corresponds to a relative displacement of the arm in space, as illustrated on the right side of figure 1.5.

An easy way to make the system robust to changes in the location of the target (the ball, in this example) is to place the origin of the system on the ball. This means that $x^* = 0$, and hence, the entire flow is asymptotically stable to the origin. The robot's motion is then computed relative to the ball. If the ball moves farther from the robot's arm, this is equivalent to have moved the robot arm to a new location, x'. As explained before, one then only needs to query for f at the new location and one is ensured to eventually reach the ball. An implementation of this, to control the arm of a humanoid robot reaching for a ball moving

Figure 1.6
Snapshot of the humanoid robot iCub reaching for a ball rolling down a slope. The path of the ball is intersected as it hits an obstacle along the way. The robot's arm adapts its path to the new ball's trajectory. It speeds up and slows down as necessary to meet the ball on time and at the right location once the ball reaches the end of the table.

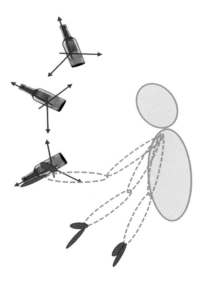

Figure 1.7
To be robust to changes in the orientation and location of an object, one can place the origin and frame of reference on the object.

rolling down a slope, is shown in figure 1.6. Obstacles in the ball's path lead the ball to go away from its original trajectory.

The same principle can be extended to reach an asymmetrical object, such as catching a falling bottle. If in addition to setting the origin of the DS on the object, we also set the frame of reference along the object's main axes, that leads the flow of motion generated by the DS to align with the axes of the object's. As the object falls and rotates, the desired path will align the robot's path in both translation and orientation to meet the object with the correct posture, as illustrated in figure 1.7.

Exercise 1.2   *Consider a linear, one-dimensional (1D) control law $\dot{x} = \alpha(x - x^*)$ around an attractor $x^* = 1$. To study the stability of the system at its attractor, draw a 2D plot of the velocity $\dot{x}$ against position x of this system for $\alpha = 1$ and $\alpha = -1$. For which values of α does the system converge and stabilize at x^*?*

Exercise 1.3   *Consider a linear, 2D control law $\dot{x} = A(x - x^*)$ with $x^* = \begin{bmatrix} 0 & 0 \end{bmatrix}^{\mathrm{T}}$.*

1. Draw the phase-plot for $A = diag\{-1, -2\}$ and for $A = diag\{1, -2\}$.

2. For which values of A does the system converge and stabilize at x^*?

1.4 Learning a Control Law to Plan Paths Automatically

We started this chapter by saying that there are an infinite number of ways to reach a target. We continued by stating that with a system following equation 1.4, there is a single path from a point in space to the target. We moved from an infinite number of solutions to unicity. This transition must be done with care, and one must be certain to embed in our control law the "correct" paths. The correct path may depend on a number of external factors. For instance, if the robot wishes to throw a ball on the floor, as when playing tennis, it will want to approach the ball from above. The arm would be in the correct configuration to go down. If, in contrast, it wishes to catch the ball before it falls, it may opt for a path that comes from below. The two sets of dynamics follow different functions, $f(x)$. To determine which function f to use, one can learn function f from examples of good trajectories.

In chapter 2, we present three ways in which we can obtain samples of good trajectories. One option is to be trained by a human expert who demonstrates examples of optimal paths. An alternative is to generate these trajectories offline from optimal control, following the approach described in the introduction of this chapter. In this way, we can embed all constraints known to the developers regarding the kinematic and dynamics limitations of the robot, as well as other constraints related to the task at hand. As it takes time to generate these solutions through optimal control, this can be done offline. Once the control law is embedded in a single function, $f(x)$, there is no need to optimize any more at run time. This, however, assumes that the optimal paths are known once and for all and that a human expert can somehow transcribe this knowledge. However, the optimal paths may not be known by the designer or may change as time passes. The third option to get data is to have the robot learn on its own, through trial and error. This is known as *reinforcement learning*.

In chapter 3, we will see how we can learn the control law given in equation 1.4. Here, we illustrate why this requires a dedicated approach and cannot rely on simply applying one's favorite machine learning technique. Recall that when we wrote equation 1.4, we require that the function should be asymptotically stable at the target. Assume you are provided with a set of M training sample pairs $X = \{x^i, \dot{x}^i\}_{i=1}^M$ of positions and velocities. In machine learning, you can parameterize your function through a set of parameters Θ. The function becomes $f(x; \Theta)$ and its parameters θ can be learned using a variety of methods for regression, such as locally weighted projection regression (LWPR) [127], Gaussian processes (GPRs) [133], Gaussian mixture regression (GMR) [26], or neural networks [146]. These methods, however, cannot guarantee that the learned DS will not diverge from the attractor, as criteria for stability are not part of the optimization (see chapter 3 for a detailed discussion of these aspects). In figure 1.8, we illustrate the vector field and solutions of the estimation of $f(x)$ when using support vector regression and neural networks. Observe that while the flow accurately follows the data, the vector field does not vanish at the attractor. The predicted trajectory in pink overshoots the attractor and continues its path.

None of these algorithms are constructed to enforce stability at a point. Their objective functions enforce that the data be fitted as closely as possible, typically through a mean-square loss. To enforce stability, the velocity should be exactly zero at the attractor. A mean-square loss function will rarely enforce that the velocity be exactly zero at a specific point. In chapter 3, we show how we can modify the original optimization framework of

Figure 1.8
Learning an estimate of the control law $\dot{x} = f(x)$ using machine learning techniques such as support vector regression (*top*) or neural networks (*bottom*) ensures a tight fit with the data, but it does not guarantee convergence at the attractor. Training data are illustrated in dark (red) lines. The learned flow is illustrated in grayscale. The pink trajectory illustrates the prediction of the model when starting from one of the training points. In both cases, the trajectory drifts once it reaches the attractor.

traditional machine learning techniques with explicit constraints to ensure that the learned dynamics be stable.

1.5   Learning How to Combine Control Laws

Up to now, we assumed that a single control law would be sufficient to control the robot. But the control law may need to change as the task changes. This will require the robot to switch across control laws at run time, as the task changes. There are many possibilities to combine the various control laws encoded in DSs. If each control law is active only in a distinct region of the state space, one can learn a partition of the state space and model only the DS valid in each partition. In chapter 4, we present an approach called augmented support vector machine (SVM), illustrated in figure 1.9. The approach learns a partition of the space. Each partition contains a single DS. The approach learns a model of the flow in each partition that leads to the partition's attractor.

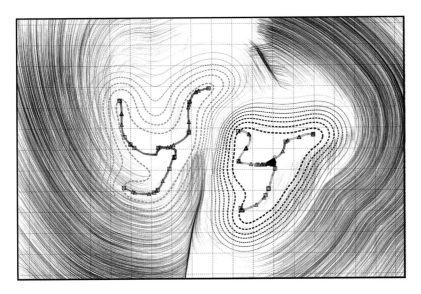

Figure 1.9
Examples of a system composed of two dynamical systems (DS) with two separate attractors. For each DS, a set of three sample trajectories is generated to train the model, delineated by the dark lines. The augmented SVM approach (see section 4.1) learns a partition of the two regions and the dynamics in each region. The local basin of attraction for each dynamics can be reconstructed by following the isolines of the learned energy function.

Consider K functions, $f_k, k = 1..K$, each of which represents one DS, augmented SVM combines these K functions in a single function. It decides which applies where through an *argmax* vote at run time:

$$f(x) = \operatorname*{argmax}_k f_k(x). \tag{1.5}$$

A robotic implementation of this technique is illustrated in [134], see excerpt in figure 1.10. The system is trained so that two separate DSs catch either the neck or the tail of a champagne glass (in simulation) and a plastic bottle (in real robotic implementation). The object is allowed to fall. The robot is tasked to grasp the object at an intercept point in front of it. A slight change in the initial rotational velocity when the object is sent off balance will lead to a different trajectory. Depending on the initial condition, hence, the object will have to be caught at its neck or tail. As the object falls very rapidly (taking less than 0.2 s for the real object), there is no time to plan. The robot must switch immediately across the tail neck DS in order to catch the object on time. Indeed, using DS for the neck when the object presents its tail would lead to failure, as the robot's gripper would close too soon. Conversely, using DS for the tail when the object presents its neck would let the object fly through the robot's fingers, as the finger aperture would be too large.

1.6 Modifying a Control Law through Learning

It is often useful to modify a control law after it was learned. For instance, one may start with a nominal linear DS and then choose to modulate it locally to avoid obstacles.

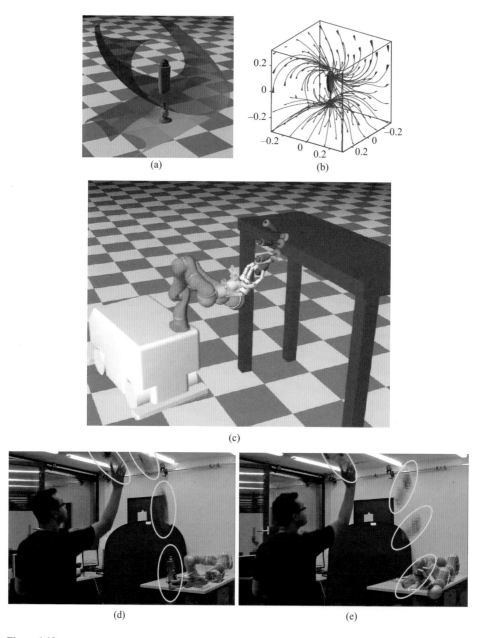

Figure 1.10
(*top*) Encoding of two DSs to catch an object at the neck or at the tail. (*Middle and bottom*) At run time, the robot switches across the two DSs to catch either the neck or the tail of the falling glass or bottle.

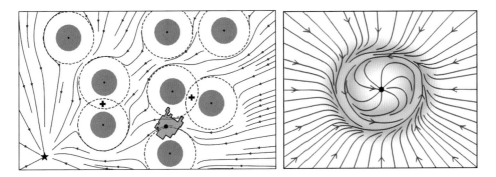

Figure 1.11
A modulation of a nominal linear DS allows for avoiding obstacles while preserving stability guarantees at the attractor [57] (*left*), or generating a limit cycle [85] (*right*) by learning a local rotation from data points provided in the orange region.

As we will see in chapter 8, it is possible to modify an original DS $f(x)$ while preserving its stability properties at an attractor. This is done by adding a multiplicative modulation term $M(x) \in \mathbb{R}^{N \times N}$, a continuous matrix function, to the original dynamics as follows:

$f\colon \mathbb{R}^N \to \mathbb{R}^N,$

$M\colon \mathbb{R}^N \to \mathbb{R}^N,$

$$\dot{x} = g(x) = M(x)f(x), \tag{1.6}$$

$\dot{x^*} = g(x^*) = f(x^*) = 0,$

$\lim_{t \to \infty} x = x^*.$

M must be full rank to prevent the flow from vanishing at a different point from the original attractor. This modulation through a multiplicative term makes the approach very flexible. In this book, we show various applications of this property, starting with examples of how it can be used to modulate the flow around one or multiple moving obstacles. We further show how it can be used to generate local nonlinear dynamics or to generate a limit cycle, as illustrated in figure 1.11.

Exercise 1.4 *Start again with the linear 2D control law created in exercise 1.3* $\dot{x} = A(x - x^*)$
with $x^* = \begin{pmatrix} 0 \\ 0 \end{pmatrix}$.

Consider the matrix $M = \begin{pmatrix} 1 & -2 \\ 0 & 1 \end{pmatrix}$.

Draw the phase-plot.
For which values of A *does the system converge and stabilize at* x^*?

One can set this modulation explicitly by hand at times, such as when avoiding obstacles. However, it is often interesting to learn this modulation. For instance, we can parameterize

$M(x)$ to locally generate a rotation in space. We set that $M(x) = e^{-\gamma \|x - x^0\|} R(\theta)$, where R is a rotation of an angle θ (in 2D). Learning such a modulation requires one to estimate how much the flow rotates θ and where the rotation applies. Assume that we are provided a set of data points locally, illustrated in orange in figure 1.11. The center of the rotation x^0 and the locality of the modulation conveyed through γ can be estimated by computing the parameters of a Gauss function over the data through expectation maximization (see exercise 1.5).

Exercise 1.5 *You are provided with a set of* M *training sample pairs* $X = \{x^i, \dot{x}^i\}_{i=1}^{M}$ *of positions and velocities, in order to estimate the modulation matrix* $M(x) = e^{-\gamma \|x - x^0\|} R(\theta)$.

1. Which value will you obtain for x^0 *and* γ *if you fit a Gauss function using maximum likelihood?*

2. How would you estimate θ?

1.7   Coupling DSs

A core property of DSs is the notion of *coupling*. Two DSs are said to be coupled if one DS explicitly depends on a second DS. Consider two variables x and y, whose dynamics are described by $\dot{x} = f_x(x)$ and $\dot{y} = f_y(y, x)$. Here, y is coupled to x through f_y.

Coupling different DSs is very useful in robotics. It allows, for instance, for enforcing that different limbs or different robots move in synch. Figure 1.12 [94] shows an example of coupling three limbs of a humanoid robot. The dynamics of the robot's arm and hand are coupled to the movement of its eyes. This coupling is inspired from similar couplings found in nature. For instance, when we reach for an object, our eyes precede and lead the arm movement, and similarly, the fingers accompany the arm movement and close in synch once the hand has reached the target. In robotics, this eye-arm-hand coupling can be modeled through a two-by-two coupling among three DSs controlling for the robot's eyes, arm, and fingers. Here, x, y, and z denote the variables controlling the eyes, arm, and hand's fingers, respectively. The eyes lead all the movements, and their dynamics is independent of the other limbs [i.e., $\dot{x} = f_x(x)$]. The arm moves with the eyes [$\dot{y} = f_y(y, x)$], and the fingers move with the arm [$\dot{z} = f_z(z, y)$]. The systems can be coupled in such a way that they converge simultaneously on the same point (e.g., at the target x^*); see exercise 1.6 for an example of how to create such a coupling.

This provides a natural robustness to perturbation. For instance, when the object is moved, the eyes track and lock on the new object's location. As the arm is coupled to the eyes, it moves in synch to the new location of the object. If the object keeps moving, the eye and arm move in synch with the object. The hand is also coupled to the arm movement in such a way that the fingers close on the object only once the arm is also at the target. As the object moves, the hand will remain open until the arm eventually locks on the object's new location (see chapter 6 for more detail).

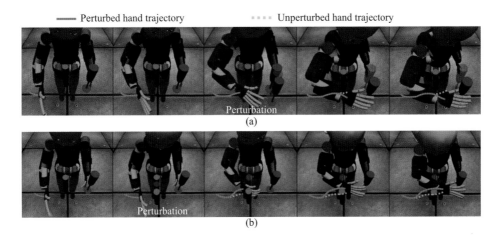

——— Perturbed hand trajectory   · · · · Unperturbed hand trajectory

Perturbation
(a)

Perturbation
(b)

Figure 1.12
Coupling eye, arm, and hand movement through explicit dependency on each limb's DS simplifies the control and ensures that the eye, arm, and hand move in synch if the object suddenly moves, so as to close simultaneously on the object in its new location (a). It also allows to react rapidly to obstacles moving on the way (b).

Exercise 1.6 *Consider two variables* $x \in \mathbb{R}$, *and* $y \in \mathbb{R}$ *coupled with the following dynamics:*

$$\dot{x} = -x,$$

$$\dot{y} = -y + \alpha x, \tag{1.7}$$

$$\alpha \in \mathbb{R}.$$

1. Draw the integral paths for the system x, y *when initialized at points* $(1,1)^T$ *and* $(-1,-1)^T$ *for different values of* α.

2. Does the system accept a fixed point?

3. Is it stable at this fixed point, and, if so, does this depend on the value of α?

Coupling DSs can also be used to synchronize the robot's movements with the dynamics of an external object over which we have no control. Assume that we have a model of the dynamics of the external object, $\dot{y} = f_y(y)$. We can set to meet this object at a point on the integral path of the object y^*. One can then couple the dynamics of the robot arm to that of the object, $\dot{x} = f_x(x, y)$, in such a way that the dynamics of the robot's arm stops at the desired interception point y^* [i.e., $lim_{t \to \infty} x = x^* = y^*$ and $f_x(x^*, y^*) = 0$]. Such coupling can be extended to continue moving with the object once the arm has met the object at y^*. To do so, one must switch smoothly from a system that stabilizes at y^* to a system that stabilizes with the velocity of the system. To do this, we can move to a second-order system and set that $lim_{t \to \infty} \dot{x} = \dot{y}$. Such a system is illustrated in figure 1.13, where a robot arm catches a flying object in midair, and is described in detail in chapter 7.

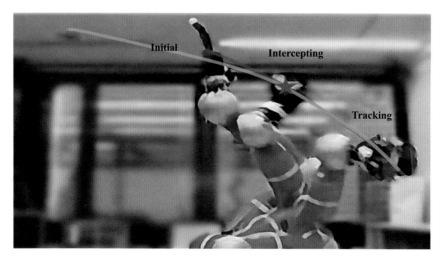

Figure 1.13
To reach and follow the dynamics of the object, we combine two DSs. The first brings the arm toward the object while the other aligns the velocity vector with that of the object.

1.8   Generating and Learning Compliant Control with DSs

Safety has become a major concern for robots these days due to their deployment in human-inhabited environments. The industry strives to have robots collaborate with humans. The advantages of this approach are numerous, ranging from gaining space to increasing productivity. It may also facilitate humans' work, with robots helping humans carry heavy loads. Yet this new trend brings with it multiple dangers. With humans in the vicinity of robots, the chances that the two will collide are high (figure 1.14). As robots move fast, the impact generated by such collisions will be great and may cause serious injuries.

To mitigate these risks and ensure that the collision results in minimal harm, one possibility is to make the robot *compliant* [35]. This can be achieved by modifying the hardware, providing the robot with soft joints or soft materials to absorb the shock. An alternative is to create this compliance through software, using impedance control [55]. Impedance control computes the torque generated by a robot through a virtual spring and damper control law. Similar to real physical springs, virtual springs are associated with a stiffness parameter, which can be regulated. Increasing or decreasing the stiffness determines how much the robot motion absorbs or resists external forces [152].

In part III of this book, we present methods by which we can make the robot compliant, either by controlling forces at contact with second-order DSs or by combining DS-based control laws with impedance control, using the former as a generator of the reference velocity for the latter. We assume that the internal dynamics of the robot is compensated for.[1] We can then write (in 2D):

$$\tau = J(q)^T D(x) \, (\dot{x} - f(x)) \tag{1.8}$$

with $D(x)$ being a positive, semidefinite matrix function of the state space; $\tau \in \mathbb{R}^{\shortmid\shortmid}$ the robot's torques at each q joint; $f(x)$ the output of the DS; and $\dot{x}$ the real velocity of the robot. Here,

(a) Stiff robot: collision   (b) Compliant robot   (c) Compliant robot: collision risk during change of direction

Figure 1.14
Robots working side by side with humans, as envisioned in a collaborative robotics paradigm, presents dangers to their human coworkers (a). To mitigate these dangers, one can render the robot compliant (b). A compliant robot would absorb the force resulting from undesired contact with humans (c), thereby mitigating the risk of injury.

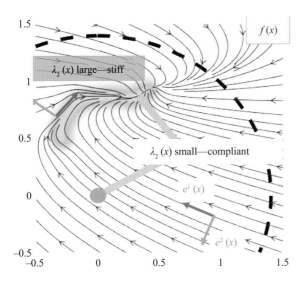

Figure 1.15
Impedance control combined with DSs. The stiffness is controlled locally through an eigenvalue and eigenvector decomposition of the stiffness matrix. The first eigenvector aligns with the DS flow, whereas the second eigenvector and eigenvalue control for the compliance to external disturbances.

$f(x)$ acts as a reference velocity, tracked through equation (1.8). How accurately the reference velocity is tracked is regulated through matrix D. Using the same approach as that introduced with the modulation of a DS, we can generate matrix D through an eigenvalue/eigenvector composition, aligning the first direction with the flow of the nominal DS. The eigenvalues can then be used to modulate the stiffness along the path, as illustrated in figure 1.15. The shaded region corresponds to a region with high stiffness. In such a region, the system rejects disturbances and follows the flow indicated by the DS. In the rest of the area, the DS accepts disturbances. A disturbance may send the system to a different location. Once the disturbance disappears, the system resumes its movement, following the new trajectory generated by the DS at the new location.

Combining DSs with impedance control is advantageous, in that it inherits both the natural ability of DSs to replan on the fly and the adaptivity of compliant control.

Figure 1.16
Force control can be achieved with DSs through decomposition of the flow with the normal component against the surface of the object generating the desired force [5]. Coupling two force-control DSs enables two robot arms to move in synch and balance the forces to lift and hold the box in the air (*left figure*), even with disturbances, such as when a human rotates the two robot arms (*middle figure*). The system remains compliant and allows the human to completely break the contact at all times (*right figure*).

The combination yields robust control approaches capable of recovering from unexpected collisions by replanning the desired trajectory and absorbing the impact.

Matrix $D(x)$ can be learned easily using a variety of commercially available machine learning techniques. However, as in the case of learning DSs, using such a technique may not guarantee that the system remains stable. Other constraints on the shape of the matrix are required to ensure that the system converges on its fixed point. Because matrix D varies over the state space, it cannot be set through a standard approach for a critically damped system. In chapter 10, we present several techniques to learn various impedance terms while preserving stability of the system.

Finally, in chapter 11, we show how one can control for force via DSs. The idea is to generate a normal component against the surface of the object. This is akin to generating force through an impedance term. If, in addition, one combines this with coupling across DSs, one can allow a pair of robot arms to pick up and lift an object in the air [5], illustrated in figure 16.

1.9   Control Architectures

To control robots with the DS-based control laws presented thus far requires a control architecture. DSs can be used either at the trajectory-planning level or directly to control the torques sent to the joints. When the DS is used at the trajectory planning level, the control architecture relies on a low-level controller to track the dynamics (i.e., speed, acceleration) prescribed by the DS, and to generate the required torques. This is usually done by a built-in position or velocity controller provided by the manufacturer. The DS also often generates a trajectory in Cartesian space. To follow this trajectory requires one to use an inverse kinematics and inverse dynamics solver.

The control architectures used in this book can be placed into two main broad categories: open-loop and closed-loop joint motion architectures. These are illustrated in figures 1.17 and 1.18. Note that the term *open loop* does not mean that the entire control system of the robot is played in an open-loop fashion. The robot's behavior remains reactive to changes in its environment at all times. Indeed, in the open-loop architecture used in this book, the robot continues to track the state of the environment through external sensors. For instance, the robot would receive information from cameras and lidars in real time to track the position of obstacles and of the target. Should either move, the DS would generate

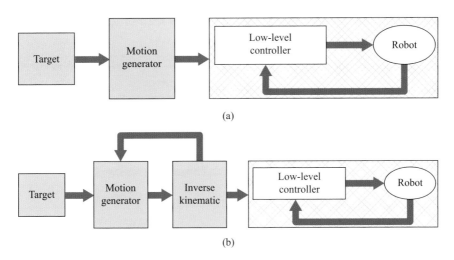

Figure 1.17
Schematic of the typical open-loop, joint motion-generator architectures used in this book. When controlling in joint space, no need to have an inverse kinematics block (a). When the DS controls end-effector path, one requires a separate inverse kinematics processing block (b). The desired motion for the robot is generated by the DS upon sensing the current position of the target. Tracking of the desired velocity generated by the DS is achieved by a low-level controller. The DS closes the loop only on the tracking of the external target, including avoiding obstacles, and not on the current position of the joints.

a new trajectory to avoid the obstacles and to adapt to the new position of the target in real time.

Note that unless explicitly defined otherwise, in all these robotic experiments (and throughout the book as well), the origin of the reference coordinates system is attached to the target. The motion is hence controlled with respect to this frame of reference. As explained earlier, such representation makes the parameters of a DS invariant to changes in the target position. Furthermore, unless it says otherwise, the Cartesian coordinate system is used to represent the robots' motions.

In the joint open-loop architectures, the state of the motion generator is not updated with respect to the sensory information, but rather with the previous desired state of the robot (see figure 1.17). These architectures have two advantages, but they also have drawbacks. The main advantage is that they are robust to delays along the robot's internal communication channel from the main central processing unit (CPU) to motor boards and back. Even though the delays might seem infinitesimal, they are not constant. Hence, their effects can be significant if the robot moves extremely rapidly, such as when catching objects in flight. Another advantage of separating the control of the joint from that of the main trajectory is that the stability and the performance of the DS motion-generator can be studied separately and do not depend on the particular dynamics and kinematics of the robot. However, as the motion generator is not updated with the real-time state of the robot, it cannot compensate for internal disturbances (e.g., the joints do not move because of internal friction, or they move imprecisely because of too much backlash). Moreover, unless the robot is provided with tactile sensors or external force-torque sensors to close the loop upon contact, or the robot is provided with a built-in impedance control law, it may not be able to react to contact. Robots driven by these architectures, therefore, may be as stiff as their low-level controller

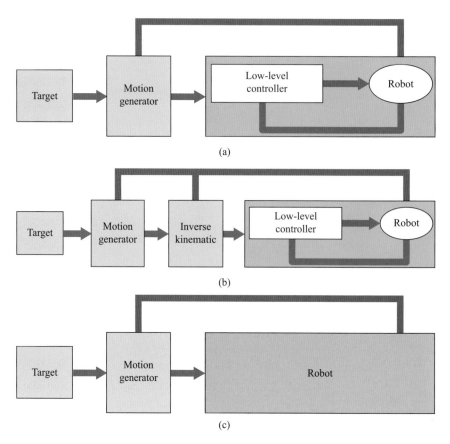

Figure 1.18
Schematic of closed-loop, joint motion-generator architectures, with joint control (a), Cartesian path control (b) or direct torque control (c). The desired motion for the robot is generated based on the current state of the robot. The delays in the communication channels highlighted in red might cause instability or deteriorate the performance. Stability of the entire control loop, therefore, must be taken into account when studying stability of the DS motion generator.

in the absence of additional sensors. These architectures are best used in scenarios that require very high speed and reactivity for the robot's motions and in which the operational space is fully safe and observable.

In contrast, closed-loop, joint motion generators update the control law at each time step based on measurements of the current state of the robot (see figure 1.18). Hence, any perturbation is transmitted into the motion generator and can be compensated for on the fly by the learned DS. In this book, we show how this can be used to train the robot at run time. This immediate and natural reactivity of the system allows the user to interrupt the robot's motion at all times and in a safe manner. The user can then use this interruption to show the robot a different path. When combined with an incremental learning update, this mechanism can be used to dynamically shape the stiffness/compliance of the robot. One drawback of closed-loop, joint motion generators is that the architecture is prone to instability caused by the delays in the communication channels. The stability of the motion generator must then be studied, taking into account the low-level controller. Figure 1.18c

Table 1.1
A summary of the implemented control architectures for each chapter

	Robotic Platform	Control Architecture	Low Level Controller	Inverse Kinematic
section 3.3.5	Katana-T	Open-loop; figure 1.17b	Position controller	Damped least square, section C.2.3
section 3.3.5	iCub	Closed-loop; figure 1.18b	Position controller	Damped least square, section C.2.3
section 3.4.5	KUKA LWR 4+	Closed-loop; figure 1.18a	Passive DS, section 10.4	NA
section 3.4.5	Wheelchair	Open-loop; figure 1.17b	Derivative controller, section C.2	Damped least squares, section C.2.3
section 3.4.5	iCub	Open-loop; figure 1.17b	Position controller	Time-projection foot-stepping [38], [37]
section 4.1.5	KUKA LWR 4+	Closed-loop; figure 1.18b	Position controller	Damped least square, section C.2.3
section 4.2.5	KUKA LWR 4+	Closed-loop; figure 1.18a	Passive DS, section 10.4	NA
section 4.2.5	iCub	Open-loop; figure 1.17b	Position controller	Time-projection foot-stepping [38], [37]
section 5.1.4	KUKA LWR 4+	Closed-loop; figure 1.18a	Passive DS, section 10.4	NA
section 5.1.4	iCub	Open-loop; figure 1.17b	Position controller	Time-projection foot-stepping [38], [37]
section 6.3.3	iCub	Closed-loop; figure 1.18b	Position controller	Damped least square
section 7.2.1	iCub	Closed-loop; figure 1.18b	Position controller	Damped least square
section 7.2.1	KUKA IIWA	Closed-loop; figure 1.18b	Position controller, low gain	Damped least square, section C.2.3
section 7.4	KUKA IIWA	Open-loop; figure 1.17b	Position controller, high gain	Damped least square, section C.2.3
section 7.6	KUKA IIWA	Closed-loop; figure 1.18b	Position controller, low gain	Damped least square, section C.2.3
section 8.2.3	KUKA LWR 4+	Closed-loop; figure 1.18a	Cartesian impedance controller 10.2	NA
section 8.2.3	Barrett WAM Arm	Closed-loop; figure 1.18a	Cartesian impedance controller 10.2	NA
section 8.3.3	KUKA LWR 4+	Closed-loop; figure 1.18a	Passive DS, section 10.4	NA
section 8.4.3	KUKA IIWA	Closed-loop; figure 1.18b	Position controller, low gain	Damped least square, section C.2.3
section 10.4.2	KUKA LWR 4+	Closed-loop; figure 1.18c	Passive DS, section 10.4	NA
section 11.1.2	KUKA LWR 4+	Closed-loop; figure 1.18b	Passive DS, section 10.4	NA

The position controllers are built-in controllers provided by the manufacturers. Hence, the details are either unpublished or partially published.

illustrates a control architecture where the motion generator and the low-level controller are integrated into one block to address the latter problem.

Various forms of closed-loop control architectures are used in this book, depending on the type of robot used for the implementation of the controller. Some robots, such as Katana, does not allow one to access the joint motion while the robot moves. Therefore, this type of robot is used only in the open-loop, joint motion model. Closed-loop control varies for other robots, such as the Barrett Arm and Wheelchair, which can be controlled only in position, or for the KUKA robot arms, which can be controlled in torques. Table 1.1 provides a summary of the control architectures presented in this book.

2 Gathering Data for Learning

This book assumes that the robot has data at its disposal to learn the controllers. This chapter presents techniques and interfaces that can be used to gather data that can then be used to train the robot's controller. It also discusses caveats to bear in mind when gathering data.

2.1 Approaches to Generate Data

One popular approach to generate data to train robots is to have an expert provide examples of the task we want the robot to learn. This is known as *learning from demonstration (LfD)* or *programming by demonstration (PbD)*[16]. LfD/PbD, sometimes referred to as *imitation learning* and *apprenticeship learning*, is a paradigm for enabling robots to learn new tasks from observing experts (usually humans) performing the task. The main principle of robot learning from demonstration is that end users can teach robots new tasks without programming. Hence, for a long time, LfD/PbD assumed that the demonstrations were provided by a human on site. This was limiting, as it required having an expert available to produce the data, and hence it restricted learning to tasks doable by humans. More recent approaches use simulations or optimal control to generate solutions that would be exhaustive or impossible to generate with a human. There are a wealth of methods to train the robot; see [120, 16, 7] for reviews.

An alternative to LfD/PbD is to let the robot learn on its own, through trial and error. This is broadly known as *reinforcement learning (RL)* [77]. An expert is still needed, but only to provide a reward function, not a demonstration of the entire task. One drawback of RL is that it takes a long time before it converges to an optimal solution. To tackle this issue, two main trends are followed. The first combines learning from demonstration with RL. The demonstrations are used to bootstrap the search, providing a feasible solution or to bound the search space (see figure 2.1). The second trend uses realistic simulators to perform the search offline and use the real world only to refine the initial search. Another generic criticism of RL is that it requires one to carefully craft the reward. *Inverse reinforcement learning* contours this problem by allowing the robot to automatically infer the reward and the optimal controller [157]. To do so, however, the robot must have access to examples of good and bad solutions to the problem. These are usually provided by a human, which

Figure 2.1
Data for training robots can either be provided by human demonstrations or gathered by the robot on its own through trial and error (known as *reinforcement learning*, or RL). The two modes can be combined. Demonstrations can be used to bootstrap the search, providing a good example of how to move the tool. Trial and error can then be conducted in a restricted search space around the demonstrated trajectory to gather more information on the force and impedance to apply when interacting with gravel.

again raises the problem of limiting the number of required demonstrations so it would be bearable to the human [23].

2.1.1   Which Method Should Be Used, and When?

The methods presented here have different requirements. Some need the user to have prior knowledge about the robot, task and environment, while others need extensive time to generate data. To help the reader determine which method is most suitable for a particular application, we provide a summary of what each method needs and provides next; see also the schematic of the requirements in figure 2.2.

Human demonstrations: Data can be gathered from a human showing the task to the robot. This has the advantage that the robot immediately obtains examples of feasible solutions. The teacher does not need to be an expert in robot control and can be the end user of the platform. This approach allows the end user to customize the robot's behavior to their preferred way of moving. This is particularly useful for robots that are meant to work in coordination with the user, such as co-bots, and for control of prostheses and exoskeletons. The disadvantage is that the number of examples is limited to only about twenty trajectories for training to be bearable to the end user, which may provide too few statistics. Finally, the dynamics of motion performed by the human sometimes may not be doable for the robot's hardware.

Optimal control: If we can write a model of the task and of the dynamics of the robot, one can use optimal control to generate a large range of feasible trajectories that satisfy all the task requirements and the robot's constraints. Optimal control depends on solvers for non-convex optimization or integer and constraint programming. Solvers may take minutes or hours to find solutions, and they are not certain to converge. There is, hence, an advantage to converting all the solutions found offline through optimal control into closed-form expressions, with guarantees to retrieve always a feasible solution in real time. This embedding is done using machine learning techniques, as presented in the rest of the book. Using optimal

Method to generate the data	Online mode	Need model of robot or world	Trainer	Number of training examples
Learning from human demonstrations	YES	NO	Anyone	<20
Optimal control	NO	YES	Skilled programmer	>100
RL (live)	NO	YES (model-based RL) NO (model-free RL)	Anyone (reward)	>100
RL (simulation)	YES	YES	Skilled programmer	>1,000

Figure 2.2
Data can be gathered through three main channels: via human demonstrations, offline through optimal control or online with the robot exploring via RL

control to generate data for learning is advantageous, as it can provide a large number of trajectories (by randomizing the initial conditions) and the trajectories are by construction feasible for the robot's dynamics.

Reinforcement learning: Data can be gathered by the robot through trial and error. At each trial, the robot receives a reward that informs it on whether this is a good or poor solution. This technique is RL. The advantage of using this technique over the previous two is that it provides a set of both feasible and unfeasible trajectories. Unfeasible trajectories can then be used during learning to delimit the range of doable motion. The disadvantage is that this is fairly slow, and it may take many trials before a feasible solution is found. It may not be doable to run this on board the robot, and it may require an accurate simulator.

In section 2.4, we present an example of using these various techniques to teach a robot to play golf. More examples of applying these methods to perform different robotic tasks will be presented in other chapters.

2.2 Interfaces for Teaching Robots

When one chooses to train a robot through human demonstrations, the choice of interface is crucial, as the interace plays a key role in the way that the information is gathered and transmitted. For the purpose of this book, we primarily consider interfaces that provide information about the kinematic of the motion (speed, position) and haptic information (force, torque, touch location). In this context, the user must hence choose an interface following the three possibilities given here.

2.2.1 Motion-Tracking Systems

To record the kinematic of human motion, one may use any of the existing motion-tracking systems, whether these are based on vision, exoskeleton, or other types of wearable motion sensors; see [15] for reviews and videos of examples. Figure 2.3 shows an example of using motion tracking to monitor a human catching a ball. The motion-capture system consist of a

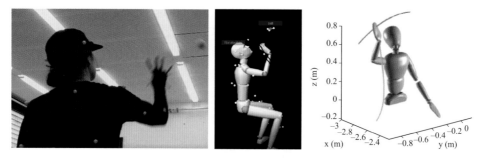

Figure 2.3
Demonstration of catching a ball recorded with a motion-capture system. The subject wears a suit covered with markers to track all upper-body joints. The reconstructed trajectories are highly accurate (c).

set of infrared cameras that track at high speed the markers worn by the subject and attached to the ball at a high frame rate of 250 frames per second within millimeters precision. Given the speed at which the motion happens, it is necessary to sample this rapidly and to have this level of precision. This can be done only with motion-capture systems or time-of-flight cameras. When one considers teaching a motion that requires a much slower pace, using standard cameras that sample at 50 Hz is an appropriate alternative. There are several systems, commercial but also open source, that can reconstruct the human motion from a camera; see [95, 119] for a few of them.

2.2.2   Correspondence Problem

These external means of tracking human motion return precise measurements of the angular displacement of the limbs and joints. They have been used in various instances for LfD of full-body motion [88, 47, 76]. These methods allow the human to move freely and to perform the task naturally. However, they require solutions to the correspondence problem (i.e., the problem of how to transfer motion from human to robot when both differ in the kinematic and dynamics of their bodies), or, in other words, if the configuration spaces are of different dimensions and sizes. This is typically done when mapping the motion of the joints that are tracked visually to a model of the human body that matches closely with that of the robot. Such mapping is always challenging to do. For instance, a robotic arm with 7 degrees of freedom (DOF) does not have a joint alignment that resembles the human arm's joint alignment. In other words, the two kinematic chains differ significantly. They also differ in terms of the range of motion. A simple scaling is not sufficient, and one must often perform double inverse kinematics, transferring displacement of the human's end point into the corresponding robot's joint angles. Given the current human's joint measurement q^h, one computes the human's end point position x^h, using Jacobian $J(q)^h$, and similarly for the robot's end point's position x^r, Jacobian J^r, and joint angles q^r, by solving the following:

$$\min_{\theta^r} \|x^h - x^r\|$$

$$x^h = J^h \theta^h$$

$$x^r = J^r \theta^r$$

The problem is particularly difficult when the two systems differ dramatically in the number of degrees of freedom. For instance, the human hand is said to have between 22 and 28 DOF, but most robotic hands have much fewer degrees of freedom. A traditional prosthetic hand would have only 5 DOF (1 for each finger), while more sophisticated hands would have 9 (Humanoid iCub hand),[1] 13 (DLR hand), 16 (Gifu hand [67] and Allegro hand[2]), and 24 (Shadow hand[3]).

Example Imagine you wish to teleoperate a robotic hand. As the robotic hand has fewer degrees of freedom than the human hand, you must find a way to map the human hand's joints onto the robotic hand's joints. Clearly, several of the human hand's joints will map to the motion of a single robotic joint. The mapping may depend on the application. In most daily hand movements, the true number of control parameters is much less than the number of degrees of freedom in the hand as joints move in coordination. This coordinated motion of the joints is often referred to as a *synergy* [126]. Synergies across joints can be constructed by projecting joint angles onto a subspace of lower dimensions, using principal component analysis (PCA). If $q^h \in \mathbb{R}^D$ denotes the D human hand joints, we construct $y \in \mathbb{R}^p$ $p < D$, with $y = Aq^h$. A is the projection matrix found after performing PCA. Each row of A embeds a particular combination of joints q^h, and y can be used to control each combination of joints separately. Consider that y is composed of binary entries, with $y_i \in 0, 1, i = 1 : p$. If we set $y_1 = 1$ and all other entries to zero, this will activate the first combination of joints, corresponding to the first row of A.

Similar mapping can be used to control the joint of a robot hand q^r for teleoperation. To use synergies to map human motion to robot motion, one must find a common map projection between the two spaces. One way to do so is to determine explicitly the mapping on the robot's fingers. As the rows of A are composed of eigenvectors and the entries of the eigenvectors correspond to the contribution of each joint to the synergy, setting these entries will determine the mapping [22]. Examples of such synergistic constructions are given in exercise 2.1.

Exercise 2.1   *Consider a robotic hand made of 5 DOF, $q \in \mathbb{R}^5$, one for each finger.*

1. Assume that the thumb is one-third shorter than the other four fingers. Which minimal synergy, $y = Aq$, do you need to close all the fingers simultaneously?

2. You wish to teleoperate this robotic hand using a data glove worn on your hand, controlling each finger individually. The glove has 2 DOF for the thumb and 3 for all the other fingers, for a total of $D = 14$ DOF. Which map, $q = Ax$, $x \in \mathbb{R}^D$, should you use?

3. You now want to control the thumb and index finger in synch and keep the other fingers static. How will the previous maps change?

2.2.3   Kinesthetic Teaching

The problem becomes even more difficult when one tries to control a full humanoid robot, or when the robot's body bears little resemblance to the human's body (e.g., an hexapod differs significantly from the human body). One way to overcome the correspondence problem is to put the human in the robot's shoes, such as by having the human guide the robot

Figure 2.4
Left figure: Demonstration of a task requiring force control through teleoperation via a haptic interface (adapted from [36]). Middle figure: Example of kinesthetic teaching, in which the human puts herself in the robot's shoes to teach the robot how to peel vegetables. Forces are measured through tactile sensors at the robot's end points. Right figure: Example of kinesthetic teaching to teach how to manipulate an object. Forces are measured through tactile sensors at the robot's fingertips.

through kinesthetic teaching, as illustrated in figure 2.4. *Kinesthetic teaching* is a technique in which the teacher embodies the robot using the mechanical backdrivibility of its joint to move it passively.[4] Kinesthetic teaching also enables the transmission of information about forces, which is not readily available from motion-tracking systems. In addition, it provides a natural teaching interface to correct a skill reproduced by the robot. One main drawback of kinesthetic teaching is that the human must often use more degrees of freedom to move the robot than the number of degrees of freedom moved on the robot. This is visible in the example on the right side of figure 2.4. To move the fingers of one hand of the robot, the teacher must use both hands. This limits the type of tasks that can be taught through kinesthetic teaching. Typically, tasks that would require moving the two hands simultaneously could not be taught in this way. One could either proceed incrementally, teaching first the task for the right hand and then, while the robot replays the motion with its right hand, teach the motion of the left hand. However, this may prove to be cumbersome. The use of external trackers, as reviewed previously, is more amenable to teaching coordinated motion among several limbs.

2.2.4   Teleoperation

Teleoperation, via joystick and remote control, can also be used to transmit the kinematics of motion. Such immersive teleoperation scenarios, where a human operator uses the robot's own sensors and effectors to perform the task, are powerful techniques to train a robot from its own viewpoint. For instance, in [31], acrobatic trajectories of an helicopter are learned by recording the motion of the helicopter when teleoperated by an expert pilot. In [51], a robot dog is taught to play soccer via a human guiding it via a joystick. Teleoperation, however, is often limited in the field of view rendered, which prevents the teacher from observing all the sensorial information required to perform the task. Teleoperation is advantageous compared to external motion tracking systems, as it entirely solves the correspondence problem because the system records directly the perception and action from the robot's configuration space. Teleoperation using a simple joystick allows one to guide only a subset of degrees of freedom. To control for all degrees of freedom, very complex, exoskeleton-type devices must be used, which is cumbersome.

2.2.5 Interface to Transfer Forces

All of the methods described thus far provide only kinematic information. This is sufficient when we want to teach tasks that can be completely described using position and velocity, such as gestures. However, when the task depends on modulating forces at contact and this is done without large movements (e.g., when inserting a peg into a hole), then one needs to use interfaces that can measure and transmit these forces.

In place of using a simple joystick, one can employ haptic devices that can render and transmit forces (see figure 2.4 left). It is advantageous compared to kinesthetic training, as it allows to train robots from a distance and hence is particularly well suited for teaching navigation and locomotion patterns. The teacher no longer needs to share the same space with the robot. This method was used, for instance, to teach balancing techniques to a humanoid robot [115]. As the teacher moves, a haptic interface attached to the torso of the demonstrator measures the interaction forces. The kinematics of motion of the demonstrator are directly transmitted to the robot through teleoperation and are combined with haptic information to train a model of motion conditioned on perceived forces. The disadvantage of teleoperation techniques is that the teacher often needs training in order to learn to use the remote control device. Also, haptic devices can sometimes poorly render the contacts perceived at the robot's end-effector, or do so with a delay. To adapt to this, one may provide the teacher with visualization interfaces to simulate the interaction forces.

To measure forces, one can also use the robot's onboard tactile or force/torque sensors. This can be used in conjunction with either kinesthetic teaching (see figure 2.4), or through the use of nonhaptic devices. While with kinesthetic teaching, the teacher can perceive the forces and modulate their gestures in response, when using a nonhaptic joystick, this perception is lost and may negatively affect the teaching.

2.2.6 Combining Interfaces

Each teaching interface has pros and cons. For that reason, nowadays people use several of these interfaces in conjunction. Figure 2.5 shows one example, where a robot is trained to scoop melons. The task is challenging, as the melon is made of soft material and the robot must learn to vary the force and stiffness as it scoops, in response to the resistance that it perceives from the melon. When the melon is hard, a stiff controller may be appropriate, whereas a more compliant controller may be needed when the melon is softer. Melons are not homogeneous, and modifying compliance on the go may be necessary. To teach the robot, the user shown in the figure exploits a combination of interfaces. She uses kinesthetic teaching to embody the robot, which allows her to better perceive and transfer the forces. Forces and torques are measured through two force/torque sensors mounted on the tool and the robot's end-effector, and also through tactile sensors covering the glove worn by the user. These sensors measure the evolution of the forces applied as a response to the change in resistance of the melon. This evolution of force measurement provides key information to allow one to adapt the impedance to changes in the fruit's resistance. In addition, marker-based vision and a data glove track kinematic information on arm and hand motions, so as to detect changes in posture and the grasp of the user's hand as she adapts to the task. More details on how this information is used to train the robot to perform the same task is documented in [149].

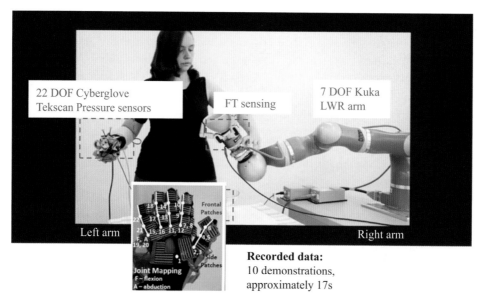

Figure 2.5
Use of multiple interfaces (i.e., vision, haptic) for training a robot to scoop melons (adapted from [149]).

2.3   Desiderata for the Data

Before we explore an example of how to gather and use data to train a robot, a few general considerations are in order.

Gathering enough data: Data are inherently noisy. Because all machine learning techniques used in this book are based on statistics, the data must include enough statistics for the algorithm to tell noise from signal. A good rule in statistics is to have at least ten samples for each dimension [54].[5] If you are teaching a task that requires control of position and orientation ($N = 6$), you would want to have at least 10^6 data points. If you sample trajectories at 50 Hz, the data will include enough statistics after as few as ten demonstrations, each one lasting less than a few seconds. However, while this may be statistically sufficient, it is not certain that this will be enough to teach the robot well. We need to make sure that the data are relevant for the task.

Consider, for instance, that you want the robot to learn how to play golf (see figure 2.6 left). To teach the robot how to putt a ball into a hole, you must present examples illustrative of how to do this. While showing ten examples may be sufficient to generate the required statistics, showing quasi-identical ones will provide little insight into the task. It is useful to generate multiple examples, starting from different configurations and approaching the target with different orientations. It may also be useful to change the experimental setup and change the position and height of the hole.

Data must be well chosen: For the focus of this book, data must be composed of samples of dynamics of movement that can be realized by the robot. Speed, acceleration profiles, torques, and impedance parameters must be within the robot's capacities. Also, the data must be representative of what we want the robot to learn. They must entail useful examples

 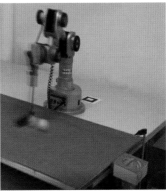

Figure 2.6
Left: The teacher makes a few demonstrations of the task of putting a golf ball by changing the orientation of the club depending on the position of the hole. *Right:* Once the robot has learned the task, it can reproduce it even when the hole is located in positions not seen in the demonstrations.

of the task at hand. While one often provides only examples of what the robot must do, it is often quite useful to provide examples of what *not* to do. This is akin to providing examples of feasible and unfeasible dynamics—dynamics that would lead the task to succeed or fail. For instance, when teaching a robot to play golf, you may want to show both examples when the ball goes into the hole and examples when the ball misses or bounces off the edge of the hole. This additional information will help the algorithm learn that tiny changes in orientation and speed may lead the task to shift from success to failure.

When teaching is performed by a human, one aims at reducing the number of demonstrations to its bare minimum to lighten the burden on the teacher. A rule of thumb is generally not to exceed twenty demonstrations for each task. Thus, it is important to choose well what you demonstrate to keep from wasting demonstrations. If you work in simulation and use optimization to generate example trajectories, you are limited only by your patience and the speed of your solver. Yet we should still be sure to collect a wide enough sample of solutions and to generate solutions that are both feasible and infeasible for the robot to learn well. Be aware that the number of feasible and infeasible solutions should be similar (or of the same order of magnitude) to avoid unbalanced datasets. The concept of an *unbalanced data set* refers to a situation when one has much more samples of one type than of another (e.g., more samples of noncancer cells than of cancer cells). The imbalance can be detrimental to learning, as the group with the largest number of samples will have more influence and be better represented in the model than the group with fewer examples.

Generalization: One major goal of machine learning is to ensure that the model *generalizes* outside the examples used for training. This is crucial, as we are quite aware of the fact that the training examples provide only a subset of all what exists. The usual practice in machine learning is to decompose the data set into a training set, a validation set, and a testing set. Training and validation are used in conjunction to determine the best hyperparameters through grid search and cross-validation. The testing set is used to assess how good the algorithm is at generalizing to unseen data. The training/test set ratio is a good metric of generalization. The smaller the number of training data required to ensure good performance of the testing set, the stronger our confidence in the model's generalization.

However, for a robot to generalize, it is important that training and testing sets be chosen in a way that leads to this generalization. Generalization can take many forms. In the golf task presented previously, limited generalization would be needed for the robot to putt the ball correctly only if the initial ball locations differ by a few centimeters from those seen during training. But proper generalization would be demonstrated if the system can putt the ball for a variety of configurations of both ball and hole. Impressive generalization capabilities would be demonstrated if the robot were capable of sinking the ball on more complex terrain than seen during training. This is rarer, however, and transferring knowledge to a new context often requires the system to be trained again. This is known as *transfer learning*. The testing set to measure the generalization capabilities of robots hence must not be the result of random sampling across the data set, but rather be chosen with care to demonstrate the expected generalization capabilities.

To conclude, the data for training robots must be chosen with care. They must be representative of what you would like the robot to achieve. They must include examples of what is a good solution to the task and what is not. Data must not be split randomly for training and testing. Rather, the data used at testing must be chosen to demonstrate the extent to which the robot generalizes the knowledge involved in the training examples.

2.4   Teaching a Robot How to Play Golf

This section illustrates how to use human demonstration, trial and error, and optimal control to train a robot to play golf.[6] Using a mock-up golf terrain, (see figure 2.6), we aim to teach the robot to move a golf club and hit a ball with the correct orientation and speed to sink the ball in the hole.

First, human demonstrations are sufficient to teach the robot the task on a flat ground. However, when the robot is tasked to sink the ball on a more complex, hilly terrain, it becomes difficult to obtain successful human demonstrations. As the task requires high precision, examples of feasible and infeasible trajectories are equally important. Trial and error thus is necessary to complement human demonstrations. Finally, the task could also have been encoded through optimal control to learn offline the desired dynamics, which would exceed in speed and precision what could be demonstrated by a human.

2.4.1   Teaching the Task with Human Demonstrations

To train the robot to putt the golf ball, one can opt for showing the movement through kinesthetic teaching (adapted from [86]), as shown in figure 2.6. As the teacher moves the robot's joints passively, the robot records the state $x \in \mathbb{R}^N$ and velocity $\dot{x} \in \mathbb{R}^N$ of its end-effector at each time step. The set of demonstrations, then, consist in a set of M trajectories composed of pairs $\{X, \dot{X}\} = \{X^m, \dot{X}^m\}_{m=1}^M = \{\{x_t^m, \dot{x}_t^m\}_{t=1}^{T_m}\}_{m=1}^M$, where T_m is the length of each mth trajectory. Each pair represents a $2N$-dimensional vector point.

We follow the approach presented in chapter 1 and embed the dynamics of movement in a dynamical system (DS) of the form $\dot{x} = g(x)$. To do so, we use the stable estimator of dynamical systems (SEDS) method, which guarantees that the flow asymptotically reaches

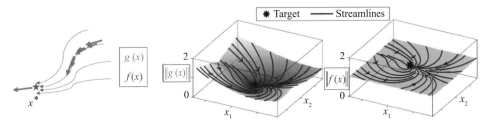

Figure 2.7
To generate a flow whose velocity does not vanish at the attractor, one can modulate two dynamics: a unitary flow asymptotically stable at the attractor $g(x)$ and a nonstationary flow that embeds a velocity profile $M(x)$ that is not necessarily stable at the attractor. The final flow $f(x) = M(x)g(x)$ is stable while also having a nonzero velocity at the attractor [86].

and stabilizes at the ball's location, such that $\lim_{t\to\infty} x = x^*$, with x^* being the ball location. This method is presented in detail in chapter 3. This approach generates a flow of motion for the end-effector that ends up correctly for hitting the ball. However, this model is not sufficient for our task because with such a system, the robot would stop once it reaches the ball. To hit the ball so that it goes into the hole, the robot needs to reach the ball with a very specific velocity. To achieve this, we modulate the initial stable DS flow so that the velocity vector at the attractor is oriented and matches the amplitude seen during the demonstrations (see the illustration in figure 2.7 left).

To achieve this, we use our demonstration data points twice: first, we learn a stable flow at the ball that embeds the demonstrated orientation when approaching it. As this flow vanishes at the ball, we store only the unitary vector field $\|g(x)\| = 1, \forall x$. Second, we learn a (not necessarily stable) function $M(x)$ to represent the amplitude of the demonstrated velocity flow: $M(x) = \|\dot{x}\|$. $M(x)$ can be learned using any machine learning technique for regression (e.g., support vector regression, Gaussian process regression, or neural networks).

The final dynamics is generated as a combination of the learned functions: $\dot{x} = f(x) = M(x)g(x)$. Then $f(x)$ converges to the ball as it follows the flow indicated by $g(x)$ with the correct amplitude specified by $M(x^*)$.[7] The convergence of the flow can be visualized by looking at the streamlines and amplitude of the energy functions of $g(x)$ and $f(x)$, as shown in figure 2.7.

2.4.1.1 Generalization and adaptation

One desideratum for this task is that the robot be capable of generalizing knowledge to various ball and hole configurations. Such a generalization can be achieved by expressing this problem in a well-chosen frame of reference. We now compute the position of the ball and of the end-effector in a frame of reference relative to the hole. When doing so, the learned orientation of the end-effector becomes independent of the world's coordinate frame, which allows the robot to generate the correct motion for different hole positions (see figure 2.8). This form of generalization assumes that the terrain is perfectly flat in all directions.

Because its motion is driven primarily by a DS, the robot can adapt to changes in the environment. For instance, if the robot is pushed away from its initial trajectory while reaching toward the ball, the system will generate a new flow that is also certain to approach the ball

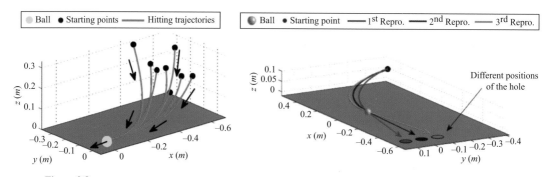

Figure 2.8
Left: Representation through DSs of trajectories to sink a golf ball. *Right:* Expressing the dynamics from the relative position of the ball to the sink allows one to nicely generalize the orientation toward the ball without further demonstrations.

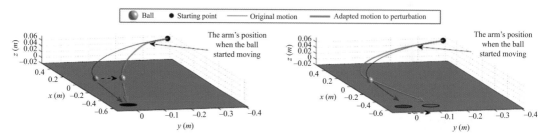

Figure 2.9
The learned DS can adapt at run time to perturbations, such as pushing the robot away from its trajectory (left) and moving the hole (right) by generating a new trajectory that reaches the target correctly [86].

with the correct orientation and amplitude. The hole's location can also be moved at run time and the robot will similarly adapt to the new location (see figure 2.9 and associated videos on the book's website, https://www.epfl.ch/labs/lasa/mit-press-book-learning/).

2.4.2 Learning from Failed and Good Demonstrations

We now task our robot to learn to sink the ball into the hole on an uneven, hilly terrain (see figure 2.10). The task is more complex than the one discussed previously. The difficulty lies in determining the precise speed and angle of attack. If one hits the ball with too little velocity, the ball runs down the slope, but too much velocity may make the ball jump over the hole. The hilly terrain is not homogeneous. A small error in the angle of attack may lead the ball to drift away from the hole. The task becomes even more difficult if we again stipulate that we want the robot to generalize to different ball and hole locations (in the setup shown in figure 2.10, the user can change the location of the hole, by sliding it along the rail located at the end of the field). To successfully sink the ball when the hole is moved, the robot must adapt its end-effector's orientation and desired speed at run time.

In the previous example, we showed that encoding the task in a relative frame of reference would allow the robot to adapt to modifications, of the hole's location. This is not possible here. In the hilly terrain of this example, many physical factors may lead the robot to be unsuccessful. This may be due to uneven friction along the terrain, which is hard to know

Figure 2.10
Left: The teacher demonstrates the task of sinking a golf ball on hilly terrain. The teacher shows both successful and failed examples so that the robot can learn a representation of the distribution of feasible and unfeasible sets of parameters. These parameters are the angle and speed when hitting the ball. *Right:* Once the robot has learned, it can reproduce the task even when the hole is located in positions not seen in the demonstrations.

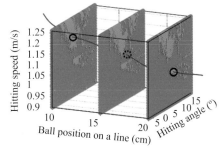

Figure 2.11
Left: The user moves the location of the hole, and the robot must adapt to this new location. *Right:* The robot learns a distribution of feasible and unfeasible parameters (hitting angle and speed) to withstand noise [86].

in advance. In addition, the task is no longer deterministic because even when adopting the theoretically correct angle and velocity, the robot may sometimes be unsuccessful.

To take into account the nonlinearity and stochasticity of the task, we opt to learn a distribution of feasible orientations and velocities. However, because the task's success is influenced by small changes in orientation and speed, one should learn not only the parameters that lead to success, but also those that lead to failures. It is not difficult to get examples of failure as humans find it difficult to generate good example. This is due to both the complexity of the terrain and the fact that it is cumbersome to move such a large robot arm. In the end, most of the demonstrations ended up being failures, and we were short of successful examples. To augment the statistics of the successful data, we let the robot learn by trial and error by systematically testing different parameters for speed and orientation. Using these demonstrations to guide the search helped to reduce the range of parameters to test. As the robot tried different parameters, it ended up gathering many examples of both successful and unsuccessful trials.

Figure 2.11 shows the distribution of parameters that lead to success versus failure in this task. Observe that the two distributions are mixed. This confirms how important it was to model the distribution of both feasible and infeasible parameters, given that the feasible regions are quite small and scattered within the infeasible region. Simple interpolation from observing only good examples could lead to predict parameters that are infeasible.

The distributions could be learned using probability density function modeling from machine learning (here, the Gaussian mixture model). At testing time, one can determine if the current configuration is deemed to fail or succeed by comparing the likelihood of the parameters under the feasible and infeasible distributions. At run time, when the environment changes (e.g., the goal location changes), one can determine a new parameter set by drawing from the closest combination of the distribution of feasible parameters. Examples of successful reproduction are shown in the video on the book's website. See also [86] for more detail.

2.5 Gathering Data from Optimal Control

In the two examples presented previously, we assumed that the speed and trajectories demonstrated by the human could be reproduced by the robot. In other words, we assumed that the demonstrations satisfy the kinematics and dynamics constraints of the robot. Kinematics constraints are satisfied when using kinesthetic teaching, as it consists of moving the robot's joint. However, the dynamics displayed in the human demonstration may be too fast, requiring larger accelerations than are possible on the robot. Conversely, some human motions are sometimes too slow and when converted into velocities for the robot, they cannot be generated if the acceleration is too small and does not overcome the robot's friction.[8]

An alternative to using human demonstrations is to search for possible solutions through RL, as described earlier in this chapter. Reinforcement learning is closely related to optimal control. In order to generate trajectories that are feasible for the robot, one must have a model of the robot. Not having such a model may be risky when running the search in the real world, as it may lead to hardware damage. In the golf example, generating trajectories that would be feasible and lead the robot to move as fast as possible could be expressed through the following iterative optimization:

Algorithm 2.1 Generate Kinematically Feasible Data Trajectories

---

Initialization:

    $x(0)$ and x^* *Randomly define the initial robot's position and final target position.*

    $q(0) = F^{-1}(x(0))$ *Inverse kinematics for the initial joint position.*

    $t = 0$. *Initialize time.*

Main loop:

    While $\epsilon \leq \|F(q(t)) - x^*\|$:

        $\dot{q}(t+dt) = \arg\max_{\dot{q}} \|\dot{q}\|$. *Maximizing the velocity of the joints.*

        subject to:

          $\frac{\dot{q}}{\|\dot{q}\|} = J^+(q(t)) \frac{x^* - F(q(t))}{\|x^* - F(q(t))\|}$. *Moving on the straight line toward the target at the task space.*

          $\dot{q}_{min}[i] \leq \dot{q}[i] \leq \dot{q}_{max}[i]$ $\forall i \in \{1, \ldots, D\}$. *Kinematic feasibility at the velocity level.*

          $q_{min} \leq q(t)[i] + \dot{q}[i]dt \leq q_{max}[i]$ $\forall i \in \{1, \ldots, D\}$. *Kinematic feasibility at the position level.*

        $q(t+dt) = q(t) + \dot{q}(t+dt)\,dt$

        $t = t + dt$

$F(.)$ and $J(.)$ are the forward kinematics and the Jacobin matrices, respectively. $+$ is Moore–Penrose inverse.

Algorithm 2.1 makes sure that the trajectory is feasible using both bounds on maximal acceleration and velocity for each joint and model of the forward and inverse kinematics.

This optimization can be run offline. If one samples different initial and final end-effector and joint locations $(x(0), q(0), x^*)$, this enables one to populate a database of feasible trajectories. All the configurations that lead to an infeasible solutions from the solver may be used as examples of a failed set of parameters. Running this in simulation is advantageous, in that it will offer many examples (much more than are doable from human demonstration) and hence have a fine-grained model. Yet this is just a simulation, remaining a proxy of reality. In the golf task, for instance, one can certainly run some of the trials in simulation using a good model of the robot and the environment. This would allow one to generate a set of initial values for feasible regions of parameters. However, real implementation will be needed to fine-tune the parameters and add more data to overcome non-linearities of the terrain.

II LEARNING A CONTROLLER

3 Learning a Control Law

This chapter presents methods to learn a control law for robot motion generation using time-invariant dynamical systems (DSs). We assume that the motion of the robotic system is fully defined by its state $x \in \mathbb{R}^N$ and characterized by a system of ordinary differential equations (ODEs). Let $f(x)$ be a first-order, autonomous DS describing a nominal motion plan for a robot, such that

$$f \colon \mathbb{R}^N \to \mathbb{R}^N,$$
$$\dot{x} = f(x), \tag{3.1}$$

where $f(\cdot) \colon \mathbb{R}^N \to \mathbb{R}^N$ is a continuous differentiable, vector-valued function representing. Learning consists of estimating the function f, which is a map from the N-dimensional input state $x \in \mathbb{R}^{\mathbb{N}}$ to its time-derivative $\dot{x} \in \mathbb{R}^{\mathbb{N}}$.

Training data consist of sample trajectories, which we assume to represent path integrals of the underlying DS. These trajectories cover a limited portion of the state space. The goal of the learning algorithm is to ensure that the learned DS reproduces the training data well, while generalizing over regions not covered by the data. We will, in particular, want to ensure that the system does not diverge from the training data points. To tackle this problem, we will embed constraints explicitly into the learning algorithm so that the learned dynamics offers guarantees from control theory. One desirable property is stability at an attractor, x^*. Hence, f must be such that

$$\dot{x}^* = f(x^*) = 0,$$
$$\lim_{t \to \infty} x = x^*. \tag{3.2}$$

Here, $f(\cdot) \colon \mathbb{R}^N \to \mathbb{R}^N$ is a continuous differentiable, vector-valued function representing a DS that converges on a single stable equilibrium point x^*, which is also called the *target* or *attractor*.

To learn the function, we choose an expression for the function to be learned, which is parameterized through a set of parameters Θ such that $f(x; \Theta)$. Learning consists of updating the parameters Θ until the function fits as accurately as possible the reference trajectories, which can be measured through a loss function $L(X, f, \Theta)$.

The problem at stake can then be rephrased as follows:

Given a set of M reference trajectories $\{\mathbf{X}, \dot{\mathbf{X}}\} = \{X^m, \dot{X}^m\}_{m=1}^{M} = \{\{x^{t,m}, \dot{x}^{t,m}\}_{t=1}^{T_m}\}_{m=1}^{M}$, where T_m is the length of each mth trajectory, estimate the parameters Θ of the function $f(x; \Theta)$, to represent at best the reference trajectories according to a loss $L(X, f, \Theta)$. Furthermore, function f must be capable of generating motion that retains some of the characteristics of the reference trajectories in the state space not covered by the training data. This can be assessed by measuring the loss incurred on a testing set composed of reference trajectories not used at training. Finally, the function f must ensure that the target x^* is reached from any point in space.

This can be summarized as two objectives:

1. Reproduce the reference dynamics.
2. Converge to the attractor (target).

From a machine learning perspective, estimating $\dot{x} = f(x)$ from data can be framed as a regression problem, where the inputs are the state variables x and the outputs are their first-order derivatives $\dot{x}$. As we will show, standard machine learning techniques can ensure objective 1. but not objective 2. In this chapter, we present techniques derived from known machine learning algorithms to learn DS that are capable of accurately reproducing the demonstrated behavior (objective 1.) and ensure convergence to the target from anywhere in the state-space (objective 2.).

We begin this chapter with a brief introduction of fundamental concepts of stability of DS, which will be reused throughout the book, in section 3.1. The reader is referred to appendix B for a review of machine learning techniques used in this chapter and throughout the book. We continue, in sections 3.3 and 3.4, by presenting two techniques to learn first-order, single attractor, nonlinear DS from data. We end with section 3.5, where we describe how these techniques can be extended to learn second-order dynamics with a single attractor.

3.1   Preliminaries

3.1.1   Multivariate Regression for DS Learning

In this section, we introduce three methods from machine learning for estimating f—namely Gaussian mixture regression (GMR), support vector regression and Gaussian process regression (GPR), and show that they generate unstable DSs.

For the reader who is unfamiliar with standard regression techniques, we provide a summary of each approach presented in this section in appendix B. Specifically, Gaussian mixture modeling and regression (GMM and GMR) can be found in section B.3, SVR is described in section B.4.2.2, and GPR is summarized in section B.5.

3.1.1.1   Gaussian Mixture Regression
With GMR, a first-order DS, $\dot{x} = f(x)$, is estimated by learning about a joint density of position and velocity measurements through a K-component Gaussian mixture model (GMM) as follows:

$$p(x, \dot{x}|\Theta_{\text{GMR}}) = \sum_{k=1}^{K} \pi_k \underbrace{p(x, \dot{x}|\mu^k, \Sigma^k)}_{\mathcal{N}(\cdot|\mu, \Sigma)}, \qquad\qquad (3.3)$$

with $\mathcal{N}(\cdot|\mu, \Sigma)$ (B.16) being the multivariate Gaussian (or normal) distribution; and π_k represent the priors of each Gaussian component, with $\sum_{k=1}^{K} \pi_k = 1$. Each kth Gaussian distribution is parameterized by $\theta_k = \{\mu^k, \Sigma^k\}$, where $\mu^k = \begin{bmatrix} \mu_x^k \\ \mu_{\dot{x}}^k \end{bmatrix}$ and $\Sigma^k = \begin{bmatrix} \Sigma_x^k & \Sigma_{x\dot{x}}^k \\ \Sigma_{\dot{x}x}^k & \Sigma_{\dot{x}}^k \end{bmatrix}$. The parameters $\Theta_{\mathrm{GMR}} = \{\pi_k, \theta_k\}_{k=1}^{K}$ of (3.3) can be estimated through either of the following:

- Maximum likelihood (ML) estimation via an iterative expectation-maximization (EM) algorithm, as described in section B.3.1 in appendix B. The number of components K can be selected via either model selection or cross-validation approaches. For the former, model selection metrics (section B.2.1 of appendix B) are used to find the optimal number of parameters that trade off the likelihood of the model with the model complexity, as described in section B.3.1.4 in appendix B. For the latter, regression metrics, such as mean squared error (MSE) variants (section B.2.4 in appendix B), are used to minimize the error between the estimated velocities and the observed ones; that is, minimize $\frac{1}{L} \sum_{i=1}^{L} \|f(x_i) - \dot{x}_i\|$ for L data points.

- Maximum a posteriori (MAP) estimation via sampling schemes, as described in sections B.3.2 and B.3.3 in appendix B, where K can be inferred either through model selection or cross-validation with regression metrics as discussed previously, or via Bayesian nonparametric estimates, which maximize the posterior distribution.

Once Θ_{GMR} and K are inferred, the *learned DS* is obtained by computing the expectation over the conditional density $\mathbb{E}\{p(\dot{x}|x)\}$ as follows:

$$\dot{x} = f(x; \Theta_{\mathrm{GMR}})$$

$$= \mathbb{E}\{p(\dot{x}|x)\} = \sum_{k=1}^{K} \gamma_k(x)\tilde{\mu}^k(x), \tag{3.4}$$

with

$$\gamma_k(x) = \frac{\pi_k p(x|\mu_x^k, \Sigma_x^k)}{\sum_{i=1}^{K} \pi_i p(x|\mu_x^i, \Sigma_x^i)}, \qquad \tilde{\mu}^k(x) = \mu_{\dot{x}}^k + \Sigma_{\dot{x}x}^k (\Sigma_x^k)^{-1}(x - \mu_x^k). \tag{3.5}$$

The terms $\tilde{\mu}^k(x)$ can be interpreted as local linear regressive functions whose slope is determined by Σ_{xx}^k (the variance of x) and $\Sigma_{\dot{x}x}^k$ (the covariance of $\dot{x}$ and x). Further, $\gamma_k(x) = p(k|x, \Theta_{\mathrm{GMR}})$ is the a posteriori probability for a data point x belonging to the kth Gaussian component, where $p(x|\mu_*^*, \Sigma_*^*) = \mathcal{N}(x|\mu_*^*, \Sigma_*^*)$ is the Gaussian density function equation (B.16). We refer herein to $\gamma_k(x)$ as the mixing weights (or functions) for the linear regressors. Further, note that the mixing functions, $\gamma_k(x)$, hold the following properties that will prove useful in the next subsections:

$$\begin{cases} 0 < \gamma_k(x) \leq 1 & \forall\, k = 1, \ldots, K \\ \sum_{k=1}^{K} \gamma_k(x) = 1. \end{cases} \tag{3.6}$$

The learned DS, $\dot{x} = f(x)$, is thus a weighted combination of K linear regressive models. For the derivation of equations (3.4) and (3.5), the reader is referred to section B.3.4.3 in

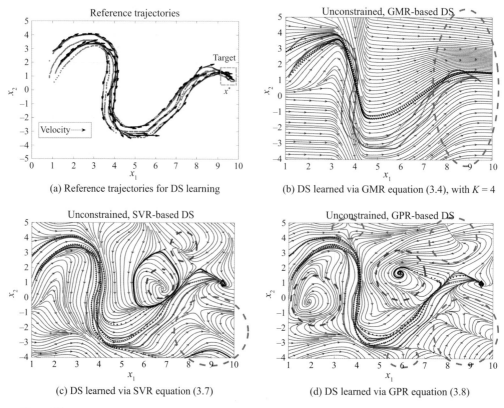

Figure 3.1
Illustrative 2D reference trajectories (a) for DS learning with standard regression algorithms (b–d) and with the proposed stable DS learning scheme (d) introduced in section 3.3. **Red** trajectories are demonstrations, and **black** trajectories are reproductions of the learned DS from the same initial state as demonstrations. Notice that for all the standard regression techniques, the DS have **spurious attractors** and **diverging regions**.

appendix B. The DS illustrated in figure 3.1b is generated by equation (3.4), with Θ_{GMR} learned via ML estimation.

3.1.1.2 Support Vector Regression

With SVR, a first-order DS, $\dot{x} = f(x)$ with $x, \dot{x} \in \mathbb{R}^N$, is estimated by learning N regressive functions, one for each nth output dimension (i.e., $f(x) = [f_1(x), \ldots, f_n(x), \ldots, f_N(x)]^T$) as follows:

$$\dot{x}_n = f_n(x; \ \theta_{\text{SVR}}^n)$$

$$= \sum_{i=1}^{L} (\alpha_i - \alpha_i^*)_n k_n(x, x^i) + b_n, \tag{3.7}$$

with $L = \sum_{m=1}^{M} T_m$ being the number of all input/output pairs (position/velocity measurements) in the collection of reference trajectories $\{\mathbf{X}, \dot{\mathbf{X}}\} = \{X^m, \dot{X}^m\}_{m=1}^{M} = \{\{x^{t,m}, \dot{x}^{t,m}\}_{t=1}^{T_m}\}_{m=1}^{M}$. Here, $\alpha_i, \alpha_i^* \in \mathbb{R}$ are the weights that indicate which ith input data point is a support vector—that is, when $(\alpha_i - \alpha_i^*)_n > 0$, x^i is a support vector for the nth regressor. Here, $b_n \in \mathbb{R}$ is the bias term. The term $k_n(x, x^i)$ represents the kernel function used in each nth regressive function. As described in section B.4 in appendix B, one has to choose different sets of hyperparameters depending on the choice of kernel function. Hence, the nth regressor may have different kernel hyperparameters. Throughout this book, we favor the radial basis function (RBF) (i.e., $k_n(x, x^i) = \exp\left(-\frac{1}{2\sigma_n^2}||x - x^i||^2\right)$), as described in appendix b, equation (B.72). This kernel has a hyperparameter σ_n, which represents the variance or width of the RBF around the ith support vector. Thus, each nth SVR is parameterized by $\theta_{\text{SVR}}^n = \{\{(\alpha_i - \alpha_i^*)_n\}_{i=1}^L, \sigma_n, b_n\}$ and the full set of parameters is determined as $\Theta_{\text{SVR}} = \{\theta_{\text{SVR}}^n\}_{n=1}^N$, with the addition of the input data points that one must keep as support vectors.

Given the chosen kernel hyperparameter to estimate the support vector weights $(\alpha_i - \alpha_i^*)_n$ and the bias b_n, one must solve a convex optimization problem that minimizes the prediction error; that is, $||f_n(x) - \dot{x}_n||_2$, while allowing an ϵ_n-deviation error in the output estimation. This is achieved by maximizing the so-called ϵ-insensitive loss function equation (B.82) via a maximum margin optimization problem equation (B.85). This requires the selection of not only the kernel hyperparameters, but also the allowed error $\epsilon_n \in \mathbb{R}_+$ and a penalty term $C_n \in \mathbb{R}_+$ associated with errors larger than ϵ_n. See equation (B.85) in section B.4.2.2 in appendix B for the definition of this optimization problem. As with the GMR approach, one must select the optimal combination of hyperparameters—in this case, $\{C_n, \epsilon_n, \sigma_n\}$ $\forall n = 1, \ldots, N$ for each nth regressor. While the optimal kernel hyperparameter σ_n might be similar for each nth regressor, the hyperparameters related to the output estimation, C_n, ϵ_n, might differ per each nth output dimension. To estimate the best hyperparameters for each regressive function, one can perform cross-validation on a set of informed hyperparameter ranges as described in section B.4.3 in appendix B, with the MSE variants as described in section B.2.4 in appendix B. An example of a two-dimensional (2D) DS learned via equation (3.7) is illustrated in figure 3.1.

3.1.1.3 Gaussian Process Regression

As with SVR, when using GPR to estimate a first-order DS, $\dot{x} = f(x)$ with $x, \dot{x} \in \mathbb{R}^N$, one must estimate N regressive functions, one for each nth output dimension (i.e., $f(x) = [f_1(x), \ldots, f_n(x), \ldots, f_N(x)]^T$) as follows:

$$\dot{x}_n = f_n(x; \theta_{\text{GPR}}^n)$$

$$= E\{p(y_n | x, \mathbf{X}, \dot{\mathbf{X}})\} \tag{3.8}$$

$$= \mathbf{k}_n(x)^T [K_n + \epsilon_{\sigma_n^2} \mathbb{I}_M]^{-1} \dot{\mathbf{X}},$$

where $\mathbf{k}_n(x^*) = \mathbf{k}_n(x, \mathbf{X}) \in \mathbb{R}^L$ denotes the vector of covariances between the query input data point and the training data set; and $K_n = K_n(\mathbf{X}, \mathbf{X}) \in \mathbb{R}^{L \times L}$, as in SVR $L = \sum_{m=1}^M T_m$, is the total number of data points in the training set. Further, $\mathbf{k}_n(x, \mathbf{X}) = [k_n(x, x^l)]_{l=1}^L$ is an L-dimensional vector that evaluates the kernel function $k_n(x, x^l)$ (i.e., covariance function) between the query point x and each lth data point from the training set. The kernel matrix K_n is built similarly (i.e., $[K_n(x^i, x^j)]_{ij} = k_n(x^i, x^j)$). Throughout the book, we use the

RBF, also known as the squared exponential (SE) function equation (B.110), as our kernel function (i.e., $k_n(x, x') = \sigma_y^2 \exp\left(-\frac{1}{2l_n^2} \sum_{i=1}^{N} (x_i - x_i')^2\right)$). As described in section B.5 in appendix B, we set the output variance to σ_y^2 and must find the optimal length scale l, which represents the width of the RBF function. As in the SVR case, the nth index on the kernel function indicates the nth regressor, which can have different hyperparameters, in the case of GPR with the RBF kernel, including $\Theta_{\text{GPR}} = \{\epsilon_{\sigma_n^2}, l_n\}_{n=1}^{N}$. Regarding hyperparameter optimization, one can perform cross-validation as described in section B.4.3 in appendix B or Bayesian model selection via ML maximization (i.e., type II ML) techniques [118]. Note that, apart from the hyperparameters Θ_{GPR}, one must also keep the training data set $\{\mathbf{X}, \dot{\mathbf{X}}\}$ to perform regression with equation (3.8).

3.1.1.4 Unstable DS regressors

As illustrated in figure 3.1, by employing the standard regression algorithms described here (i.e., GMR equation (3.4), SVR equation (3.7) and GPR equation (3.8)), we obtain an unstable DS. Specifically, the resulting DSs (figure 3.1) are unstable, as they:

1. Diverge in regions of the state space where no data were collected.

2. Do not guarantee that the motion generated by the DS will stop at the desired target.

Such instabilities arise as the learned regressive function parameters (i.e., $\Theta_{\text{GMR}}, \Theta_{\text{SVR}}$ or Θ_{GPR}), are not constrained to guarantee the absence of 1 and 2. In other words, they are not forced to ensure the global asymptotic stability (GAS) constraints expressed in equation (3.2).

In this chapter, we enforce the stability constraints expressed in equation (3.2) for a DS mapping function $f(x; \Theta_*): \mathbb{R}^N \to \mathbb{R}^N$ by imposing conditions on the learned parameters Θ_* derived via Lyapunov's second method for GAS [137].

Programming Exercise 3.1 *The aim of this instructional exercise is to familiarize readers with the standard regression algorithms used for DS learning. Readers will be able to reproduce the plots illustrated in figure 3.1 and generate new DS from hand-drawn or pre-collected datasets.*

Instructions: *Open MATLAB and set the directory to the folder corresponding to chapter 3 exercises. In this folder, you will find, among others, the following MATLAB scripts:*

```
1  ch3_ex1_gmrDS.m
2  ch3_ex1_svrDS.m
3  ch3_ex1_gprDS.m
```

In these scripts, you will find instructions in the block comments that will enable you to:

- *Draw trajectories via a graphical user interface (GUI) or from pre-collected 2D and three-dimensional (3D) data sets.*

- *Learn the first-order DS from the drawn/loaded trajectories with GMR, SVR, and GPR.*

Readers should modify the code in the scripts corresponding to each regression technique (GMR, SVR, GPR) to achieve the following:

1. Find the best optimization technique (GMR/SVP?GPR) and set of hyperparameters that minimize the $\mathrm{RMSE} = \sqrt{\frac{1}{L} \sum_{i=1}^{L} ||f(x^i) - \dot{x}^i||_2}$ *between the observed velocities and the predicted ones from the learned DS.*

2. Find the set (or sets) of hyperparameters that yield the best trade-off between RMSE and stability. This can be achieved by manual exploration or a post-hoc verification technique.

3. Modify the number of datapoints L *by increasing/decreasing the number of trajectories and samples per trajectory. Analyze the sensitive of each technique when* L *changes wrt. RMSE, model complexity and stability.*

3.1.2   Lyapunov Theory for Stable DSs

Theorem 3.1 (GAS [137])   *A function* $\dot{x} = f(x)$ *is a DS that is GAS at the attractor* $x^* \in \mathbb{R}^N$, *if there exists a continuous and continuously differentiable Lyapunov candidate function* $V(x) : \mathbb{R}^N \to \mathbb{R}$ *that is radially unbounded; that is,* $||x|| \to \infty \Rightarrow V(x) \to \infty$ *and satisfies the following conditions:*

$$(I)\ V(x^*) = 0, \quad (II)\ V(x) > 0 \ \forall \ x \in \mathbb{R}^N \setminus x = x^*$$
$$(III)\ \dot{V}(x^*) = 0, \quad (IV)\ \dot{V}(x) < 0 \ \forall \ x \in \mathbb{R}^N \setminus x = x^* \tag{3.9}$$

Intuitively, theorem 3.1 states that for any DS of the form given in equation (3.2) to be GAS there should be a corresponding energy-like function $V(x)$, as the one depicted in figure 3.2 (left), which should be nonincreasing along all trajectories of $f(x)$. In other words, the DS should always be dissipating energy, with zero energy at target x^*. Condition (IV) can also be interpretated geometrically. Namely, the vector field produced by $f(x)$ should always point toward the lower-level sets of the Lyapunov function $V(x)$. In other words, if

$$\dot{V}(x) = \frac{\partial V(x)}{\partial x} f(x) < 0, \tag{3.10}$$

then the angle between the gradient of the Lyapunov function $\nabla V(x) = \frac{\partial V(x)}{\partial x}$ and the direction of motion of the vector field $f(x)$ should be obtuse, as depicted in figure 3.2 (right). This guarantees that the trajectories of $f(x)$ are directed toward lower values of the Lyapunov function $V(x)$ at all times. Hence, a GAS DS must have a corresponding positive definite function $V(x)$ ensuring that conditions (I–IV) are met.

In control theory, the most common Lyapunov function used for linear and/or linearized DS [137, 70] is the quadratic Lyapunov function, (QLF)—that is,

$$V(x) = \frac{1}{2}(x - x^*)^T (x - x^*). \tag{3.11}$$

A 2D example of equation (3.11) is depicted in figure 3.2 (left).

Next, we showcase the derivation of sufficient stability conditions from equation (3.11) for a linear time-invariant (LTI) DS of the following form:

$$\dot{x} = f(x)$$
$$= Ax + b, \tag{3.12}$$

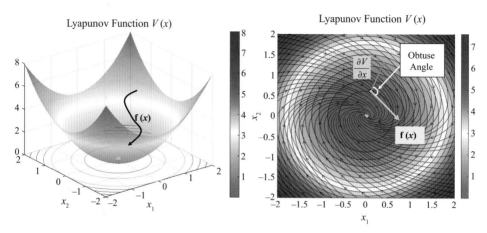

Figure 3.2
Left: Illustration of a Lyapunov function. *Right:* Geometrical intuition of theorem 3.1.

where $b \in \mathbb{R}^N$ can be seen as an offset or translation of the origin of the DS, and $A \in \mathbb{R}^{N \times N}$ is the linear system matrix that defines the dynamics of the LTI system.

Proposition 3.1   *The linear DS defined in equation* (3.2) *is GAS at the attractor* x^* *if,*

$$\begin{cases} b = -Ax^* \\ A^T + A \prec 0, \end{cases} \tag{3.13}$$

where $\prec$ refers to negative definiteness of a matrix, respectively. A matrix A *is deemed negative definite if its symmetric part* $\tilde{A} = \frac{1}{2}(A^T + A)$ *has all negative eigenvalues.*[1]

Proof Proposition 3.1 can be proved if there is a continuous and continuously differentiable Lyapunov function $V(x) : \mathbb{R}^N \to \mathbb{R}$ such that $V(x) > 0$, $\dot{V}(x) < 0 \ \forall \ x \neq x^*$ and $V(x^*) = 0, \dot{V}(x^*) = 0$. By considering equation (3.11) as the candidate Lyapunov function, we can ensure that $V(x) > 0$ due to its quadratic form. The negative definite condition follows by taking the derivative of $V(x)$ with respect to time and expanding it as shown here:

$$\dot{V}(x) = \frac{\partial V(x)}{\partial x} \frac{\partial x}{\partial t} = \frac{1}{2} \frac{\partial}{\partial x} \left((x - x^*)^T (x - x^*) \right) \dot{x}$$

$$= (x - x^*)^T f(x)$$

$$= (x - x^*)^T \underbrace{(Ax + b)}_{\text{Via } (3.12)} \tag{3.14}$$

$$= (x - x^*)^T \underbrace{A}_{\prec 0 \ (3.13)} \underbrace{(x - x^*)}_{\text{Via } (3.13)} < 0 \quad \forall x \in \mathbb{R}^N \quad \& \quad x \neq x^*$$

By substituting $x = x^*$ into equations (3.11) and (3.14):

$$V(x^*) = \frac{1}{2}(x - x^*)^T (x - x^*)\big|_{x = x^*} = 0 \tag{3.15}$$

Stable Linear DS $\dot{x} = Ax + b$

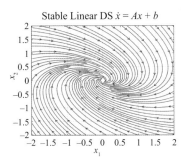

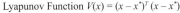

Lyapunov Function $V(x) = (x - x^*)^T (x - x^*)$

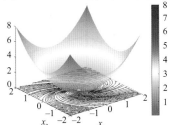

Lyapunov Function Derivative $\dot{V}(x)$

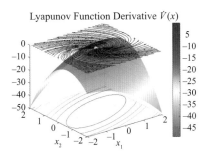

Figure 3.3
Top: GAS linear DS as equation (3.12); *middle:* corresponding Lyapunov function $V(x) = (3.11)$; *bottom:* derivative of Lyapunov function $\dot{V}(x)$ with regard to equation (3.12)

$$\dot{V}\left(x^*\right) = \left(x - x^*\right)^T A \left(x - x^*\right)\big|_{x=x^*} = 0. \tag{3.16}$$

Therefore, the DS given by (3.12) is GAS at the attractor x^*, if the conditions stated in (3.13) are satisfied. $\qquad\square$

Notice that by substituting the first condition from equation (3.13) into equation (3.12) reduces it to

$$\begin{aligned} \dot{x} &= f(x) \\ &= A(x - x^*). \end{aligned} \tag{3.17}$$

Hence, for a linear DS to be GAS in the sense of Lyapunov (theorem 3.1) with regard to a QLF [equation (3.11)], one solely requires the real eigenvalues of the symmetric part of matrix A to be negative; that is, $\tilde{A} = \frac{1}{2}(A + A^T)$ must be Hurwitz.

Often, however, using a QLF such as equation (3.11) to ensure the stability of a DS might result in overly conservative constraints that restrict the complexity of the dynamics

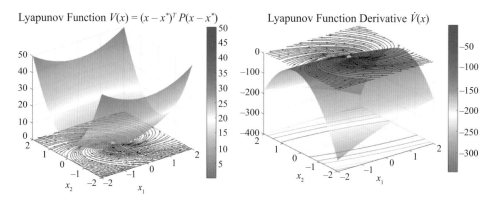

Figure 3.4
(*left*) P-QLF Lyapunov function V(x) = equation (3.18); (*right*) derivative of Lyapunov function $\dot{V}$(x) wrt. (3.12)

of $f(x)$. To alleviate this, another common Lyapunov function can be used, the *parametrized quadratic Lypaunov function (P-QLF)*, which has the following form:

$$V(x) = (x - x^*)^T P(x - x^*), \tag{3.18}$$

where $P \in \mathbb{S}_N^+ \subset \mathbb{R}^{N \times N}$ is a symmetric positive definite matrix that reshapes a QLF [equation (3.18)]; see figure 3.4 (left). Such reshaping offers a less strict stability condition that permits stronger nonlinearities in the dynamics of the corresponding DS $f(x)$. In this case, rather than enforcing the eigenvalues of the symmetric part of matrix A to be negative, the sufficient conditions described next must hold.

Proposition 3.2   *The linear DS defined in equation* (3.12) *is GAS at the attractor* x* *if*

$$\begin{cases} b = -Ax^* \\ A^T P + PA = Q, \quad P = P^T \succ 0, \quad\quad Q = Q^T \prec 0, \end{cases} \tag{3.19}$$

where $\prec$ *(*$\succ$*) refer to negative (positive) definiteness of a matrix, respectively.*

Proof *Following the proof of proposition 3.1, by considering equation (3.18) as the candidate Lyapunov function, we can ensure that* V(x) > 0. *The negative definite condition is obtained by taking the derivative of* V(x) *with regard to time and expanding it as follows:*

$$\begin{aligned} \dot{V}(x) = \frac{\partial V(x)}{\partial x}\frac{\partial x}{\partial t} &= \dot{x}^T P(x - x^*) + (x - x^*)^T P^T \dot{x} \\ &= f(x)^T P(x - x^*) + (x - x^*)^T P f(x) \\ &= \underbrace{(Ax + b)^T}_{\text{Via equation (3.12)}} P(x - x^*) + (x - x^*)^T P \underbrace{(Ax + b)}_{\text{Via equation (3.12)}} \\ &= \underbrace{(x - x^*)^T A^T}_{\text{Via equation (3.19)}} P(x - x^*) + (x - x^*)^T P \underbrace{A(x - x^*)}_{\text{Via equation (3.19)}} \\ &= (x - x^*)^T \underbrace{\left(A^T P + PA\right)}_{Q \prec 0 \text{ via equation (3.19)}} (x - x^*) < 0 \quad \forall x \in \mathbb{R}^N \quad \& \quad x \neq x^*. \end{aligned} \tag{3.20}$$

Following the proof from proposition 3.1, by substituting $x = x^*$ *into equations (3.12) and (3.20), we ensure that* $V(x^*) = 0$, $\dot{V}(x^*) = 0$. *Therefore, the linear DS given by equation (3.12) is GAS with respect to an attractor* x^* *if the conditions stated in equation (3.19) are satisfied.* □

Notice that the conditions expressed in proposition 3.2 require both $P \in \mathbb{R}^{N \times N}$ and $Q \in \mathbb{R}^{N \times N}$ matrices to be symmetric. Further, given $P \succ 0$ and $Q \prec 0$, the linear DS will be GAS if the real eigenvalues of the matrix A are negative; that is, A must be Hurwitz, yet $\tilde{A} = \frac{1}{2}(A + A^T)$ need not be. This allows more flexibility for the design of the dynamics matrix A. Yet finding the matrices P and Q that ensure equation (3.19) for a desired A is not a trivial task, even for a simple linear DS such as equation (3.12).

In sections 3.3 and 3.4, we will present approaches to design matrices $A, P, Q \in \mathbb{R}^{N \times N}$ with different optimization techniques. The next sub-section will introduce the nonlinear DS formulation used in this chapter.

Exercise 3.1 *Find the combination of matrices* $P \in \mathbb{R}^{N \times N}$ *and* $Q \in \mathbb{R}^{N \times N}$ *that reduce proposition 3.2 to proposition 3.1.*

Exercise 3.2 *Design a matrix* $A \in \mathbb{R}^{2 \times 2}$ *and Lyapunov function shaping matrix* $P \in \mathbb{R}^{2 \times 2}$ *to ensure that a linear DS,* $\dot{x} = f(x) = A(x - x^*)$, *with attractor at the origin* $x^* = [0\ \ 0]^T$ *to be GAS wrt. the conditions stated using either:*

(a) a matrix $Q \in \mathbb{R}^{2 \times 2}$ *with the following form:*

$Q = q\mathbb{I}_2$, $q \in \mathbb{R}$

(b) a matrix $Q \in \mathbb{R}^{2 \times 2}$ *with the following form:*

$$Q = \begin{bmatrix} q_1 & q_2 \\ q_2 & q_1 \end{bmatrix}, \quad q_1, q_2 \in \mathbb{R}$$

HINT: What conditions should the entries of the Q matrices satisfy?

3.2 Nonlinear DSs as a Mixture of Linear Systems

The Lyapunov stability conditions derived from a QLF or P-QLF in propositions 3.1 or 3.2, respectively, solely hold for linear DS as equation (3.12). In robotics, we often require the robots to perform nonlinear motions that cannot be represented by a single linear DS matrix $A \in \mathbb{R}^{N \times N}$. Stability analysis of nonlinear DS is still an open question, and theoretical solutions exist only for particular cases. In this chapter, we offer several solutions to this problem by formulating a nonlinear DS as a mixture of linear systems of the following form:

$$\dot{x} = f(x)$$

$$= \sum_{k=1}^{K} \gamma_k(x) \left(A^k x + b^k \right), \tag{3.21}$$

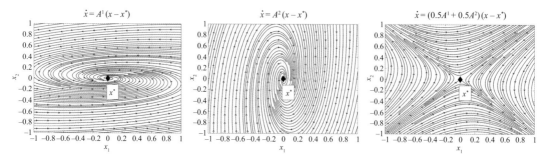

Figure 3.5
While each subsystem (left) $\dot{x} = A^1(x - x^*)$ and (center) $\dot{x} = A^2(x - x^*)$ is GAS, a weighted sum of these systems $x = \gamma_1(x)A^1(x - x^*) + \gamma_2(x)A^2(x - x^*)$ may become unstable (right). Here, the system remains stable only for points on the line $x_2 = x_1$.

where $A^k \in \mathbb{R}^{N \times N}$ and $b^k \in \mathbb{R}^N$ are the kth linear system parameters. Here, $\gamma_k(x) : \mathbb{R}^N \to \mathbb{R}_+$ is the *mixing* function, also referred to as the *activation* function. Such state-dependent function defines in which regions of the state-space the kth linear DS must be activated. Observe that $f(x)$ is now expressed as a nonlinear sum of linear DS. By formulating a nonlinear DS as equation (3.21), we can derive Lyapunov stability conditions akin to those derived in section 3.1.2 with QLFs and P-QLFs. In doing so, one might be tempted to directly apply proposition 3.2 to equation (3.21) by following the intuition that the nonlinear DS $f(x)$ should be stable if all eigenvalues of matrices $A^k \; \forall k = 1..K$ have strictly negative real parts. This is not true, however, as will be illustrated next.

Example 3.1 *Consider a 2D, nonlinear DS expressed as equation (3.21) composed of $K = 2$ linear systems, with the attractor at the origin—that is, $\mathrm{x}^* = [0 \; 0]^\mathrm{T}$, and*

$$\begin{cases} A^1 = \begin{bmatrix} -1 & -10 \\ 1 & -1 \end{bmatrix}, \quad A^2 = \begin{bmatrix} -1 & 1 \\ -10 & -1 \end{bmatrix} \\ b^1 = b^2 = 0 \\ \gamma_1(x) = \gamma_2(x) = 0.5 \end{cases} \tag{3.22}$$

The eigenvalues of the matrices A^1 and A^2 are $-1 \pm 3.16i$. Hence, given a $P \succ 0$ and $Q \prec 0$ each matrix determines a stable linear system according to proposition 3.2. However, the combination of the two matrices is stable only when $\mathrm{x}_2 = \mathrm{x}_1$, and is unstable everywhere else in the state-space—that is, $\mathbb{R}^N \setminus \{(\mathrm{x}_2, \mathrm{x}_1) \,|\, \mathrm{x}_2 = \mathrm{x}_1\}$. See figure 3.5 for illustrations.

Exercise 3.3 *Consider a DS of the form $\dot{x} = \sum_{k=1}^{K} \gamma_k(x)(A^k x + b^k)$ with $N = 2$, $K = 2$, and $\gamma_1(\mathrm{x}) = 0.7$ and $\gamma_2(\mathrm{x}) = 0.3$. Design nondiagonal matrices $A^1 \in \mathbb{R}^{2 \times 2}$ and $A^2 \in \mathbb{R}^{2 \times 2}$ such that the DS is GAS at the origin—that is, $\mathrm{x}^* = [0 \; 0]^\mathrm{T}$, with a QLF.*

Exercise 3.4   *Consider a DS of the form* $\dot{x} = \sum_{k=1}^{K} \gamma_k(x)(A^k x + b^k)$ *with* N = 2, K = 2, *and* $\gamma_1(x) = 0.5$ *and* $\gamma_2(x) = 0.5$. *Design nondiagonal matrices* $A^1 \in \mathbb{R}^{2\times2}$ *and* $A^2 \in \mathbb{R}^{2\times2}$ *such that it is GAS at the origin—that is,* $x^* = [0 \ 0]^T$, *with a P-QLF with*

$$P = \begin{bmatrix} 2 & 0 \\ 0 & 2 \end{bmatrix}, Q = \begin{bmatrix} -1 & 0 \\ 0 & -1 \end{bmatrix}.$$

Programming Exercise 3.2   *This exercise will allow the reader to understand the complexity in designing globally asymotically stable DS as equation (3.21) by designing and visualizing DS as those defined in example 3.1.*

Instructions: *Open MATLAB, set the directory to to the chapter 3 exercises folder. By following the instructions in the following MATLAB script:*

```
1   ch3_ex2_designDS.m
```

The reader will be able to reproduce the illustrations from figure 3.5. Further, we recommend the following:

- *Design values of* $\gamma_1, A^1, \gamma_2, A^2$ *such that a globally asympotically stable system is generated. One can use this code as support to solve exercises 3.3 and 3.4.*
- *Add a third linear DS, K = 3, can you find the full set of parameters to ensure GAS? How about simply stability? (See appendix A.)*

Next, we introduce learning algorithms, based on GMM and GMR, to automatically design stable parameters of a first-order, nonlinear DS formulated as equation (3.21) with QLF (section 3.3) and with P-QLF (section 3.4).

3.3   Learning Stable, Nonlinear DSs

It is straightforward to draw parallels between the nonlinear DS formulation [equation (3.21)] and the Gaussian mixture regressor equation presented in equation (3.4). Namely, the nonlinear DS formulation is a *mixture* of linear DS, while the Gaussian mixture regressor is a *mixture* of linear regressors. In this section, we present an approach for learning nonlinear DS that exploits this resemblance by formulating a constrained GMR learning algorithm referred to as the *stable estimator of DSs (SEDS)*, originally published in [72]).

3.3.1   Constrained Gaussian Mixture Regression

Let us begin by defining the nonlinear DS parameters of equation (3.21) [i.e., $\Theta_{f(x)} = \{\gamma_k(x), A^k, b^k\}_{k=1}^{K}$], via GMR. Next, we restate and expand the GMR regressor equation [equation (3.4)]; that is, the posterior mean of the conditional distribution $p(\dot{x}|x)$ as follows:

$$\dot{x} = f(x; \; \Theta_{\text{GMR}})$$

$$= \mathbb{E}\{p(\dot{x}|x)\} = \sum_{k=1}^{K} \frac{\pi_k p(x|\mu_x^k, \Sigma_x^k)}{\sum_{i=1}^{K} \pi_i p(x|\mu_x^i, \Sigma_x^i)} \left(\mu_{\dot{x}}^k + \Sigma_{\dot{x}x}^k \left(\Sigma_x^k \right)^{-1} \left(x - \mu_x^k \right) \right). \tag{3.23}$$

This expanded GMR regressor equation can be simplified through a change of variables. Let us define this as follows:

$$\begin{cases} A^k = \Sigma_{\dot{x}x}^k \left(\Sigma_x^k \right)^{-1} \\ b^k = \mu_{\dot{x}}^k - A^k \mu_x^k \\ \gamma_k(x) = \frac{\pi_k p(x|\mu_x^k, \Sigma_x^k)}{\sum_{i=1}^{K} \pi_i p(x|\mu_x^i, \Sigma_x^i)}. \end{cases} \tag{3.24}$$

Substituting equation (3.24) into equation (3.23) yields

$$\dot{x} = f(x; \; \Theta_{\text{GMR}})$$

$$= \mathbb{E}\{p(\dot{x}|x)\} = \sum_{k=1}^{K} \gamma_k(x) \left(A^k x + b^k \right). \tag{3.25}$$

As can be seen, equation (3.25) is identical to the mixture of linear DS formulations introduced in equation (3.21), but with each DS parameter encoded via equation (3.24). Using such reparametrization, we can analyze the influence of the GMR parameters, $\Theta_{GMR} = \{\pi_k, \mu^k, \Sigma^k\}_{k=1}^{K}$, on the resulting DS. As illustrated in figure 3.6, for a one-dimensional (1D) model constructed with three Gaussians, each linear dynamics $A^k x + b^k$ corresponds to a line that passes through the centers of the Gaussian μ^k with slope A^k that is determined by Σ_{xx}^k (the variance of x) and $\Sigma_{\dot{x}x}^k$ (the covariance of x and $\dot{x}$). This interpretation scales to higher dimensions. Further, the nonlinear weighting terms, $\gamma_k(x)$, where $0 < \gamma_k(x) \leq 1$ and $\sum_{k=1}^{K} \gamma_k(x) = 1$, give a measure of the relative influence of each Gaussian locally. In other words, they define the regions in the state space in which a kth linear DS has more weight or importance.

Until now, we have reparametrized the GMR regressor equation [equation (3.4)] such that it is written in the form of equation (3.21). This does not guarantee that the learned DS will be GAS at a target x^*. To guarantee such a convergence property, we propose the following sufficient conditions derived from a QLF equation (3.11).

Theorem 3.2   *The nonlinear DS defined in equation (3.25), parametrized by $\Theta_{\text{GMR}} = \{\pi_k, \mu^k, \Sigma^k\}_{k=1}^{K}$ via equation (3.24), is GAS at the target $x^* \in \mathbb{R}^N$ if*

$$\begin{cases} b^k = -A^k x^* \\ A^k + \left(A^k \right)^T \prec 0 \end{cases} \qquad \forall k = 1 \ldots K, \tag{3.26}$$

where $\left(A^k \right)^T$ is the transpose of A^k, and $\prec 0$ refers to the negative definiteness of a matrix.

Proof *Theorem 3.2 can be proved if there exists a continuous and continuously differentiable Lyapunov function $V(x) : \mathbb{R}^N \to \mathbb{R}$ such that $V(x) > 0$, $\dot{V}(x) < 0 \; \forall \; x \neq x^*$ and*

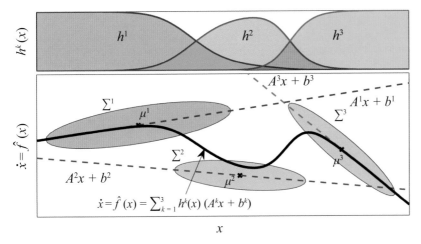

Figure 3.6
Illustration of parameters defined in equation (3.24) and their effects on $f(x)$ for a 1D model constructed with three Gaussians [72]. Please refer to the text for further information.

$V(x^*) = 0$, $\dot{V}(x^*) = 0$ *(from theorem 3.1). By considering a QLF [equation (3.11)] as the candidate Lyapunov function, we can ensure* $V(x) > 0$ *due to its quadratic form. The negative definite condition follows by taking the derivative of* $V(x)$ *with regard to time and expanding it as shown here:*

$$\dot{V}(x) = \frac{\partial V(x)}{\partial x} \frac{\partial x}{\partial t} = \frac{1}{2} \frac{\partial}{\partial x} \left(\left(x - x^* \right)^T \left(x - x^* \right) \right) \dot{x}$$

$$= \left(x - x^* \right)^T f(x)$$

$$= \left(x - x^* \right)^T \underbrace{\sum_{k=1}^{K} \gamma_k(x) \left(A^k x + b^k \right)}_{\text{via (3.21)}}$$

$$= \left(x - x^* \right)^T \sum_{k=1}^{K} \gamma_k(x) \underbrace{A^k \left(x - x^* \right)}_{\text{via (3.26)}}.$$

(3.27)

Rearranging terms in equation (3.27) and using the conditions defined in equation (3.6) for the GMR mixing functions $\gamma_k(x)$ *and equation (3.26) for the* A^k *matrices yields*

$$\dot{V}(x) = \sum_{k=1}^{K} \underbrace{\gamma_k(x)}_{>0 \text{ via (3.6)}} \left(x - x^* \right)^T \underbrace{A^k}_{<0 \text{ via (3.26)}} \left(x - x^* \right)$$

$$= \sum_{k=1}^{K} \underbrace{\gamma_k(x)}_{\gamma_k > 0} \underbrace{\left(x - x^* \right)^T A^k \left(x - x^* \right)}_{<0} \quad < 0 \quad \forall x \in \mathbb{R}^N \quad \& \quad x \neq x^*$$

(3.28)

Furthermore, following the proof from proposition 3.1, by substituting $x = x^*$ *into the candidate Lyapunov function* $V(x)$ *[equation (3.11)] and its time derivative* $\dot{V}(x)$ *[equation (3.28)] yields*

$$V\left(x^*\right) = \frac{1}{2}\left(x - x^*\right)^T \left(x - x^*\right)\Big|_{x=x^*} = 0 \tag{3.29}$$

$$\dot{V}\left(x^*\right) = \sum_{k=1}^{K} \gamma_k(x)\left(x - x^*\right)^T A^k \left(x - x^*\right)\Big|_{x=x^*} = 0. \tag{3.30}$$

Therefore, the nonlinear DS given by equation (3.25) is GAS at the attractor x^*, *if the conditions stated in equation (3.26) are satisfied.* $\qquad\square$

Exercise 3.5 *Consider the nonlinear DS* $\dot{x} = \mathbb{E}\{p(\dot{x}|x)\} = \sum_{k=1}^{K} \gamma_k(x)$ $\left(A^k x + b^k\right)$ *with* $N = 1$, $K = 4$. *Design the GMR parameters* $\Theta_{\text{GMR}} = \{\pi_k, \mu^k, \Sigma^k\}_{k=1}^{2}$ *that will produce a GAS DS at the target* $x^* = 0$ *such that:*

- *If* $x = -1$, *then* $\dot{x} = 0.5$; *and when* $x = -2$, *then* $\dot{x} = 2$.
- *If* $x = +1$, *then* $\dot{x} = -0.5$; *and when* $x = +2$, *then* $\dot{x} = -2$.

Exercise 3.6 *Considering the nonlinear DS* $\dot{x} = \mathbb{E}\{p(\dot{x}|x)\} = \sum_{k=1}^{K} \gamma_k(x)$ $\left(A^k x + b^k\right)$. *Derive sufficient conditions for GAS with a P-QLF.*

3.3.2 Stable Estimator of DSs

Section 3.3.1 provided sufficient conditions for $f(x)$, defined by equation (3.25), to be GAS at a target x^*. We now introduce a procedure for computing the unknown parameters of equation (3.25); that is, $\Theta_{\text{GMR}} = \{\pi_k, \mu^k, \Sigma^k\}_{k=1}^{K}$, which we use to describe the Stable estimator of Dynamical Systems (SEDS). SEDS is a learning algorithm that computes the optimal values of Θ_{GMR} by solving an optimization problem under the GAS constraints defined in equation (3.26).

Due to its probabilistic nature, the parameters for a GMR regressor, Θ_{GMR}, are generally estimated either by maximizing the likelihood or the posterior probability of the GMM representing the joint distribution of inputs/outputs, $p(x, \dot{x})$. In other words, they seek to estimate the parameters of the joint probability distribution that best describes the observed data. Other regression algorithms, on the other hand, estimate their parameters by minimizing the error between the observed and the estimated outputs (e.g., SVR as described in section 3.1.1.2). To estimate the hyperparameters of nonparametric algorithms like GPR (see section 3.1.1.3), one can either follow the probabilistic route (i.e., maximizing likelihood/posterior) or use an external metric that minimizes the regression error. As summarized in section B.2 of appendix B, the MSE and its variants are the most commonly used metrics to evaluate regression algorithms. We, thus, consider two candidates for the objective function of the SEDS algorithm: log-likelihood, and MSE.

3.3.2.1 SEDS [Likelihood]

The SEDS [Likelihood] objective function follows the ML parameter estimation approach for GMM. As previously discussed, the parameters of a GMM are typically estimated by an EM algorithm that finds the optimimum of the likelihood of the GMM. This is done by maximizing the log-likelihood, defined in equation (B.22), via an iterative scheme (see section B.3.1 in appendix B for EM descriptions). Hence, in this SEDS variant, we maximize the log-likelihood of our reference trajectories $\{\mathbf{X}, \dot{\mathbf{X}}\} = \{X^m, \dot{X}^m\}_{m=1}^M = \{\{x^{t,m}, \dot{x}^{t,m}\}_{t=1}^{T_m}\}_{m=1}^M$ being described by a GMM with parameter Θ_{GMR} subject to the stability constraints defined in equation (3.26) as follows:

$$\min_{\Theta_{\text{GMR}}} J(\Theta_{\text{GMR}}) = -\frac{1}{L} \sum_{m=1}^M \sum_{t=0}^{T_m} \log p\left(x^{t,m}, \dot{x}^{t,m} | \Theta_{\text{GMR}}\right) \tag{3.31}$$

subject to

$$\begin{cases}
\textbf{(a)} \ b^k = -A^k x^* \\
\textbf{(b)} \ A^k + \left(A^k\right)^T \prec 0 \\
\textbf{(c)} \ \Sigma^k \succ 0 \qquad \forall \, k = 1, \ldots, K \\
\textbf{(d)} \ 0 < \pi_k \leq 1 \\
\textbf{(e)} \ \sum_{k=1}^K \pi_k = 1,
\end{cases} \tag{3.32}$$

where $p\left(x^{t,m}, \dot{x}^{t,m} | \Theta_{\text{GMR}}\right)$ is the probability distribution function (PDF) of the GMM given by equation (3.3) and $L = \sum_{m=1}^M T_m$ is the total number of training data points from the reference trajectories $\{\mathbf{X}, \dot{\mathbf{X}}\}$. Constraints (a–b) in equation (3.32) are stability conditions from theorem 3.2. Constraints (c–e) are imposed by the nature of the GMM to ensure that all Σ^k are positive definite matrices, prior π^k values are positive scalars smaller than or equal to 1, and the sum of all the priors is equal to 1. This last constraint is necessary for the PDF of the GMM to be statistically sound (i.e., the probability of a point belonging to the kth Gaussian cannot be > 1). Thanks to this property, the mixing/activation function $\gamma(x)$ yields a weighted sum of linear DS for equation (3.25).

3.3.2.2 SEDS [MSE]

The SEDS [Likelihood] variant can be seen as an alternative to the standard EM algorithm, but subject to constraints. To follow the parameter estimation approaches for other regression algorithms (i.e., GPR/SVR), we propose an alternative objective function based on the MSE:

$$\min_{\Theta_{\text{GMR}}} J(\Theta_{\text{GMR}}) = \frac{1}{2L} \sum_{m=1}^M \sum_{t=0}^{T_m} \left\| f(x^{t,m}) - \dot{x}^{t,m} \right\|^2 . \tag{3.33}$$

This optimization problem seeks to minimize equation (3.33) subject to the same constraints given by equation (3.32). Furthermore, the values of $f\left(x^{t,m}\right)$ in equation (3.33) are computed directly from equation (3.25).

3.3.2.3   SEDS parameter optimization

Both SEDS [Likelihood] and SEDS [MSE] can be formulated as nonlinear programming (NLP) problems [10], and thus can be solved using standard constrained optimization techniques. In this section, we propose to use a successive quadratic programming (SQP) approach that relies on a quasi-Newton method to solve the optimization problems defined in equations (3.31) and (3.33) and constrained by equation (3.32).

Implementation details: SQPs minimize a quadratic approximation of the Lagrangian of an NLP with a linear approximation of the constraints. Given the derivative of the constraints and the first- (gradient) and second-order (Hessian matrix) derivatives of the cost function with regard to to the optimization parameters, SQP methods find a proper descent direction (if any) that minimizes the cost function while not violating the constraints. Next, we provide details on the proposed method to solve the SEDS optimization problems:

• *Equality constraints:* The SQP finds a descent direction that minimizes the cost function by varying the parameters on the hypersurface satisfying the equality constraints.

• *Inequality constraints:* The SQP follows the gradient direction of the cost function whenever the inequality holds (inactive constraints). Only at the hypersurface (where the inequality constraints becomes active) does SQP search for a descent direction that either minimizes the cost function by varying the parameters on the hypersurface or points it toward the inactive constraint domain.

• *Cost function derivatives:* Rather than computing the Hessian matrix of second-order derivatives explicitly, for SEDS, we use a quasi-Newton method that computes an estimate of the Hessian matrix via the gradients. We use the Broyden-Fletcher-Goldfard-Shanno (BFGS) algorithm [10] as the Hessian approximation method.

The SQP implementation proposed for SEDS has several advantages over general-purpose solvers. First, we have an analytic expression of the gradients that improves performance significantly, allowing us to employ the BFGS algorithm. Second, the implemented code is tailored to solve the specific problem at hand. Namely, one can simplify the SEDS optimization paramaters, $\Theta_{\mathrm{GMR}} = \{\pi_k, \mu^k, \Sigma^k\}_{k=1}^{K}$, via a change of variables in which the Cholesky decomposition of the covariance matrices is used rather than the full covariance matrices. Such reformulation guarantees that the optimization constraints (**a**), (**c**), (**d**), and (**e**), as defined in equation (3.32), are satisfied. Hence, only constraint (**b**) from equation (3.32) must be enforced in this optimization. The analytical derivation of the gradients and the mathematical reformulation to satisfy the optimization constraints are decribed in detail in [74].

While a feasible solution to the SEDS NLP problem always exists, it is nonconvex; hence, one cannot be sure of finding the globally optimal solution. Yet with a good initialization of the parameters, the SQP solver will converge to some local minima of the objective function. One way of ensuring this would be to run the EM algorithm to obtain an initial guess of the GMM parameters to optimize, $\Theta_{\mathrm{GMR}}^{\mathrm{init}} = \{\pi_k, \mu^k, \Sigma^k\}_{k=1}^{K}$. Often, however, this initial guess might not be suitable, as it does not take into account the constraints defined in equation (3.32). To ensure this, we offer two procedures (algorithms 3.1 and 3.2) that provide means to compute a good initial guess that is fed to the SEDS solver.

Algorithm 3.1 Initialization algorithm for SEDS solver (by parameter deformation)

Require: Demonstrations $\{\mathbf{X}, \dot{\mathbf{X}}\} = \{\{x^{t,m}, \dot{x}^{t,m}\}_{t=1}^{T_m}\}_{m=1}^{M}$ and hyperparameter K

1: Run EM (section B.3.1 in Appendix B) over $\{\mathbf{X}, \dot{\mathbf{X}}\}$ to find an intial estimate of $\{\pi_k, \mu^k, \Sigma^k\}_{k=1}^{K}$.

2: Define $\tilde{\pi}_k = \pi_k$ and $\tilde{\mu}_x^k = \mu_x^k$

3: Transform the covariance matrices such that they satisfy the stability constraints given by (3.32) **(b)** and **(c)**:

$$
\begin{cases}
\tilde{\Sigma}_x^k = \mathbb{I}_N \circ \mathrm{abs}\left(\Sigma_x^k\right) \\[6pt]
\tilde{\Sigma}_{\dot{x}x}^k = -\mathbb{I}_N \circ \mathrm{abs}\left(\Sigma_{\dot{x}x}^k\right) \\[6pt]
\tilde{\Sigma}_{\dot{x}}^k = \mathbb{I}_N \circ \mathrm{abs}\left(\Sigma_{\dot{x}}^k\right) \\[6pt]
\tilde{\Sigma}_{x\dot{x}}^k = -\mathbb{I}_N \circ \mathrm{abs}\left(\Sigma_{x\dot{x}}^k\right)
\end{cases}
\qquad \forall k = 1..K,
\tag{3.34}
$$

where $\circ$ and abs(.) corresponds to entrywise product and absolute value function.

4: Compute $\tilde{\mu}_{\dot{x}}^k$ by solving the optimization constraint given by (3.32) **(a)**:

$$
\tilde{\mu}_{\dot{x}}^k = \tilde{\Sigma}_{\dot{x}x}^k \left(\tilde{\Sigma}_x^k\right)^{-1} \left(\tilde{\mu}_x^k - x^*\right)
\tag{3.35}
$$

Ensure: $\Theta_{\mathrm{GMR}}^{\mathrm{init}} = \left\{\tilde{\pi}_k, \tilde{\mu}^k, \tilde{\Sigma}^k\right\}_{k=1}^{K}$.

Algorithm 3.2 Initialization algorithm for SEDS solver (by kth parameter estimation)

Require: Demonstrations $\{\mathbf{X}, \dot{\mathbf{X}}\} = \{\{x^{t,m}, \dot{x}^{t,m}\}_{t=1}^{T_m}\}_{m=1}^{M}$ and hyperparameter K

1: Run EM (section B.3.1 in Appendix B) over $\{\mathbf{X}, \dot{\mathbf{X}}\}$ to find an intial estimate of $\{\pi_k, \mu^k, \Sigma^k\}_{k=1}^{K}$.

2: Define $\tilde{\pi}_k = \pi_k$

3: **for** k = 1 to K **do**

4:    - Extract $\{\mathbf{X}^k, \dot{\mathbf{X}}^k\}$: the data-points from the reference trajectories belonging to the kth Gaussian function via $p(k|x^i, \{\mu_x^k, \Sigma_x^k\})$ (see section B.3.4.1 in Appendix B).

5:    - Run the SEDS solver on the kth reference trajectories, $\{\mathbf{X}^k, \dot{\mathbf{X}}^k\}$, assuming $K = 1$ in equation (3.25) with initial parameters $\{\pi_1 = 1, \mu^1 = \mu^k, \Sigma^1 = \Sigma^k\}$.

6:    Define $\tilde{\mu}^k = \mu_{SEDS}^1$ and $\tilde{\Sigma}^k = \Sigma_{SEDS}^1$.

7: **end for**

Ensure: $\Theta_{\mathrm{GMR}}^{\mathrm{init}} = \left\{\tilde{\pi}_k, \tilde{\mu}^k, \tilde{\Sigma}^k\right\}_{k=1}^{K}$.

Hyperparameters and prespecifications: Notice that the SEDS initialization algorithm, and consequently, the SEDS learning algorithm, require the prespecification of the number of Gaussian functions K and the attractor of the DS $x^* \in \mathbb{R}^N$. While the latter is defined by the user (see the note at the end of this subsection), the optimal number of K Gaussians is identified using the standard GMM model selection scheme (see the section B.3.1.4 in appendix B). To recall, when selecting the optimal K for a GMM with model selection, we employ the Bayesian information criterion (BIC) metric to trade off a model's likelihood with the number of parameters needed to encode the data. The equation for BIC is provided in equation (B.2) and relies on the computation of the total number of learned parameters, which for a GMM fitted onto N-dimensional position and velocity pairs would be $B = K\left(1 + 3N + 2N^2\right)$ parameters—that is, the priors π_k, means μ^k, and covariance matrices Σ^k are of size $1, 2N$, and $N(2N + 1)$, respectively.

- For SEDS [Likelihood], however, the number of parameters can be reduced because the constraints given by equation (3.32) (**a**) provide an explicit formulation to compute $\mu_{\dot{x}}^k$ from other parameters (i.e., μ_x^k, Σ_x^k, and Σ_{xx}^k). Thus, the total number of parameters needed to construct a GMM with K Gaussians is $B = K(1 + 2N(N + 1))$.

- For SEDS [MSE], the number of parameters can be further reduced because when constructing $f(x)$, the term $\Sigma_{\dot{x}}^k$ is not used and is omitted during the optimization. Hence, the total number of learning parameters reduces to $B = K\left(1 + \frac{3}{2}N(N + 1)\right)$.

For both SEDS variants, learning grows linearly with K and quadratically with N.
Note: The means to automatically identify attractors, $x^* \in \mathbb{R}^N$, from demonstrations involving sequences, dynamics/attractors is provided in chapter 5.

3.3.3   Evaluating the Learning of Nonlinear DS

As with any machine learning problem, it is crucial to evaluate the performance of the learning method when estimating nonlinear DS. To evaluate this performance, we need to go back to the original objectives set forth for our learning problem. Recall that we had two objectives. Objective 1 was to provide an accurate model of the dynamics, and objective 2 was to ensure that the learned DS would be globally stable at an attractor. This latter objective does not need to be assess numerically, as it is already enforced by construction of the learned model (it was hence assessed through a formal proof). Objective 1, however, needs to be assessed numerically. The most natural means to assess that the learned function reproduces the dynamics well is through MSE as a loss function. MSE must assess the proper reproduction of velocity flow. As we will see in the rest of this chapter, learning approaches to estimating nonlinear DS differ primarily in their ability to fit the natural curvature of the nonlinear dynamics. The more curvy, the more difficult it is to fit the data while preserving the stability of the attractor. To highlight this trade-off between accuracy and stability, we will use a 2D data set of nonlinear dynamics that embeds a different level of complexity. This is known as the *LASA Handwriting Dataset*.

3.3.4   LASA Handwriting Dataset: Benchmark for Evaluating
the Learning of Stable DS

The LASA Handwriting Dataset consists of a library of 2D handwriting motions recorded from a Tablet-PC. For each motion, the user was asked to draw seven demonstrations of a desired pattern by starting from different initial positions (but fairly close to each other) and ending at the same final point. The demonstrations may intersect each other. In total, a library of thirty human handwriting motions were collected, of which twenty-six corresponded to a single dynamic behavior/motion pattern, and the remaining four each included more than one dynamic behavior/motion pattern (named Multi-Models). Figure 3.7 shows the full set of motions from the LASA Handwriting Dataset. In the accompanying MAT-LAB source code, we provide scripts to load the demonstrations from this data set, as well as means to draw your own data sets and learn N-dimensional, GAS DS models via the SEDS variants.

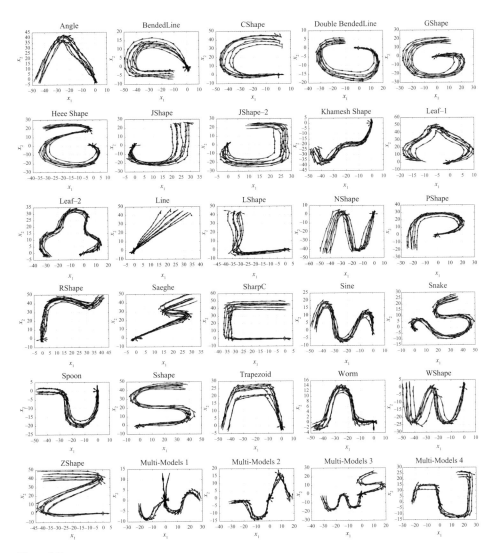

Figure 3.7
The full set of motions from the LASA Handwriting Dataset [72].

The LASA Handwriting Dataset has now become a benchmark for evaluating and comparing different approaches to learn nonlinear DSs. In this chapter, we use the data set to ease the comparison across the various approaches. We start here with an evaluation when using SEDS (extensive evaluation of the method can be found in [72]).

SEDS [Likelihood] versus SEDS [MSE] SEDS can be trained using either likelihood or MSE as the objective function. A qualitative appraisal of the performance of SEDS at modeling dynamics of a subset of the LASA Handwriting Dataset using either the [Likelihood] or [MSE] objective function can be seen in figures 3.8 and 3.9, respectively. In these two sets of plots, we show handwriting motions numbered 1 (Angle) to 15 (PShape). Further, in figures 3.10 and 3.11, we illustrate the remaining motions; that is, motions 16 (RShape) to 30 (Multi-Models 4).

Recall that the approach has one hyperparameter, the number of K Gaussian functions, which needs to be set prior to running the estimation. For each motion, the optimal K was computed using a model selection approach via the BIC criterion. Model selection with BIC consists in running the method (here SEDS) multiple times, changing the parameter K in each case. The BIC criterion is a metric that offers a trade-off between accuracy and precision of the estimate. For the sake of reproducibility, we list the obtained K for each DS used to generate the illustrations in figures 3.8 to 3.11.

Number of K Gaussians for the LASA Handwriting Dataset: Angle ($K=4$), BendedLine ($K=4$), Cshape ($K=5$), Double BendedLine ($K=5$), Gshape ($K=4$), Heee ($K=4$), Jshape ($K=5$), Jshape-2 ($K=5$), Khamesh ($K=5$), Leaf-1 ($K=4$), Leaf-2 ($K=5$), Line ($K=2$), Lshape ($K=3$), Nshape ($K=4$), Pshape ($K=4$), Rshape ($K=4$), Saeghe ($K=4$), SharpC ($K=5$), Sine ($K=4$), Snake ($K=4$), Spoon ($K=4$), Sshape ($K=6$), Trapezoid ($K=4$), Worm ($K=4$), Wshape ($K=6$), Zshape ($K=5$), Multi-Models 1 ($K=5$), Multi-Models 2 ($K=6$), Multi-Models 3 ($K=7$), and Multi-Models 4 ($K=6$).

Parameter initialization: From our experiments, the choice of initialization procedure for the optimization parameters, Θ_{GMR}, depends greatly on three factors: (1) the shape of the motion, (2) the number of K Gaussian functions and (3) the initial estimate of the GMM parameters given by the standard EM algorithm. For example, let us assume that the number of Gaussian functions K was correctly chosen, and that the standard EM algorithm perfectly fits the Gaussians on the reference trajectories such that the ML of the model is achieved. In such a case, the two initialization algorithms will enforce the stability constraints on the covariance matrices, but not drastically modify the overall set of parameters. On the other hand, when the GMM is ill fitted to the reference trajectories with the standard EM algorithm, the initialization algorithms might enforce constraints on parameters that are a bad guess and hence yield worse results. Hence, we recommend running SEDS with multiple initializations until good reproduction is achieved.

Performance analysis: The plots in figures 3.8 to 3.11 illustrate the demonstrated reference trajectories, $\{X, \dot{X}\}$, with **red** trajectories, the learned DS [equation (3.25)] represented as a **light black** vector field, and the reproduced trajectories by the learned DS with **dark black** trajectories. These reproduced trajectories are generated by simulating the forward integration of the DS, excluding a robot controller to avoid modeling errors. The simulations begin at the same initial points as all the demonstrations. Hence, to qualitatively evaluate the *accuracy* of the DS, we aim for the reproductions (the **dark black** trajectories) to be as close as possible to the demonstrated reference trajectories (the **red** trajectories). To evaluate the *stability* of the DS, we want all trajectories to lead the robot to the target/attractor. As can be seen, for all motions of the LASA Handwriting Dataset, the learned DS, be it via SEDS [Likelihood] or SEDS [MSE], is GAS.

Reproduction *accuracy*, on the other hand, depends on the motion at hand. From the thirty motions in the LASA Handwriting Dataset, the SEDS learning variants are capable of accurately reproducing the reference trajectories and/or producing similar motion

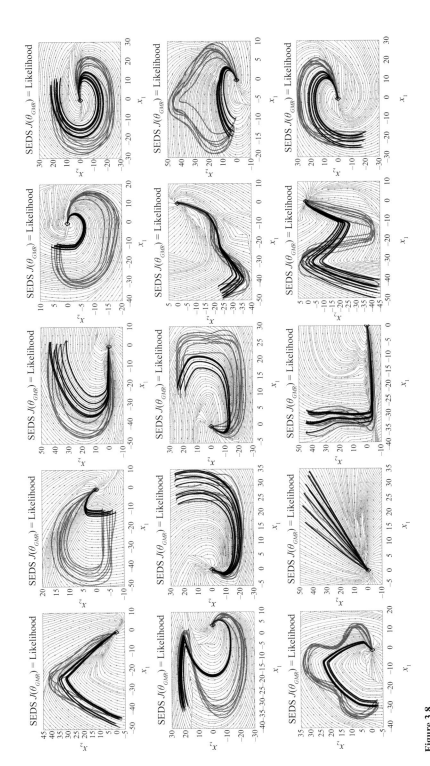

Figure 3.8
Performance of **SEDS [Likelihood]** on the LASA Handwriting Dataset (motions 1–15).

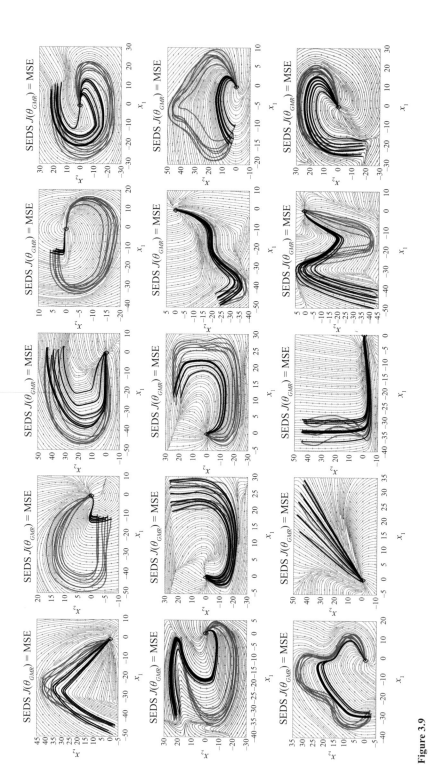

Figure 3.9
Performance of **SEDS [MSE]** on the LASA Handwriting Dataset (motions 1–15).

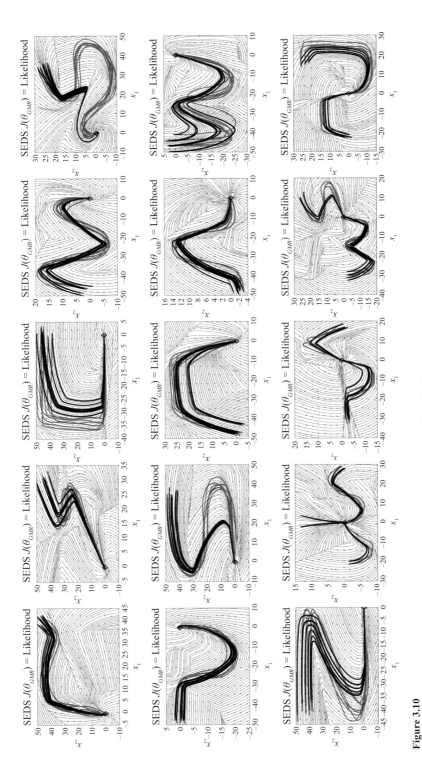

Figure 3.10
Performance of **SEDS [Likelihood]** on the LASA Handwriting Dataset (motions 16–30).

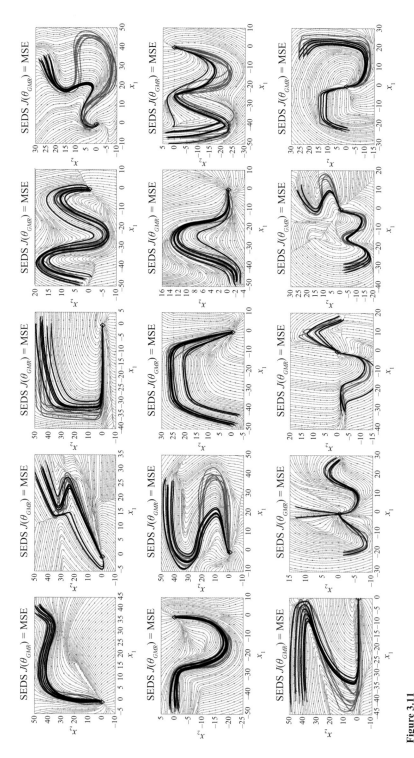

Figure 3.11
Performance of **SEDS [MSE]** on the LASA Handwriting Dataset (Motions 16–30).

patterns for the majority of motions (i.e., twenty-two), while drastically failing for eight of them (BendedLine, DoubleBendedLine, Heee, Leaf, Leaf-2, Nshape, Snake, Sshape). One can see that the subset of motions where SEDS fails exhibit (1) high curvatures, (2) motions that move away from the attractor, and (3) trajectories where the initial and end points are close to each other. The stability constraints and DS formulation presented in this section do not account for these characteristics in the motions. These will be addressed in the following sections of this chapter. Hence, performance of the SEDS variants should not be evaluated based on these motions; we simply present them here to show the limitations of the algorithm.

Considering the twenty-two motions that are accurately reproduced by the SEDS algorithm, when comparing the performance of SEDS [Likelihood] to SEDS [MSE], there are only a few differences between the two methods. These differences are in direct relation to the objective function. For example, one can see that in many cases of the LASA Handwriting Dataset, the motions generated with SEDS [MSE] are smoother than the ones generated by SEDS [Likelihood] *(Sine, Worm, Spoon, Saeghe, and Rshape)*; that is, the latter yields motions with sharper edges and linearized motion segments. SEDS [MSE] seeks to minimize the error between the velocities of the trajectories and those generated by the DS model, without considering the probabilistic or geometrical properties of the GMM parameters. In other words, the geometric intuition of the DS being a mixture of linear regressors *placed precisely on linear segments* of the nonlinear trajectory (see figure 3.6) is lost. This is because the means μ^k and covariances Σ^k are optimized to minimize the [MSE] loss function, not to maximize the model's likelihood. Nevertheless, although the placement of the Gaussian functions is drastically changed by SEDS [MSE], we are capable of achieving smooth, nonlinear motions. On the other hand, because SEDS [Likelihood] seeks to maximize the log-likelihood of the GMM parameters, the Gaussian function placement is not drastically changed, and yet nonlinear motions that are less smooth are achieved. It was reported in the original published work [72] that the SEDS [Likelihood] approach was slightly more accurate quantitatively than the SEDS [MSE]. This result was obtained from a subset of the full LASA Handwriting Dataset presented in this book. For further quantitative comparisons, we refer the reader to [72].

Model and computational complexity: SEDS [MSE] is better than SEDS [Likelihood], in that it requires fewer parameters [this number is reduced by a factor of $\frac{1}{2}KN(N+1)$]. On the other hand, SEDS [MSE] has a more complex cost function that requires computing GMR at each iteration over all training data points. As a result, the use of MSE makes the algorithm computationally more expensive and causes a slightly longer training time. See [72] for further details.

Programming Exercise 3.3 *The aim of this instructional exercise is to familiarize the readers with the SEDS learning algorithm. We provide code to learn DSs (and visualize them) from 2D drawn data sets, as well as the LASA Handwriting Dataset, using the SEDS algorithm.*

Instructions: *Open MATLAB and set the directory to the folder corresponding to chapter 3 exercises; within this folder, you will find the following scripts:*

```
1  ch3_ex0_drawData.m
2  ch3_ex3_seDS.m
```

In these scripts, you will find instructions in the block comments that will enable you to:

- *Draw 2D trajectories on a GUI*
- *Load the LASA Handwriting Dataset shown in figure 3.7*
- *Learn the first-order nonlinear DS with the SEDS variants*

Note: *The MATLAB scripts provide options to load/use various types of data sets and to select various options for the learning algorithm. Please read the comments in the script and run each block separately.*

Readers should test the following with either self-drawn trajectories or motions from the LASA Handwriting Dataset:

1. Compare SEDS [Likelihood] to SEDS [MSE] in terms of reconstruction error (RMSE), model complexity and computation time. .

2. Compare the results when manually selecting K versus using model selection via BIC.

3. Compare the results when using M = 1 versus M >> 1 trajectories.

4. Compare the SEDS initialization algorithms. Which one yields the best performance in reconstruction, model complexity, and solver convergence time? Can you think of another approach to initialize the parameters?

5. The Multi-Models from the LASA Handwriting Dataset are a concatenation of two single-model motions. Could you think of a way to learn the two DS models seperately, and then merge them? If so, compare the performance of the merged DS models to learning a single DS model from the merged trajectories.

6. Draw self-intersecting trajectories. Is SEDS suitable for these? Where should the self-intersecting trajectories be disconnected/segmented in order to properly use multiple DS learned via SEDS?

7. What happens if the trajectories have distinct end points? Will the SEDS algorithm be able to encode these? Can they be encoded with standard GMR?

The reader can also use an instructional GUI, like the one shown in figure 3.12, to complement the previous exercise.

With this GUI, only the self-drawn trajectories can be used to learn DS models via SEDS variants. Yet it provides a robot simulation that allows the user to execute the motion of the learned DS and apply perturbations to the robot while it is executing the task.

Instructions: *Open MATLAB and set the directory to the folder corresponding to chapter 3 exercises; in this folder, you will find the following script:*

```
1   gui_seDS.m
```

or, typing the following alternative in the MATLAB Command window:

```
1   >> gui_seDS
```

With this GUI, the user will be able to do the following:

- *Draw trajectories within the workspace of a robot arm with 2 degrees of freedom (DOF).*
- *Manually select K, or use model selection via BIC.*
- *Learn a DS model using either SEDS [Likelihood] or SEDS [MSE].*

- *Simulate the robot motion following the learned DS.*
- *Apply perturbations to the robot while it is executing the motion.*

The initialization of GMM parameters is performed via algorithm 3.2.

Note: *An instructional movie on how to use this GUI can be found at the following YouTube video:* `https://youtu.be/fQL-tCOqCH0`.

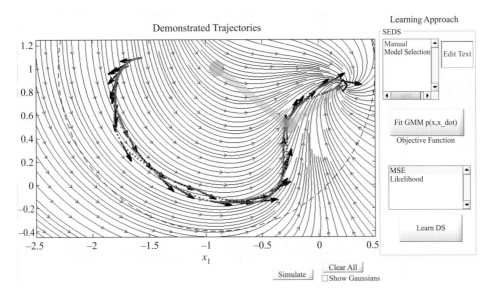

Figure 3.12
The GUI for the SEDS learning evaluation provided in MATLAB source code.

3.3.5 Robotic Implementation

From [72], we present here one example of how the nonlinear DS, $f(x)$, estimated using SEDS, can be used as a control law to generate motion in the 6-DOF arm of the industrial robot Katana-T. The robot is tasked to place a saucer on a tray (task 1) and then a cup on a saucer (task 2).

For both tasks, $M = 4$ reference trajectories were demonstrated and $K = 4$ Gaussian functions were used to learn the DS models. Figure 3.13 shows demonstrated trajectories and testing trajectories for both tasks, superimposed on one another. Testing trajectories that start from points different from those demonstrated follow well the dynamics of the demonstrations. All trajectories land at the target.

To exemplify the natural and instantaneous adaptation to the new target's position, we induce perturbation along the path by moving the target (the tray and saucer, respectively). The target (the saucer or tray, respectively, is displaced during the execution of the task at $t = 2$ s after onset of the motion. Figure 3.14 shows the trajectories generated by the learned DS to respond to the perturbations in both tasks. Note that, while task 2 is executed after

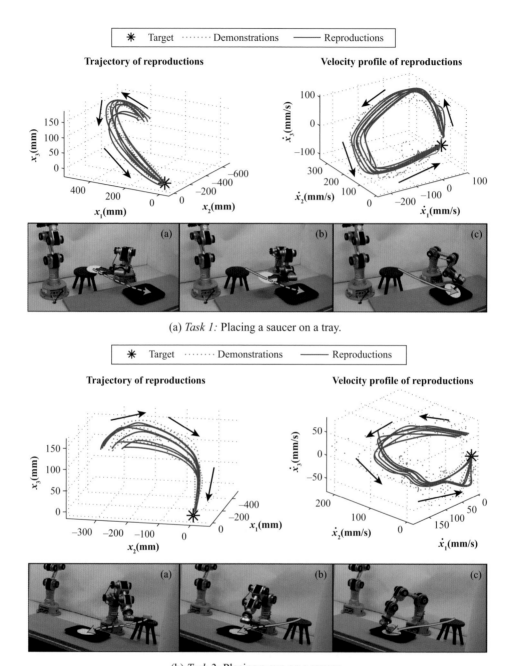

(a) *Task 1:* Placing a saucer on a tray.

(b) *Task 2:* Placing a cup on a saucer.

Figure 3.13
A 6-DOF Katana-T arm is tasked to place a saucer on a tray (a) and then a cup on a saucer (b). Training and testing trajectories follow the same dynamics and converge to their respective targets [72].

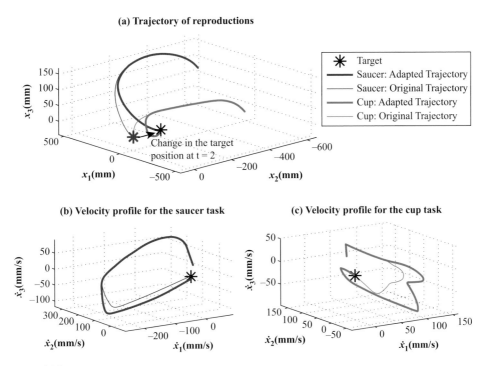

Figure 3.14
On-the-fly adaptation to a change in the target's position for task 2 and task 3 [72].

the completion of task 1, we superimpose the trajectories for both tasks in the same plot. In both experiments, the robot handles well the perturbation by regenerating a new path that lands it to the target.

3.3.6 Shortcoming of the SEDS formulation

We started this section by formulating learning of the control law using constrained GMR [equation (3.25)]. We presented one approach, SEDS, to learn the parameters of the GMR using a constrained nonconvex optimization. This formulation guarantees that the resulting control law is GAS at a single attractor. Moreover, we showed that robot motion generated with such a model provides online adaptation to changes in the target's location. However, the approach suffers from a number of limitations, discussed next, which led to an alternative approach presented subsequently.

Incremental learning Incremental learning is often crucial to allow the user to either refine the model in an interactive manner or add new demonstrations in regions not seen before. The SEDS learning algorithm does not allow incremental retraining of the model, as it is a batch offline optimization algorithm. If one added new demonstrations after training a model, one would have to either entirely retrain the model based on the combined set of old and new demonstrations, or build a new model from the new demonstrations and merge it with the previous model. For example, one can merge two GMMs with K_1 and K_2 number of Gaussian functions into a single model by simply concatenating their parameters; that is, $K = K_1 + K_2$ Gaussian functions by concatenating their parameters,

$\Theta_{\text{GMR}} = \{\pi_k, \mu^k, \Sigma^k\}_{k=1}^K \; \forall k = 1, \ldots, K_1, \ldots, K$. The resulting model is no longer locally optimal; yet it could be an accurate estimation of both models. This approach, however, is suitable only if there is no overlap between the two models/demonstrations. In section 3.4, we propose an alternative nonlinear DS formulation and learning approach that is capable of training the DS model incrementally while dealing with the overlap issue without the need of using the old demonstrations and preserving GAS.

Encoding highly nonlinear complex motions As mentioned in section 3.3.3, although the simulations show that a large portion of the LASA Handwriting Dataset motions can be modeled while satisfying the sufficient stability conditions derived in theorem 3.2, these global stability conditions might be too restrictive to accurately model highly nonlinear motions. This is evidence by the failure of the SEDS variants to encode eight of the thirty LASA Handwriting Dataset plots (i.e., *BendedLine, DoubleBendedLine, Heee, Leaf, Leaf-2, Nshape, Snake, and Sshape*). Furthermore, while the vector fields from another subset of plots, including *Cshape, Gshape, Jshape, Jshape-2, Pshape, SharpC, and Zshape*, resemble the motion patterns demonstrated in the reference trajectories, the reproductions are not necessarily accurate. The reason for this is that the SEDS learning algorithm suffers from the "accuracy versus stability" dilemma; that is, it performs poorly in highly nonlinear motions that contain high curvatures or are nonmonotic (i.e., are temporarily moving away from the attractor). This is mainly due to the choice of Lyapunov function; for example, geometrically, a QLF can allow only trajectories whose L_2-norm (i.e., $\|x - x^*\|_2$) distances decrease monotonically [72, 108]. One solution is to relax the stability conditions for SEDS by using a P-QLF, see section 3.1, to allow higher nonlinearities.

Sensitivity to hyperparameters and initializations As detailed in section 3.3.3, the SEDS learning algorithm relies heavily on two main factors: a good choice of K Gaussian functions and a good initial estimate of the GMM parameters to be optimized by the SEDS solver. This becomes a disadvantage when robot autonomy is desired, as these "good" estimates must be monitored by a human or through ad hoc heuristics.

Next, we propose an approach that tackles the limitations presented here.

3.4   Learning Stable, Highly Nonlinear DSs

This section presents an alternative approach to learning a nonlinear DS as a control law. It addresses the "accuracy versus stability" dilemma that is, core to the SEDS approach (see section 3.3), and the need to allow incremental learning and remove sensitivity to a choice of hyperparameter and initializations in the accuracy. To do so, the approach:[2]

1. Introduces an alternative P-QLF Lyapunuv candidate function [equation (3.18)] to satisfy stability constraints, and then enables it to better fit nonlinearities.

2. Preserves the locality of the Gaussian functions, allowing incremental learning.

3. Estimates the optimal number of K Gaussian functions automatically.

   To achieve these goals, we attempt a Bayesian nonparametrics GMM estimation.

3.4.1 Untied Linear Parameter Varying Formulation

Observe that the mixture of linear DS, defined in (3.21), $\dot{x} = f(x) = \sum_{k=1}^{K} \gamma_k(x)\left(A^k x + b^k\right)$, is a linear parameter varying (LPV) system [6], with each $A^k x$ an LTI system, and $\gamma_k(x)$ a state-dependent parameter vector $\boldsymbol{\gamma} = [\gamma_1, \dots, \gamma_K]$. In LTI systems, Lyapunov functions of the form $V(x) = (x - x^*)^T P(x - x^*)$ are commonly used to ensure stability. The matrix P becomes a parameter of the QLF (see section 3.1.2 and proposition 3.1 for an example using a linear DS). Replacing the QLF of SEDS with a P-QLF function yields the sufficient conditions described next to ensure GAS at x^*.

Theorem 3.3 *The nonlinear DS defined in equation (3.21) is GAS at an attractor* x^* *if* $\exists$ $P = P^T$ *and* $P \succ 0$, *with* $V(x) = (x - x^*)^T P(x - x^*)$, *such that:*

$$\begin{cases} (A^k)^T P + PA^k = Q^k, \quad Q^k = (Q^k)^T \prec 0 \\ b^k = -A^k x^* \end{cases} \quad \forall k = 1, \dots, K \quad (3.36)$$

Proof *We first show that* $V(x) = (x - x^*)^T P(x - x^*)$ *is a candidate Lyapunov function.* $P \succ 0$, $V(x) > 0 \forall x \neq x^*$ *and* $V(x^*) = 0$. *Taking the derivative of* $V(x)$ *with regard to time, we have:*

$$\dot{V}(x) = (x - x^*)^T P f(x) + f(x)^T P(x - x^*)$$

$$= (x - x^*)^T P \underbrace{\left(\sum_{k=1}^{K} \gamma_k(x)(A^k x + b^k)\right)}_{\text{via (3.21)}} + \underbrace{\left(\sum_{k=1}^{K} \gamma_k(x)(A^k x + b^k)^T\right)}_{\text{via (3.21)}} P(x - x^*)$$

$$= (x - x^*)^T P \left(\sum_{k=1}^{K} \gamma_k(x)(A^k x - \underbrace{A^k x^*}_{\text{via (3.36)}})\right) + \left(\sum_{k=1}^{K} \gamma_k(x)(A^k x - \underbrace{A^k x^*}_{\text{via (3.36)}})^T\right) P(x - x^*)$$

$$= (x - x^*)^T P \left(\sum_{k=1}^{K} \gamma_k(x) A^k\right)(x - x^*) + (x - x^*)^T \left(\sum_{k=1}^{K} \gamma_k(x)(A^k)^T\right) P(x - x^*)$$

$$= (x - x^*)^T \left(\sum_{k=1}^{K} \underbrace{\gamma_k(x)}_{>0 \text{ via (3.21)}} \underbrace{((A^k)^T P + PA^k)}_{Q_k \prec 0 \text{ via (3.36)}}\right)(x - x^*) < 0, \quad (3.37)$$

with $Q_k = Q_k^T \prec 0$. $V(x)$ *is a valid Lyapunov function for the DS given in (3.21), if conditions (3.36) are satisfied and the DS is GAS with respect to* x^*. □

Recall from section 3.1.2 that the effect of P is a reshaping of the Lyapunov function. When P is identity, the Lyapunov function offers an isotropic measure in space. Shaping the entries of P leads to an anisotropic measure. This can be particularly useful to model trajectories exhibiting high curvature and different speed of convergence the speed of convergence toward x^* along the various dimensions.

If one replaces the stability constraints of the SEDS learning algorithm equation (3.32) (**a–b**) with those defined in equation (3.36), the optimization becomes more complex, as equation (3.36) requires the joint estimation of P and A^k's. The problem is nonconvex and depends on a good initial estimate of P. Solving this problem is not infeasible, but heavy parameter tuning and an ad hoc solver must be implemented to address the problem. To alleviate this situation, we optimize the parameters of equation (3.21) with constraints derived from equation (3.36) in a decoupled way.

Let us first understand where the coupling comes from. In the SEDS formulation, the DS parameters $\Theta_{f(x)} = \{\gamma_k(x), A^k, b^k\}_{k=1}^K$ are all tied to the GMM parameters $\Theta_{\mathrm{GMR}} = \{\pi_k, \mu^k, \Sigma^k\}_{k=1}^K$, fitted on the joint distribution, $p(x, \dot{x})$, of the reference trajectories $\{\mathbf{X}, \dot{\mathbf{X}}\}$. Such parameter tying is a result of the change of variables introduced in equation (3.24) and is highlighted as follows:

$$\gamma_k(x) = \frac{\pi_k p(x|\mu_x^k, \Sigma_x^k)}{\sum_{i=1}^K \pi_i p(x|\mu_x^i, \Sigma_x^i)}, \quad A^k = \Sigma_{\dot{x}x}^k \left(\Sigma_x^k\right)^{-1}, \quad b^k = \mu_{\dot{x}}^k - A^k \mu_x^k \quad \forall k = 1, \ldots, K.$$

As can be seen, the mixing function, $\gamma_k(x)$, and the linear DS parameters, $\{A^k, b^k\}$, are tied by the means and covariance matrices, $\{\mu_x^k, \Sigma_x^k\}$, of the **position** variables within the GMM. This is what makes the SEDS learning algorithm an NLP. Even though the main stability constraint of SEDS [i.e., equation (3.32) **(a)**] is a linear matrix inequality (LMI) that could be solved as a semi definite program (SDP), it actually is not possible to do this because of this tying of the parameters. We thus propose to untie $\Theta_{f(x)}$ from Θ_{GMR} as follows:

$$\dot{x} = f(x, \Theta_{\mathrm{LPV}})$$

$$= \sum_{k=1}^K \underbrace{\gamma_k(x)}_{\Theta_\gamma} \underbrace{\left(A^k x + b^k\right)}_{\Theta_f} \tag{3.38}$$

with

$$\Theta_{\mathrm{LPV}} = \{\underbrace{\{\pi_k, \theta_\gamma^k = (\mu^k, \Sigma^k)\}_{k=1}^K}_{\Theta_\gamma}, \underbrace{\{A^k, b^k\}_{k=1}^K}_{\Theta_f}\}. \tag{3.39}$$

Due to this untying of parameters, we solely need to fit a GMM density on the position variables [i.e., $p(x|\Theta_\gamma) = \sum_{k=1}^K \pi_k \mathcal{N}(x|\mu^k, \Sigma^k)$] and the set of linear DS parameters [$\Theta_f = \{A^k, b^k\}_{k=1}^K$] remain free to be optimized with any approach, independent of Θ_γ. This allows us to use off-the-shelf SDP solvers for LMIs such as the ones defined in the stability conditions proposed in equation (3.36), which yield less conservative constraints. Furthermore, by untying/decoupling the parameters, this approach preserves the geometric representation of the GMM, as its parameters are no longer modified in the constrained optimization phase.

Caveat: While the proposed approach, which we refer to as the *LPV-DS formulation*, allows us to use a P-QLF to derive less restrictive stability constraints and estimate Θ_f via SDPs, its reproduction accuracy relies heavily on a good estimate of Θ_γ. Not only must one find the correct number of Gaussians K that best represent the reference trajectories, but they

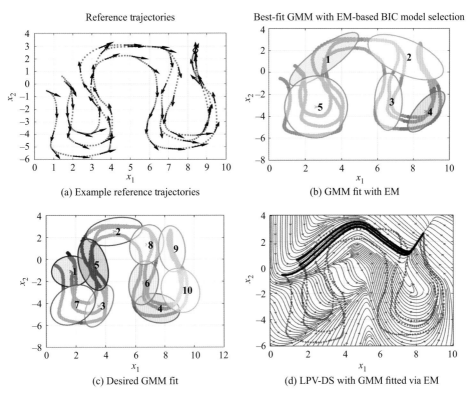

Figure 3.15
Example of a GMM fit on the position variables of a set of highly nonlinear reference trajectories and its effect on LPV-DS reproduction accuracy and motion modeling. When fitting a GMM to the reference trajectories illustrated in (a) with a standard EM and model selection scheme, we achieve an optimal $K = 5$, as illustrated in (b). As can be seen, however, the Gaussians corresponding to clusters 1 and 2 contain noncontiguous trajectories and points whose velocities are in opposite directions. Such clustering violates the initial LPV-DS assumption, which states that each Gaussian should represent a linear DS. This results in an inaccurate representation of the demonstrated motion by the LPV-DS model, as shown in (d). To alleviate this situation, we look for a GMM fit that is physically consistent with the trajectory data, as illustrated in (c).

should also be aligned with the trajectories such that each Gaussian represents a local region in the state space that follows a linear DS (see figure 3.15).

Exercise 3.7 *Consider a nonlinear DS such as* $\dot{x} = f(x, \Theta_{LPV}) = \sum_{k=1}^{K} \gamma_k(x)$ $\left(A^k x + b^k\right)$ *with* N = 1, K = 4. *Design the LPV-DS parameters* $\Theta_{LPV} = \{\pi_k, \mu^k, \Sigma^k, A^k, b^k\}_{k=1}^{2}$ *that will produce a GAS DS at the target* $x^* = 0$ *such that:*

- *If* x = −1, *then* $\dot{x} = 0.5$; *and when* x = −2, *then* $\dot{x} = 2$.
- *If* x = +1, *then* $\dot{x} = −0.5$ *and when* x = +2, *then* $\dot{x} = −2$.

Exercise 3.8 *Consider a nonlinear DS as* $\dot{x} = f(x, \Theta_{LPV}) = \sum_{k=1}^{K} \gamma_k(x)$ $\left(A^k x + b^k\right)$ *with* N = 2, K = 2. *The design matrices* $P \in \mathbb{R}^{2 \times 2}$, $A^1 \in \mathbb{R}^{2 \times 2}$, *and* $A^2 \in \mathbb{R}^{2 \times 2}$

and activation functions $\gamma_1(x), \gamma_2(x)$, such that it is GAS at the origin—that is, $x^* = [0\ \ 0]^T$, with a P-QLF and with a P-QLF and free choice of $Q \in \mathbb{S}_N^+$. The DS should exhibit the following dynamic behaviors:

- *When $x_1 < 0$, the DS should follow a linear motion toward the target.*
- *When $x_1 > 0$, the DS should follow a curved-linear motion toward the target.*

Recall that $x = [x_1, x_2]$. The code in the programming exercises can be used to design this DS.

3.4.2   Physically Consistent Bayesian Nonparametric GMM

As described in figure 3.15, for an accurate reproduction with the LPV-DS formulation, not only is the number of optimal K Gaussian functions important, but the Gaussians must be consistent with the LPV-DS assumptions (i.e, each Gaussian should represent data points that follow a quasi-linear motion). Such an estimate is empirically hard to find with either EM estimation or Bayesian nonparametric approaches. While Bayesian nonparametric modeling and estimation approaches (such as sampling and variational inference) can allow on to determine automatically an adequate number of K Gaussian functions, they cannot ensure that each Gaussian would cluster data points that represent linear motion. We refer to this property as *physical consistency* (i.e., ensuring that the location and coverage of each Gaussian component corresponds to a linear DS). In this section, we propose a Bayesian nonparametric model and estimation approach for GMMs, referred to as a *physically consistent Gaussian mixture model (PC GMM)*.

Note: In this discussion, we will provide a brief introduction to the Bayesian nonparametric treatment for GMM. However, we assume that the reader has a good understanding of clustering problems and Bayesian nonparametrics. These can be reviewed in appendix B, specifically section B.3, for the various GMM estimation approaches.

Let us interpret the GMM as a hierarchical model where each kth mixture component is viewed as a cluster, represented by a Gaussian distribution $\mathcal{N}(\cdot|\theta_\gamma^k)$ with $\theta_\gamma^k = \{\mu_\gamma^k, \theta_\gamma^k\}$ and mixing weight π_k. Each data point x_i is assigned to a cluster k via cluster assignment indicator variables $Z = \{z_1, \ldots, z_M\}$, where $i : z_i = k$ for M samples; that is,

$$z_i \in \{1, \ldots, K\}$$

$$p(z_i = k) = \pi_k \tag{3.40}$$

$$x_i | z_i = k \sim \mathcal{N}(\theta_\gamma^k).$$

Via equation (3.40), the probability density function of the mixture model is defined by

$$p(x|\Theta_\gamma) = \sum_{k=1}^{K} p(z_i = k) \cdot \mathcal{N}(x|\mu^k, \Sigma^k). \tag{3.41}$$

With the Bayesian treatment, one normally imposes a prior on the latent variables of the probabilistic model. In the GMM case, a prior distribution is imposed on the set of indicator variables Z and the set of Gaussian distribution variables $\Theta_\gamma = \{\mu^k, \Sigma^k\}_{k=1}^{K}$.

With EM estimation, K is known a priori; hence, the marginal distribution over Z is solely defined by the set of mixing weights $\boldsymbol{\pi} = \{\pi_k\}_{k=1}^{K}$. Thus, the prior probability of the cluster assignment indicator variable z_i reduces to $p(z_i = k) = \pi_k$, as discussed in the previous sections of this chapter.

With the Bayesian nonparametric treatment, not only are priors imposed on the latent variables, but the number of Gaussian functions is assumed to be unknown and infinite; $K \rightarrow +\infty$. Hence, a Bayesian nonparametric GMM is also known as as the *infinite GMM*.

Priors for latent variables Because K is assumed to be infinite, one needs an infinite distribution as a prior. The Dirichlet process (DP) is an infinite distribution over distributions that is used to evaluate an *infinite* mixture model on a *finite* set of samples, following [64]. Via inference, one can jointly estimate the optimal K and Θ_γ.

We use the normal-inverse-Wishart ($\mathcal{N}$IW) distribution as a prior on the Gaussian parameters θ_γ^k. The optimal K and Θ_γ can be estimated by maximizing the posterior probability of equation (3.41) via sampling or variational inference. While using such priors solves the problem of estimating the optimal K, they perform poorly when the distribution of the data points exhibits idiosyncrasies such as high curvatures and nonuniformities, as shown in figures 3.15 and 3.16. This is because the Gaussian exponential $\exp\left\{-\frac{1}{2}(x - \mu^k)^T(\Sigma^k)^{-1}(x - \mu^k)\right\}$ does not take into consideration the directionality of the trajectories. To enforce that the model follows the flow of the DS, we introduce a new measure of similarity, described next.

Physical consistency via $\dot{x}$-similarity To cluster trajectory data in a physically consistent way, one must consider both directionality and locality. We thus propose a similarity measure composed of a locally scaled, shifted cosine similarity kernel, which we refer to as $\dot{x}$-*similarity*:

$$\Delta_{ij}(x^i, x^j, \dot{x}^i, \dot{x}^j) = \underbrace{\left(1 + \frac{(\dot{x}^i)^T\dot{x}^j}{||\dot{x}^i||||\dot{x}^j||}\right)}_{\text{Directionality}} \underbrace{\exp\left(-l||x^i - x^j||^2\right)}_{\text{Locality}}. \tag{3.42}$$

The first term, measuring directionality, is the shifted cosine similarity of pairwise velocity measurements [i.e., $\overline{\cos}(\angle(\dot{x}^i, \dot{x}^j)) \in [0, 2]$] and it is bounded by the angle between the pairwise velocities $\phi_{ij} = \angle(\dot{x}^i, \dot{x}^j)$. When $\phi_{ij} = \pi$ (its maximum value), the velocities are in opposite directions and $\overline{\cos}(\phi_{ij}) = 0$. When $\phi_{ij} = \{\pi/2, 3\pi/2\}$ the velocities are orthogonal to each other, which yields $\overline{\cos}(\phi_{ij}) = 1$. Finally, when $\phi_{ij} = \{0, 2\pi\}$ the cosine similarity value is at its maximum $\overline{\cos}(\phi_{ij}) = 2$, as the pairwise velocities are pointing in the same direction. This would suffice as a measure of physical consistency for trajectories that do not include repeating patterns; however, for trajectories such as a sinusoidal wave, $\overline{\cos}(\phi_{ij})$ can yield its maximum value even if the trajectories are not close to each other in Euclidean space. To enforce locality we scale the $\overline{\cos}(\phi_{ij})$ with an RBF kernel on the position measurements, the second term in equation (3.42). Notably, $l = \frac{1}{2\sigma^2}$ is a hyperparameter that can be a nuisance if not set properly. We thus propose to set σ with the following data-driven heuristic: $\sigma = \sqrt{\text{Mo}(\mathbf{D})/2}$, where $\mathbf{D} \in \mathbb{R}^{M \times M}$ is a matrix of pairwise squared Euclidean distances

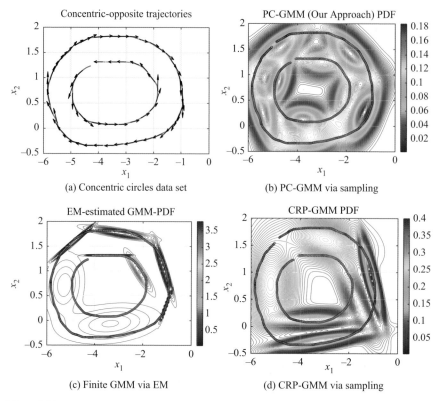

Figure 3.16
Performance of different GMM fitting strategies on a concentric trajectory data set. The arrows in (a) indicate velocity directions.

$d_{ij} = ||x^i - x^j||^2$ and Mo is the mode of all entries of **D**. Intuitively, we are approximating the length scale of the trajectories. Such an approximation is sufficient, as we only need it to scale high $\overline{\cos}(\phi_{ij})$ values that are far away in Euclidean space.

To include equation (3.42) in the estimation of a GMM while preserving the form of equation (3.41), rather than using CRP as a prior on $p(Z)$, we adopt the distance-dependent (dd)-CRP [19]. The (dd)-CRP focuses on the probability of customers sitting with other customers (i.e., observation x^i being clustered with x^j) based on an external measure of distance.

The Physically Consistent-CRP The Physically Consistent-CRP adapts (dd)-CRP in order to generate a prior distribution $p(C)$ over customer seating assignments $C = \{c_1, \ldots, c_M\}$, where $i : c_i = j$ indicates that the ith and jth customers are clustered, that is that observations x^i and x^j are clustered together based on equation (3.42). This prior is constructed by a sequence of probabilities, where the ith customer (x^i) has two choices, she/he can: sit with the jth customer (x^j) with a probability proportional to equation (3.42), or sit alone with a probability proportional to α. Such a sequence yields a prior distribution that is a multinomial over customer seating assignments C conditioned on Δ and α; that is, $p(C|\Delta, \alpha)$ and can be computed as

$$p(C \mid \Delta, \alpha) = \prod_{i=1}^{M} p(c_i = j \mid \Delta, \alpha), \quad \text{where} \quad p(c_i = j \mid \Delta, \alpha) = \begin{cases} \frac{\Delta_{ij}(\cdot)}{\sum_{j=1}^{M} \Delta_{ij}(\cdot) + \alpha} & \text{if} \quad i \neq j \\ \frac{\alpha}{M + \alpha} & \text{if} \quad i = j, \end{cases}$$

$$(3.43)$$

where $\Delta \in \mathbb{R}^{M \times M}$ is the matrix of pairwise similarities computed by equation (3.42) and α is the concentration parameter.

PC-GMM By using the *PC*-CRP prior, we no longer explicitly compute the cluster indicator variables $Z = \{z_1, \dots, z_N\}$, and yet they are recovered via the recursive mapping function $Z = \mathbf{Z}(C)$, which gathers all linked customers. Hence, using equation (3.43) and the $\mathcal{N}$IW distribution as priors, we construct the following **P**hysically-**C**onsistent GMM:

$$c_i \sim PC\text{-}CRP(\Delta, \alpha)$$

$$z_i = \mathbf{Z}(c_i)$$

$$\theta_\gamma^k \sim \mathcal{N}\text{IW}(\lambda_0) \qquad (3.44)$$

$$x^i | z_i = k \sim \mathcal{N}(\theta_\gamma^k).$$

Equation (3.44) indicates that $C = \{c_1, \dots, c_N\}$ are sampled from the *PC-CRP* prior, and for each kth Gaussian, its parameters θ_γ^k are drawn from an $\mathcal{N}$IW distribution, with hyperparameters $\lambda_0 = \{\mu_0, \kappa_0, \Lambda_0, \nu_0\}$. To estimate these parameters, one has to maximize the posterior distribution of equation (3.44) [i.e., $p(C, \theta_\gamma | \mathbf{X})$]. Due to conjugacy, however, one can integrate out the model parameters θ_γ^k from the posterior distribution and estimate only the posterior of the latent variable C—that is, $p(C | \mathbf{X}, \Delta, \alpha, \lambda) = \frac{p(C \mid \Delta, \alpha) p(\mathbf{X} | \mathbf{Z}(C), \lambda)}{\sum_C p(C \mid \Delta, \alpha) p(\mathbf{X} | \mathbf{Z}(C), \lambda)}$. As this full posterior is intractable, we approximate it via collapsed Gibbs sampling by drawing samples of c_i from the following distribution:

$$p(c_i = j \mid C_{-i}, \mathbf{X}, \Delta, \alpha, \lambda) \propto \underbrace{p(c_i = j \mid \Delta, \alpha)}_{\substack{\text{Similarities in} \\ \text{scaled velocity space}}} \underbrace{p(\mathbf{X} \mid \mathbf{Z}(c_i = j \cup C_{-i}), \lambda)}_{\substack{\text{Observations in} \\ \text{position space}}}, \qquad (3.45)$$

where the first term is given by equation (3.43) and the second term is the likelihood of the table assignments that emerge from the current seating arrangement, $\mathbf{Z}(c_i = j \cup C_{-i})$. C_{-i} indicates the customer seating assignments for all customers except the ith one. Details on the implementation of the collapsed Gibbs sampler are provided in section D.1.1 in appendix D.

Estimating the full set of GMM parameters After obtaining the MAP estimate from sampling [equation (3.45); see section D.1.1 in appendix D], we solely obtain the indicator variables C. We then estimate the optimal number of Gaussians with $K = |\mathbf{Z}(C)|$ and the observations that are assigned to each Gaussian as $\mathbf{X}_{\mathbf{Z}(C)=k}$. Hence, in order to estimate the Gaussian parameters $\theta_\gamma^k = \{\mu^k, \Sigma^k\}_{k=1}^K$, we take $\mathbf{X}_{\mathbf{Z}(C)=k}$ for each kth cluster and the set of hyperparameters λ and sample θ_γ^k from the posterior of the $\mathcal{N}$IW (refer to

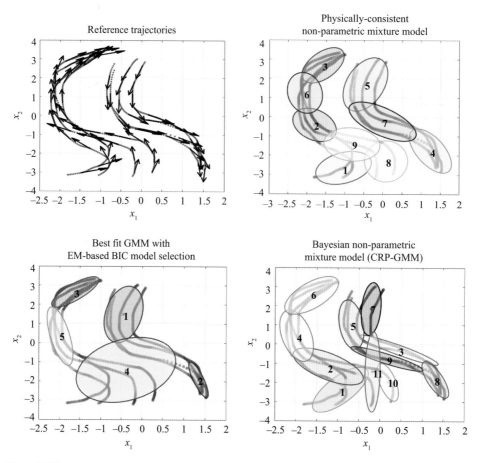

Figure 3.17
GMM fit on 2D opposing motions (different targets) data set.

section B.3.2.2 in appendix B for the exact equations). Finally, the mixing weights $\{\pi_k\}_{k=1}^{K}$ are estimated as $\pi_k = M_k/M$, where $M_k = |\mathbf{X}_{\mathbf{Z}(C)=k}|$ is the number of observations assigned to the kth cluster. Figure 3.16 shows the performance of *PC*-GMM versus a *CRP*-GMM versus EM on a data set of the concentric circular trajectories. Further examples are illustrated in figures 3.17 to 3.19.

3.4.3   Stable Estimator of LPV DSs

Given the set of reference trajectories $\{\mathbf{X}, \dot{\mathbf{X}}\} = \{X^m, \dot{X}^m\}_{m=1}^{M} = \{\{x^{t,m}, \dot{x}^{t,m}\}_{t=1}^{T_m}\}_{m=1}^{M}$, the attractor x^* (i.e., the desired target), and the GMM parameters $\Theta_\gamma = \{\pi_k, \mu^k, \Sigma^k\}_{k=1}^{K}$ estimated via the *PC*-GMM approach, we now estimate the set of linear DS parameters $\Theta_f = \{A^k, b^k\}_{k=1}^{K}$ by minimizing the MSE as in the SEDS [MSE] variant [equation (3.33)]. Due to the decoupling/untying of the parameters in the LPV-DS formulation, imposing constraints on the DS parameters is a very flexible process. Next, we propose three constraint variants derived from (O1), a QLF (like SEDS) [equation (3.32)], and (O2 and O3), from a P-QLF [equation (3.36)]:

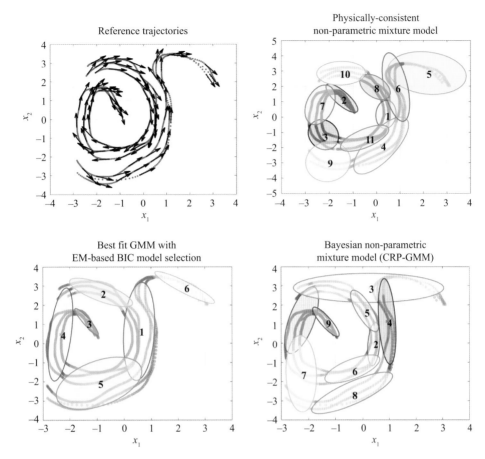

Figure 3.18
GMM fit on 2D multiple motions (different targets) data set.

$$\min_{\Theta_f} J(\Theta_f) \quad \text{subject to}$$

$$(O1) \begin{cases} (A^k)^T + A^k \prec 0, b^k = -A^k x^* \;\; \forall k = 1, \dots, K \end{cases}$$

$$(O2) \begin{cases} (A^k)^T P + PA^k \prec 0, \;\; b^k = \mathbf{0} \;\; \forall k = 1, \dots, K; \;\; P = P^T \succ 0 \end{cases} \quad (3.46)$$

$$(O3) \begin{cases} (A^k)^T P + PA^k \prec Q^k, \;\; Q^k = (Q^k)^T \prec 0, \;\; b^k = -A^k x^* \;\; \forall k = 1, \dots, K. \end{cases}$$

3.4.3.1 Offline LPV-DS Parameter Optimization

• (O1) follows the same conditions used in SEDS [72], and yet, instead of it being a non-linear, constrained optimization problem, it is a convex semidefinite optimization problem that can be solved via standard SDP solvers such as SeDuMi [142].

• (O2) has nonconvex constraints, as P is unknown, and yet it can be solved via nonlinear SDP solvers such as PENLAB [40]. Note that (O2) assumes that the attractor is at the origin, hence the constraint $b^k = \mathbf{0}$; absent this assumption, it may converge on unstable solutions. This approach was initially proposed in [100].

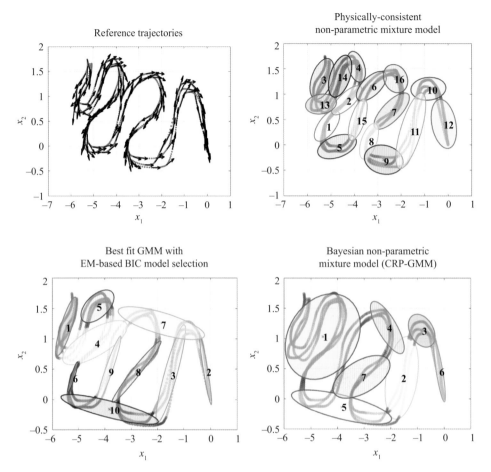

Figure 3.19
GMM fit on 2D messy snake data set.

• (*O3*) is similar to (*O2*), but to solve it, we assume that it has a prior estimate of P obtained through the data-driven approach. To find a suitable P for a given set of reference trajectories, we solve the following constrained optimization problem:

$$\min_P J(P) = \sum_{m=1}^{M} \sum_{t=1}^{T^m} \left(\frac{1+\bar{w}}{2} \text{sgn}(\psi_{t,m}) \psi_{t,m}^2 + \frac{1+\bar{w}}{2} \psi_{t,m}^2 \right)$$

(3.47)

subjectto

$$\left\{ P \succ 0, \; P = P^T, \right.$$

where $\bar{w} > 0$ is a small, positive scalar (i.e., $\bar{w} << 1$) and $\psi_{t,m}$ is a function defined as $\psi_{t,m} = \frac{\nabla_x V(x^{t,m})^T \dot{x}^{t,m}}{||\nabla_x V(x^{t,m})^T|| ||\dot{x}^{t,m}||}$. Equation (3.47) is a modified version of the one proposed in [75] for learning more complex Lyapunov candidate functions. Intuitively, equation (3.47) optimizes P such that the lowest number of data points in $\{\mathbf{X}, \dot{\mathbf{X}}\}$ violate the Lyapunov stability

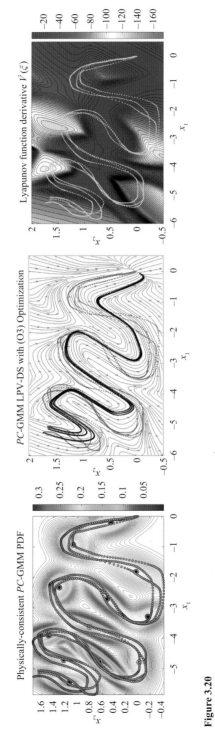

Figure 3.20
(left) *PC*-GMM. **(center)** LPV-DS estimated with (*O3*). **(right)** $\dot{V}(x)$.

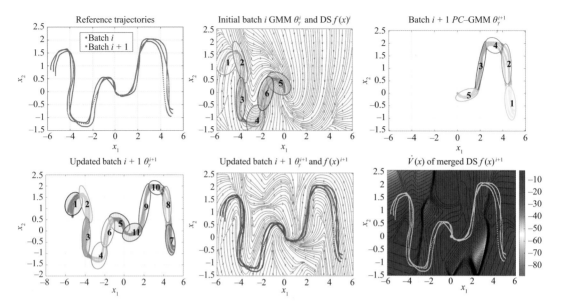

Figure 3.21
Example 1 of the incremental LPV-DS learning approach.

condition in equation (3.10). Given such prior estimates of P, we introduce the auxiliary matrices Q^k, which allow a wider exploration of the parameter space.

The aforementioned solvers (SeDuMi and PENLAB) are used with the YALMIP MATLAB toolbox [93]. ($O2$) yields similar results to ($O3$), albeit when it finds a feasible solution. Due to the nonlinearity of the problem, it might converge on a solution in which not all the constraints are met. ($O3$) always converges on a feasible solution, so long as $P = P^T$ and has well-balanced eigenvalues.

3.4.3.2 Incremental LPV-DS Parameter Optimization

Assume that we obtained an initial batch ($i = 1$) of reference trajectories $\{\mathbf{X}, \dot{\mathbf{X}}\}^i$, and we then learn a DS $f^i(x)$ on these demonstrations and use it to generate motions. After a while, we obtain a new batch of reference trajectories $\{\mathbf{X}, \dot{\mathbf{X}}\}^{i+1}$ that we wish to use to update the DS $f^i(x) \to \tilde{f}^{i+1}(x)$.

A naive approach to tackle this problem would be to simply concatenate the batches of reference trajectories $\{\mathbf{X}, \dot{\mathbf{X}}\} = \{\mathbf{X}, \dot{\mathbf{X}}\}^i \cup \{\mathbf{X}, \dot{\mathbf{X}}\}^{i+1} \cup \cdots \cup \{\mathbf{X}, \dot{\mathbf{X}}\}^{i+\infty}$ and learn a new DS every time new data arrive. Such an approach not only becomes computationally expensive as the number of batches increases, it also does not use the previous batches of data in an efficient way. For example, if a new reference trajectory arrives that is overlapping with trajectories from the old batches, learning new parameters Θ_γ and Θ_f might have no effect on the resulting model, and so there is no need to relean the DS. Another example of such data inefficiency occurs if the batches of reference trajectories are nonoverlapping, as in the example shown in figure 3.21. As demonstrated there, it is not necessary to learn the DS parameters Θ_f^i associated with the previous batch from scratch—it would suffice only to learn a new linear DS on the data from the new batch that are not overlapping the previously learned model.

Thus, we propose an incremental learning algorithm that can update a DS $f^i(x)$ learned on a data set $\{\mathbf{X}, \dot{\mathbf{X}}\}^i$ with new incoming reference trajectories $\{\mathbf{X}, \dot{\mathbf{X}}\}^{i+1}$, while reusing the parameters Θ_γ^i and Θ_f^i learned from the previous batch, solely using the newly arrived batch of data $\{\mathbf{X}, \dot{\mathbf{X}}\}^{i+1}$ and preserving GAS.

An incremental learning approach for LPV-DS Consider each batch of demonstrations $\{\mathbf{X}, \dot{\mathbf{X}}\}^i$ and $\{\mathbf{X}, \dot{\mathbf{X}}\}^{i+1}$ representing two independent DS:

$$f^i(x) = \sum_{k=1}^{K^i} \gamma_k^i(x)(A^{k^i}x + b^{k^i}) \tag{3.48}$$

$$f^{i+1}(x) = \sum_{k=1}^{K^{i+1}} \gamma_k^{i+1}(x)\left(A^{k^{i+1}}x + b^{k^{i+1}}\right). \tag{3.49}$$

The updated DS, $\tilde{f}^{i+1}(x) = f^i(x) \oplus f^{i+1}(x)$, is constructed by merging equations (3.48) and (3.49), with operator $\oplus$ denoting the merge operation of two functions or parameter spaces. The updated set of parameters is then $\tilde{\Theta}^{i+1} = \left\{ \{\Theta_\gamma^i, \Theta_f^i\} \oplus \{\Theta_\gamma^{i+1}, \Theta_f^{+1}\} \right\}$, which results not in a concatenation, but a merging of parameters. In order to preserve global stability with the updated set of parameters used in the updated DS $\tilde{f}^{i+1}(x)$, we propose the following sufficient conditions derived from a QLF.

Theorem 3.4 *The merged DS, $\tilde{f}^{i+1}(x) = \mathrm{f}^i(x) \oplus \mathrm{f}^{i+1}(x)$, composed of individual DSs [equations (3.48) and (3.49)] is globally asymptotically converging at the attractor x^* if*

$$\sum_{k=1}^{K^i} \gamma_k^i(x) + \sum_{k=1}^{K^{i+1}} \gamma_k^{i+1}(x) = 1$$

$$(A^{k^i})^T + A^{k^i} \prec 0, \quad b^{k^i} = -A^{k^i}x^* \qquad \forall\, k^i = 1, \ldots, K^i \tag{3.50}$$

$$(A^{k^{i+1}})^T + A^{k^{i+1}} \prec 0, \quad b^{k^{i+1}} = -A^{k^{i+1}}x^* \quad \forall\, k^{i+1} = 1, \ldots, K^{i+1}.$$

Proof *We first show that $\mathrm{V}(x) = (x - x^*)^T(x - x^*)$ is a candidate Lyapunov function. $\mathrm{V}(x) > 0 \forall x \neq x^*$ and $\mathrm{V}(x^*) = 0$. Taking the derivative of $\mathrm{V}(x)$ with regard to time, we have*

$$\dot{\mathrm{V}}(x) = \frac{\partial \mathrm{V}(x)}{\partial x}\frac{\partial x}{\partial t} = (x - x^*)^T \tilde{\mathrm{f}}^{i+1}(x)$$

$$= (x - x^*)^T \left(\sum_{k=1}^{K^i} \gamma_k^i(x)(A^{k^i}x + b^{k^i}) + \sum_{k=1}^{K^{i+1}} \gamma_k^{i+1}(x)(A^{k^{i+1}}x + b^{k^{i+1}}) \right)$$

$$= (x - x^*)^T \left(\sum_{k=1}^{K^i} \gamma_k^i(x) \underbrace{(A^{k^i})(x - x^*)}_{\text{via equation }(3.50)} + \sum_{k=1}^{K^{i+1}} \gamma_k^{i+1}(x) \underbrace{(A^{k^{i+1}})(x - x^*)}_{\text{via equation }(3.50)} \right)$$

$$= (x - x^*)^{\mathrm{T}} \left(\sum_{k=1}^{K^i} \gamma_k^i(x) A^{k^i} + \sum_{k=1}^{K^{i+1}} \gamma_k^{i+1}(x) A^{k^{i+1}} \right) (x - x^*)$$

$$= (x - x^*)^{\mathrm{T}} \Big(\underbrace{\sum_{k=1}^{K^i + K^{i+1}} \tilde{\gamma}_k^{i+1}(x)}_{>0 \ \textit{via equation (3.50)}} \underbrace{\left(\tilde{A}^{k^{i+1}} \right)}_{\prec 0 \ \textit{via equation (3.50)}} \Big) (x - x^*) < 0 \qquad (3.51)$$

where $\sum_{k=1}^{K^i + K^{i+1}} \tilde{\gamma}_k^{i+1}(x) = \sum_{k=1}^{K^i} \gamma_k^i(x) + \sum_{k=1}^{K^{i+1}} \gamma_k^{i+1}(x)$ and $\tilde{A}^{k^{i+1}}$ is the kth matrix in the set of concatenated matrices $\tilde{A}^{i+1} = \{A^{1^i}, \ldots, A^{K^i}, A^{1^{i+1}}, \ldots, A^{K^{i+1}}\}$. By substituting $x = x^*$ in V(x) [equation (3.11)] and $\dot{V}(x)$ [equation (3.51)], we ensure that $V(x^*) = 0$, $\dot{V}(x^*) = 0$. Therefore, $\tilde{f}^{i+1}(x)$ is GAS with respect to an attractor x^* if the conditions in equation (3.50) are satisfied.   □

Note that the stability constraints imposed on each linear system matrix A^{k^*} for its corresponding k^*th linear DS, be it from the previous (*i*th) or the current batch ($i + 1$th), are independent of each other. Hence, so long as each linear system matrix $A^{k^i} \ \forall k = 1, \ldots, K^i$ and $A^{k^{i+1}} \ \forall k = 1, \ldots, K^{i+1}$ is negative definite as defined in theorem 3.4, then the merged DS, $\tilde{f}^{i+1}(x) = f^i(x) \oplus f^{i+1}(x)$, remains GAS.

Incremental parameter update algorithm Because the constraints on the A^{k^*} matrices are independent, one can optimize for only the new set of K^{i+1} linear dynamics, without updating the old set of K^i linear dynamics parameters. An implicit constraint from the stability conditions introduced in theorem 3.4 is that the merged set of linear dynamics $\tilde{\Theta}^{i+1} = \left\{ \{\Theta_\gamma^i, \Theta_f^i\} \oplus \{\Theta_\gamma^{i+1}, \Theta_f^{+1}\} \right\}$ do not overlap. If this holds and the system dynamics matrices are negative definite, then the incremental LPV-DS preserves its original stability properties after every update. Hence, given a DS learned on the *i*th batch, $f^i(x) : \mathbb{R}^N \to \mathbb{R}^N$ with mixing function parameters Θ_γ^i and DS parameters Θ_f^i, if a new batch of reference trajectories $\{X, \dot{X}\}^{i+1}$ arrives, we perform the following steps:

1. Learn mixing function parameter (Θ_γ^{i+1}) with the PC-GMM algorithm.

2. Update the mixing function parameters $\tilde{\Theta}_\gamma^{i+1} = \Theta_\gamma^i \to \Theta_\gamma^{i+1}$ with the Kullback-Leibler divergence metric, selecting which Gaussians to merge to the existing model.

3. Update the priors of the GMM mixing function such that the stability conditions hold; that is, $\sum_{k=1}^{K^i + K^{i+1}} \pi^k = 1$ so that $\sum_{k=1}^{K^i} \gamma_k^i(x) + \sum_{k=1}^{K^{i+1}} \gamma_k^{i+1}(x) = 1$ can be ensured.

4. Learn DS parameters corresponding to the newly merged Gaussians via the SDP optimization [equation (3.46)] with constraint option (*O1*).

5. If new data arrive, repeat step 1.

Next, we describe some of the steps of the incremental parameter update algorithm in more detail.

Step 2: Updating the mixing function parameters $\tilde{\Theta}_\gamma^{i+1} = \Theta_\gamma^i \to \Theta_\gamma^{i+1}$ from new data. Given new data, we learn a new *PC*-GMM Θ_γ^{i+1}. We then follow the incremental estimation of the GMM's approach for the online data stream clustering approach

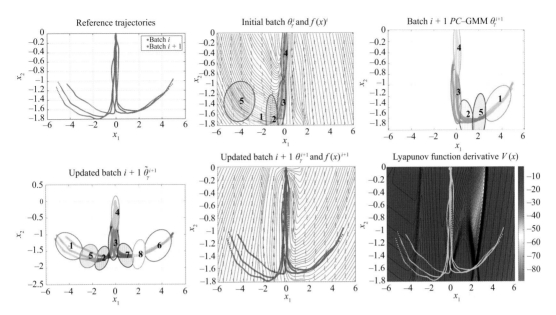

Figure 3.22
Example 2 of the incremental LPV-DS learning approach.

presented in [139]. However, instead of using multivariate statistical tests for equality of covariance and mean as a merging strategy, we use the Kullback-Leibler divergence $D_{KL}(\mathcal{N}(\mu^{k^i}, \Sigma^{k^i})||\mathcal{N}(\mu^{k^{i+1}}, \Sigma^{k^{i+1}}))$ between each kth Gaussian component of each GMM. If any bidirectional $D_{KL} < \tau$ is less than a threshold $\tau \in \Re$ (which is normally set to 1), then the Gaussian functions are deemed similar and their sufficient statistics are used to update the components. See figures 3.21 and 3.22.

Step 4: Learn DS parameters $\Theta_f^i \to \Theta_f^{i+1}$ for new local components. For the new set of Gaussian components created from step 3, we now compute the DS parameters Θ_γ^{i+1} on the newly arrived training data $\{\mathbf{X}, \dot{\mathbf{X}}\}^{i+1}$ by solving the constrained optimization problem [equation (3.46)] with variant ($O1$).

Exercise 3.9 *Derive the sufficient conditions for the merged DS $\tilde{f}^{i+1}(x) = f^i(x) \oplus f^{i+1}(x)$ to be GAS with a P-QLF.*

Exercise 3.10 *Assume that you use a P-QLF to derive the sufficient conditions for $\tilde{f}^{i+1}(x) = f^i(x) \oplus f^{i+1}(x)$ to remain GAS. Derive an algorithm to update both the LPV-DS parameters and P as new batches of data arrive.*

3.4.4 Offline Learning Algorithm Evaluation

To quantitatively evaluate the physical consistency of the stable estimator of LPV DSs (SE-LPVDS), we use the entire LASA Handwriting Dataset (excluding the multimodel

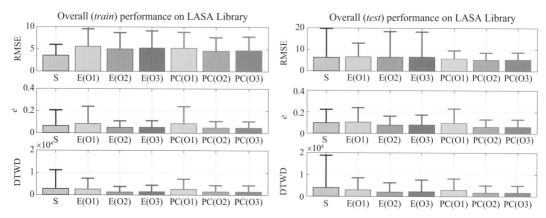

Figure 3.23
Overall performance metrics on LASA Library. **(Left)** Performance on the training data. **(Right)** Performance on the testing data. Each bar graph shows the mean (and standard) of the error for every approach. S=SEDS, $E(\cdot)$=EM-GMM, $PC(\cdot)$=PC-GMM.

motions)—that is, twenty-six motion sets. Because each motion set contains seven trajectories, we use four to train our models and evaluate reproduction accuracy, and the remaining three to test generalization accuracy (i.e., the reproduction accuracy of unseen trajectories). We employ three metrics: (1) prediction RMSE $= \frac{1}{M} \sum_{m=1}^{M} ||\dot{x}^m - f(x^m)||$ as in [99]; (2) prediction cosine similarity $\dot{e} = \frac{1}{M} \sum_{m=1}^{M} \left| 1 - \frac{f(x^m)^T \dot{x}^m}{||f(x^m)^T \dot{x}^m|| ||\dot{x}^m||} \right|$ as in [100]; and (3) dynamic time warping distance (DTWD) as in [122]. While (1) and (2) give an overall similarity of the shape of the resulting DS with regard to the demonstrations, (3) measures the dissimilarity between the shapes of the reference trajectories and their corresponding reproductions from the same initial points.

Figure 3.23 shows the performance of SEDS (S), EM-based GMM $E(\cdot)$ fitting, and PC-GMM $PC(\cdot)$ with ($O1 - 3$) LPV-DS optimization variants [equation (3.46)]. Root-mean-square error (RMSE) is comparable throughout all methods, as this metric is not representative of reproduction accuracy. Yet, if we focus on $\dot{e}$ and the DTWD on the training set, methods $E(O2 - 3)$ and $PC(O2 - 3)$ clearly outperform SEDS with a drastic gap in the DTWD. While the $E(O2 - 3)$ approaches have comparable accuracy on the training set, their error increases on the testing set. This indicates that the $E(O2 - 3)$ are overfitting and only locally shaping the DS, and yet the overall shape of the motion is not being generalized. The relative train/test errors for the LPV-DS approach tend to be in the same range.

In figure 3.24c (continued), we illustrate the PC-GMM fit and the learned LPV-DS with variant ($O3$) for the subset of motions in which SEDS fails or underperforms (see figures 3.8 to 3.11). As shown, for all these motions (except for *BendedLine*), the LPV-DS approach is capable of outperforming SEDS in terms of reproduction accuracy and motion pattern shaping. We omit illustrating the results of optimization variants ($O1$), as doing so produces results comparable to SEDS; and ($O2$), as doing so results in DSs similar to those found with ($O3$) when given a good initial P. Further quantitative comparisons are found in [43].

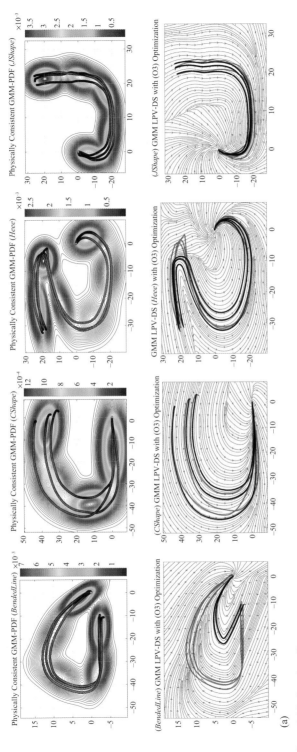

Figure 3.24a (continued)

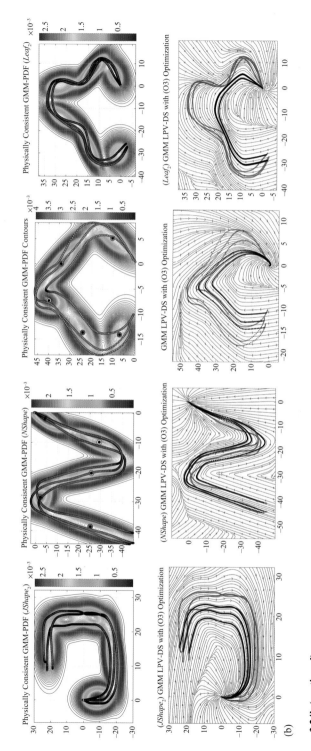

Figure 3.24b (continued)

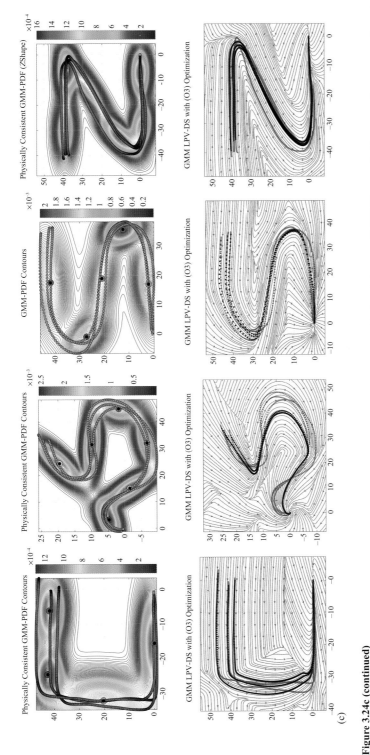

Figure 3.24c (continued)
Motion with high nonlinearities from the LASA Handwriting Dataset that are successfully learned with the PC-GMM and LPV-DS approach. From top to bottom rows, for each pair of rows, we show the PC-GMM density superposed to the training trajectories (top row) and (bottom row) the learned DS estimated with LPV-DS via Eq. (O3).

Programming Exercise 3.4   *The aim of this instructional exercise is to familiarize the readers with the SE-LPVDS learning algorithm.*

Instructions: *Open MATLAB and set the directory to the folder corresponding to chapter 3 exercises; within this folder, you will find the following scripts:*

```
1  ch3_ex0_drawData.m
2  ch3_ex4_lpvDS.m
```

In these scripts, you will find instructions in block comments that will enable you to do the following:

- *Draw 2D trajectories on a GUI*
- *Load the LASA Handwriting Dataset shown in figure 3.7*
- *Learn first-order nonlinear DS with the SE-LPVDS variants*

Note: *The MATLAB script provides options to load/use different types of data sets and to select options for the learning algorithm. Please read the comments in the script and run each block separately.*

We recommend that readers test the following with either self-drawn trajectories or motions from the LASA Handwriting Dataset:

1. Compare results qualitatively (phase plot) and quantitatively (mean-square error) with different optimization variants (O1–O3).

2. Compare results with different GMM fitting approaches:
   (i) EM with Model Selection via BIC
   (ii) Collapsed sampler for CRP-GMM
   (iii) Collapsed sampler for PC-GMM

3. Compare results when using $M = 1$ versus $M \gg 1$ trajectories.

4. Draw self-intersecting trajectories. Is SE-LPVDS suitable for these?

5. Draw trajectories where the initial and end points are close to each other. Is SE-LPVDS suitable for these?

6. Use the block of code provided in `ch3_ex4_lpvDS.m` *to implement the incremental learning algorithm. Can you think of other metrics to determine if the Gaussian functions are overlapping?*

Readers can also use an instructional GUI, such as the one shown in figure 3.25, to complement the previous exercise.

With this GUI, only the self-drawn trajectories can be used to learn DS models via SE-LPVDS variants. However, it provides a robot simulation that allows the user to execute the motion of the learned DS and apply perturbations to the robot while it is executing the task.

Instructions: *Open MATLAB and set the directory to the folder corresponding to chapter 3 exercises; in this folder, you will find the following script:*

```
1  gui_lpvDS.m
```

or, alternatively typing in the MATLAB Command window:

```
1  >> gui_lpvDS
```

With this GUI, the user will be able to do the following:

- *Draw trajectories within the workspace of a 2-DOF robot arm.*

- *Manually select* K, *or use model selection via BIC or an automatic PC-GMM fit.*
- *Learn a LPV-DS with optimization variant (O1) or (O3).*
- *Simulate the robot motion following the learned DS.*
- *Apply perturbations to the robot while it is executing the motion.*

Note: *An instructional movie on how to use this GUI can be found at the following YouTube video:* `https://youtu.be/jDcPaOUMwvI`.

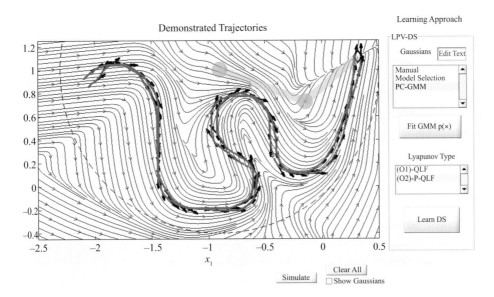

Figure 3.25
The GUI for the SE-LPVDS learning evaluation provided in MATLAB source code.

3.4.5 Robotic Implementation

Next, we present three implementations of the LPV-DS method to learn control laws for robotic systems. To showcase that the approach does not depend on the type of robot used, we consider an implementation on three very different platforms and different tasks: (1) a 7-DOF KUKA LWR 4+ robotic arm for manipulation tasks; (2) a simulated, semiautonomous wheelchair for navigation tasks; and (3) iCub humanoid robots for navigation and co-manipulation tasks.

3.4.5.1 Manipulation task

The manipulation tasks involve learning a complex 3D motion corresponding to a task that should be performed by a 7-DoF robot arm. We validate this approach for three tasks: an inspection task (figure 3.26), a branding task (figure 3.27), and a shelf-arranging task (figure 3.28).

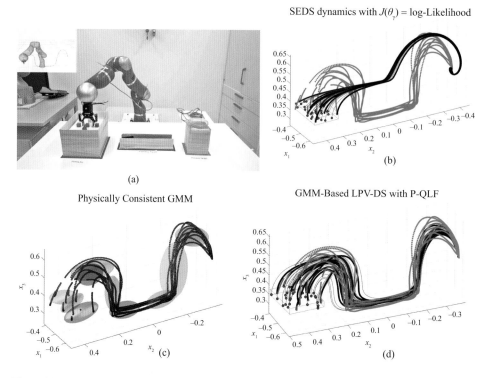

SEDS dynamics with $J(\theta_\gamma) = \log\text{-Likelihood}$

(a)

(b)

Physically Consistent GMM

GMM-Based LPV-DS with P-QLF

(c)

(d)

Figure 3.26
Inspection task: (a) Illustration of the experimental set-up. The robot is tasked to put the colored boxes from the far left box on the table to the box on the far right of the table. Its nominal DS moves in straight line. A teacher provides demonstrations to teach the robot to move through the narrow passage created by the rectangular box in the middle. (b) red trajectories are demonstrations, **black** trajectories are generated by a learned SEDS model (from Section 3.3) from initial positions used for learning, and blue trajectories are generated by the learned SEDS model from new initial positions sampled from the volume of initial positions used for learning (blue markers); (c) 3D trajectories collected from kinesthetic demonstrations, and the ellipsoids represent the fitted Gaussian with the PC-GMM algorithm; (d) 3D demonstrations and execution of the learned LPV-DS; the trajectory colors follow the same convention as for SEDS execution.

Data collection: For each tasks, we collected between nine and twelve trajectories of successful demonstrations. The trajectories are collected kinesthetically by controlling a 7-DoF KUKA LWR+ robot arm in gravity compensation mode. The data was collected at 100 Hz. Each trajectory can have a different length, and there is no need for alignment or dynamic time warping across the demonstrations. The raw trajectories contain only position measurements. After collection, we then compute the velocity measurements by numerical differentiation. While all the trajectories can be used to learn a model, as opposed to SEDS, the LPV-DS learning scheme and formulation do not rely on the joint distribution; hence, sufficient variance in the demonstrations is not necessary. In practice, an LPV-DS can be learned with the $M = 1$ trajectory, but, to ensure accurate reproduction of highly nonlinear motions, we recommend $M \geq 3$ trajectories for learning.

Learning and Evaluation: After selecting the demonstrated trajectories, we preprocess them with a Savitsky-Golay filter to smooth out the trajectories. The smoother the position and velocity profiles, the smoother the resulting DS will be. The trajectories can also be

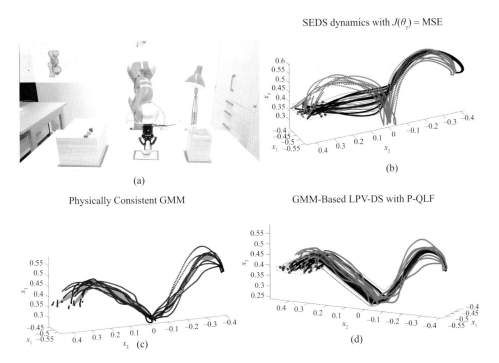

Figure 3.27
Branding task: (a) Variant from task in figure 3.26. The robot must move through one particular point located in the blue box in the middle. (b) red trajectories are demonstrations, **black** trajectories are generated by a learned SEDS model (from section 3.3) from initial positions used for learning and blue trajectories are generated by the learned SEDS model from new initial positions sampled from the volume of initial positions used for learning (blue markers); (c) 3D trajectories collected from kinesthetic demonstrations, and the ellipsoids represent the fitted Gaussian with the PC-GMM algorithm; (d) 3D demonstrations and execution of the learned LPV-DS; trajectory colors follow the same convention as for SEDS execution.

subsampled to have fewer samples for trajectory and make the learning scheme more efficient. Of course, if one subsamples the data too much, then we risk losing accuracy in reproduction. In our experiments, we subsample the trajectories by a factor of 2. Once this preprocessing is done, we perform the following optimization steps:

1. Learn mixing function parameters (Θ_γ^{i+1}) with the PC-GMM sampling algorithm.

2. Estimate P for the P-QLF candidate Lyapunov function by solving by solving the constrained optimization problem in equation (3.47).

3. Learn DS parameters corresponding to the newly merged Gaussians via the SDP optimization [equation (3.46)] with constraint option ($O3$).

In figures 3.26, 3.27, and 3.28 show the following:

• For the demonstrated preprocessed trajectories in **red**, we illustrate the fitted GMM with ellipsoids superimposed on the trajectories.

• The **black** trajectories are open-loop trajectories generated by integrating the learned LPV-DS, with initial positions $x(0) \in \mathbb{R}^N$ equivalent to those used for learning.

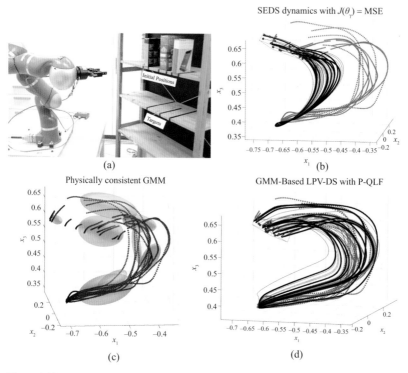

Figure 3.28
Shelf-arranging task: (a) The robot is taught to bring the books from top shelf to the middle shelf in a dedicated location for each book. (b) red trajectories are demonstrations, **black** trajectories are generated by a learned SEDS model (from Section 3.3) from initial positions used for learning, and the blue trajectories are generated by the learned SEDS model from new initial positions sampled from the volume of initial positions used for learning (blue markers); (c) 3D trajectories collected from kinesthetic demonstrations, and ellipsoids represent the fitted Gaussian with the PC-GMM algorithm; (d) 3D demonstrations and execution of learned LPV-DS; trajectory colors follow the same convention as SEDS execution.

- The **blue** trajectories are open-loop trajectories generated by integrating the learned LPV-DS, with initial positions $x(0) \in \mathbb{R}^N$, randomly sampled from a cube enclosing all the initial positions from the demonstrated trajectories.

These figures also show the open-loop trajectories generated by integrating a learned SEDS model (from section 3.3); as can be seen, SEDS fails to accurately encode this complex, highly nonlinear motions, while the LPV-DS approach succeeds.

Robot control and execution: To use the learned LPV-DS to perform each of the learned tasks, we employ the KUKA LWR 4+ arm that was utilized to collect the trajectories via kinesthetic teaching. However, during execution, we control the robot in torque mode via the closed-loop DS-based impedance control-law [82] that will be described in chapter 10. We do not need to integrate the DS in this control mode, as is it a negative feeback velocity error contro law. Due to the passivity provided by this control law, we can actively perturb the robot while executing the commanded velocities from the learned DS, showcasing the ability of our learned LPV-DS models to be robust to perturbations.

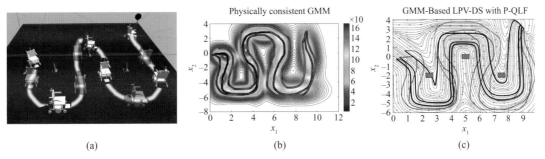

Figure 3.29
Wheelchair navigation task: (a) A wheelchair simulated in GAZEBO is trained to travel along a snake like path. (b) 2D demonstrations collected from hand-drawn trajectories via a MATLAB GUI, and ellipsoids represent the fitted Gaussian with the PC-GMM algorithm; (c) red trajectories are demonstrations and **black** trajectories are open-loop trajectories generated by integrating the learned LPV-DS model from the initial positions used for learning.

3.4.5.2 Wheelchair navigation task

We validate the LPV-DS approach on a navigation scenario for a semiautonomous wheelchair in the Gazebo physics simulator. As shown in figure 3.29, we create a constrained road navigation scenario in which the wheelchair must circle around the traffic cones before reaching the target (in this case, the stop sign).

Data collection: For this 2D road navigation task, we re-create the map from Gazebo in MATLAB and use the GUI provided in the programming exercises to draw the desired trajectories. We collected three trajectories with 300 to 500 samples each. We also scale the velocity such that the DS can be learned with $dt = 0.01$ s for the control of the wheelchair in simulation.

Learning and evaluation: We apply the same preprocessing and learning steps as in the manipulation task described in section 3.4.5.1. We show the task, trajectories, fitted GMM, and reproduced open-loop trajectories in figure 3.29, and we use the same color convention. We do not include a comparison to SEDS, as this approach will clearly fail in this highly-nonlinear motion, as evidenced in figures 3.10 and 3.11 for the w-shape, which is quite similar to our demonstrated trajectories.

Robot control and execution: The wheelchair is controlled at the velocity level, so at each time step, the desired velocity is computed directly from the LPV-DS and the angular velocity is computed such that the heading of the wheelchair aligns to the direction of motion defined by the DS. Further details on how the learned DS $f(x)$ is used to control the wheelchair are provided in [42]. The perturbations were simulated by applying external forces to the wheelchair. The wheelchair is robust to perturbations, as it will continue to follow the desired path delineated by the DS. If the perturbation is too high and the wheelchair is too close to the target, then it will simply converge it. This scenario showcases the advantage of using DS for navigation, as it is capable of replanning the entire path on the fly. Furthermore, as will be demonstrated in chapter 9, one can modulate a DS to perform online obstacle avoidance for robust, real-time navigation.

3.4.5.3   iCub navigation/co-manipulation task

We validated the LPV-DS approach for motion planning in a navigation/co-manipulation scenario between two iCub humanoid robots (see figure 3.30). The simulated task requires that the two robots grab the rectangular object and move it to the other side of the wall. In this case, the learned LPV-DS defines the desired velocity of the object; hence, the problem is treated as a 2D navigation task, as in section 3.4.5.2.

Data collection: The reference trajectories for this scenario were collected by teleoperating the robots within the Gazebo physics simulator. Specifically, the robots are being controlled in Gazebo with compliant, low-level walking controllers [38] [37] and an arm impedance control law as described in detail in [44]. After the robots have succesfully grasped the object, the user provides the desired motion of the object via keyboard inputs emulating a joystick for teleoperation. We collected six demonstrations, three of them circling the wall from the left and the other three from the right, as seen in figure 3.30. Data is recorded at 100 Hz.

Learning and evaluation: We apply the same preprocessing and learning steps as in the manipulation task described in section 3.4.5.1. We show the task, trajectories, fitted GMM, and reproduced open-loop trajectories in figure 3.29, we use the same color convention as in section 3.4.5.1. We do not include comparison to SEDS as this approach will clearly fail in this highly-nonlinear motion as evidenced in figure 3.11 and 3.10 for the w-shape, which is quite similar to our demonstrated trajectories.

Robot control and execution: The learned LPV-DS, $f(x)$, generates the desired velocities for the object, which are then transformed to velocities in the frame of each robot's base and fed to a compliant, low-level walking controller [38] [37]. Further details of the control architecture used for the iCub humanoid robots are beyond the scope of this book, but interested readers can learn more in [44].

In this subsection, we have introduced an alternative to the SEDS learning approach that is capable of learning a wider range of complex motions. Historically, DSs have been mainly used to encode simple reaching motions. With the LPV-DS formulation and SE-LPVDS learning scheme, we are capable of encoding and reproducing entire tasks in a single DS that otherwise would have needed to be discretized into simpler motions. As evidenced by our robotic experiments, the LPV-DS is capable of reproducing highly nonlinear manipulation and navigation tasks that go beyond simply reaching to a target. Further, by untying the parameters in the mixture of linear DS formulation, we are capable of learning a DS incrementally. Although we only show examples in simulation, to the best of our knowledge, a nonlinear DS that ensures GAS and allows an incremental learning approach does not exist. Finally, even though the SE-LPVDS learning scheme outperforms SEDS for highly complex nonlinear motions, it is not equipped to encode motions that exhibit self-intersecting trajectories or trajectories whose initial and end points coincide or are close to each other (resembling a loop), as the *BendedLine* or *DoubleBendedLine* motions from the LASA Handwriting Dataset.

The main issue with self-intersecting trajectories is that velocity directions will be ambiguous at the self-intersections. Because the mixing/activation function $\gamma(x)$ is solely dependent on position, the direction to follow is difficult to define. This problem can be

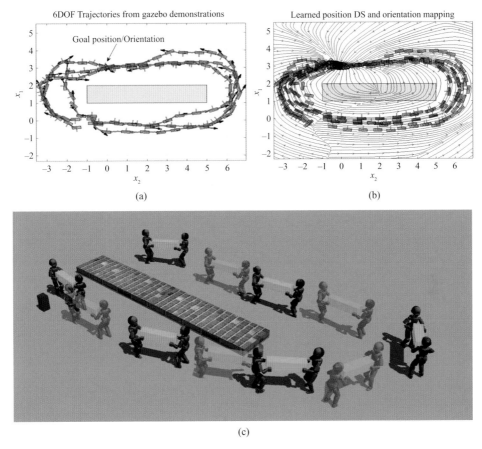

Figure 3.30
iCub navigation/co-manipulation around a wall; (a) 2D trajectories of the rectangular object being teleoperated in Gazebo, with the position and orientation of the object illustrated with red rectangles; (b) 2D demonstrated trajectories (red rectangle) and execution with the learned DS (**dark gray rectangle**); (c) set of iCubs executing the learned task using the LPV-DS model.

solved if one defines the motion in terms of position, velocity, and acceleration (i.e., a second-order dynamics as introduced in section 3.5). Furthermore, for trajectories that resemble loops, a sequence of DS would be more suitable. This topic will be addressed further in chapter 5.

3.5 Learning Stable, Second-Order DSs

In previous sections, we introduced methods to learn first-order DSs [i.e., $\dot{x} = f(x)$]. Here, we showcase how the previous approaches can be extended to learn second-order DSs [i.e., $\ddot{x} = f(x, \dot{x})$].[3] Such DSs are useful when one seeks to generate smooth and feasible motions for a robot (by constraining velocity/acceleration) or when ambiguities in the velocity direction are present in the demonstrations (such as self-intersecting motions).

3.5.1   Second-Order LPV-DS Formulation

A second-order LPV-DS can be formulated as follows:

$$
\begin{bmatrix} \dot{x} \\ \ddot{x} \end{bmatrix} = \sum_{k=1}^{K} \gamma_k(x,\dot{x}) \left(\begin{bmatrix} 0 & I \\ A_1^k & A_2^k \end{bmatrix} \begin{bmatrix} x \\ \dot{x} \end{bmatrix} + \begin{bmatrix} 0 \\ b^k \end{bmatrix} \right),
\tag{3.52}
$$

where $x, \dot{x}, \ddot{x} \in \mathbb{R}^N$ denote the position., velocity, and acceleration of the robot state, respectively, Further, $\{A_1^k, b^k\}_{k=1}^{K}$ is the set of linear DS parameters describing the first-order behavior in the system, while $\{A_2^k\}_{k=1}^{K}$ is the set of linear system matrices that describe the second-order behavior. Finally, the activation/mixing function $\gamma_k(x,\dot{x})$ now relies not only on the position variables, but also on the velocities. Hence, in order to learn a $\gamma_k(x,\dot{x})$ that is normalized as $\forall k = 1, \ldots, K$ as in the first-order cases, we parametrize it as the a posteriori probability, $p(k|[x\ \dot{x}]^T, \theta_\gamma^k)$, of the GMM density learned from the position and velocity measurements $p([x\ \dot{x}]^T | \Theta_\gamma)$ [equation (3.3)], as shown here:

$$
\gamma_k(x,\dot{x}) = \frac{\pi_k p([x\ \dot{x}]^T | \mu^k, \Sigma^k)}{\sum_{i=1}^{K} \pi_i p([x\ \dot{x}]^T | \mu^i, \Sigma^i)},
\tag{3.53}
$$

with $\mu^k = \begin{bmatrix} \mu_x^k \\ \mu_{\dot{x}}^k \end{bmatrix}$, $\Sigma^k = \begin{bmatrix} \Sigma_x^k & \Sigma_{x\dot{x}}^k \\ \Sigma_{\dot{x}x}^k & \Sigma_{\dot{x}}^k \end{bmatrix}$ and $\sum_{k=1}^{K} \pi_k = 1$ for $0 \leq \pi_k \leq 1$,   and $p([x\ \dot{x}]^T | \mu^k, \Sigma^k) = \mathcal{N}([x\ \dot{x}]^T | \mu^k, \Sigma^k)$ is the Gaussian function. Further, if we expand equation (3.53) and set $b^k = 0\ \forall k \in \{1, \ldots, K\}$, which is intuitively placing the attractor/target at the origin, we achieve the following reduced second-order LPV DS:

$$
\ddot{x} = \sum_{k=1}^{K} \gamma_k(x,\dot{x}) A_1^k x + \sum_{k=1}^{K} \gamma_k(x,\dot{x}) A_2^k \dot{x}
\tag{3.54}
$$

One can then directly apply theorem 3.3 to equation (3.52) and derive the sufficient conditions for GAS described next.

Theorem 3.5   *The second-order LPV-DS defined in equation (3.52) is GAS at the origin* $x^* = 0$—*that is,* $\lim_{t \to \infty} [x\ \dot{x}]^T = [0\ 0]^T$ *if,*

$$
\begin{cases} \begin{bmatrix} 0 & I \\ A_1^k & A_2^k \end{bmatrix}^T P + P \begin{bmatrix} 0 & I \\ A_1^k & A_2^k \end{bmatrix} \prec 0 & \forall k = 1, \ldots, K, \\ b^k = 0 \end{cases}
\tag{3.55}
$$

with a P-QLF used as the Lyapunov candidate function, i.e., $P = P^T \succ 0$. $\square$

Exercise 3.11   *Derive the sufficient stability conditions presented in equation (3.55).*

Exercise 3.12   *What if the attractor* x^* *is not placed at the origin? Derive the sufficent condition for* b^k.

3.5.1.1 Change of variables for GMR interpretation

Interestingly, the second-order LPV-DS can be interpreted as a constrained GMR regressor, as in the SEDS approach (discussed in section 3.3.1). Namely, if we learn a GMM joint density of the positions, velocities, and acceleration measurements in the following form:

$$p(\underbrace{x, \dot{x}}_{\text{Inputs}}, \underbrace{\dot{x}, \ddot{x}}_{\text{Outputs}} | \Theta_{GMM}) = \sum_{k=1}^{K} \pi_k p(x, \dot{x}, \dot{x}, \ddot{x} | \mu^k, \Sigma^k) \tag{3.56}$$

with

$$\mu^k = \begin{bmatrix} \mu_x^k \\ \mu_{\dot{x}}^k \\ \mu_{\dot{x}}^k \\ \mu_{\ddot{x}}^k \end{bmatrix}, \qquad \Sigma^k = \begin{bmatrix} \Sigma_x^k & \Sigma_{x\dot{x}}^k & \Sigma_{x\dot{x}}^k & \Sigma_{x\ddot{x}}^k \\ \Sigma_{\dot{x}x}^k & \Sigma_{\dot{x}}^k & \Sigma_{\dot{x}}^k & \Sigma_{\dot{x}\ddot{x}}^k \\ \Sigma_{\dot{x}x}^k & \Sigma_{\dot{x}}^k & \Sigma_{\dot{x}}^k & \Sigma_{\dot{x}\ddot{x}}^k \\ \Sigma_{\ddot{x}x}^k & \Sigma_{\ddot{x}}^k & \Sigma_{\ddot{x}}^k & \Sigma_{\ddot{x}}^k \end{bmatrix} \tag{3.57}$$

and consider the vector $[x \ \dot{x}]^T$ as the input and $[\dot{x} \ \ddot{x}]^T$ as the output to the regressive model, then the second-order DS can be computed as the posterior mean of the conditional distribution $p([\dot{x} \ \ddot{x}]^T | [x \ \dot{x}]^T)$ as follows:

$$[\dot{x} \ \ddot{x}]^T = f([x \ \dot{x}]^T; \ \Theta_{GMR})$$

$$= \mathbb{E}\{p([\dot{x} \ \ddot{x}]^T | [x \ \dot{x}]^T)\} = \sum_{k=1}^{K} \gamma_k(x, \dot{x}) \left(\begin{bmatrix} 0 & I \\ A_1^k & A_2^k \end{bmatrix} \begin{bmatrix} x \\ \dot{x} \end{bmatrix} + \begin{bmatrix} 0 \\ b^k \end{bmatrix} \right). \tag{3.58}$$

Equation (3.58) is derived due to the following change of variables:

$$\begin{cases} \begin{bmatrix} 0 & I \\ A_1^k & A_2^k \end{bmatrix} = \begin{bmatrix} \Sigma_{\dot{x}x}^k & \Sigma_{\dot{x}}^k \\ \Sigma_{\ddot{x}x}^k & \Sigma_{\ddot{x}}^k \end{bmatrix} \begin{bmatrix} \Sigma_x^k & \Sigma_{x\dot{x}}^k \\ \Sigma_{\dot{x}x}^k & \Sigma_{\dot{x}}^k \end{bmatrix}^{-1} \\[12pt] b^k = \mu_{\ddot{x}}^k - A_2^k \mu_{\dot{x}}^k - A_1^k \mu_x^k \\[12pt] \gamma_k(x, \dot{x}) = \dfrac{\pi_k p\left([x \ \dot{x}]^T \Big| [\mu_x^k \ \mu_{\dot{x}}^k]^T, \begin{bmatrix} \Sigma_x^k & \Sigma_{x\dot{x}}^k \\ \Sigma_{\dot{x}x}^k & \Sigma_{\dot{x}}^k \end{bmatrix} \right)}{\sum_{i=1}^{K} \pi_i p\left([x \ \dot{x}]^T \Big| [\mu_x^i \ \mu_{\dot{x}}^i]^T, \begin{bmatrix} \Sigma_x^i & \Sigma_{x\dot{x}}^i \\ \Sigma_{\dot{x}x}^i & \Sigma_{\dot{x}}^i \end{bmatrix} \right)}. \end{cases} \tag{3.59}$$

Notice that by following the parameter estimation approach of SE-LPVDS, we will estimate values for $\Theta_\gamma = \{\pi_k, [\mu_x^k \ \mu_{\dot{x}}^k]^T, \begin{bmatrix} \Sigma_x^k & \Sigma_{x\dot{x}}^k \\ \Sigma_{\dot{x}x}^k & \Sigma_{\dot{x}}^k \end{bmatrix}\}_{k=1}^{K}$ and $\Theta_f = \{A_1^k, A_2^k, b^k\}_{k=1}^{K}$, and yet not for the full set of GMM parameters of the joint density (3.57). If one seeks to compute the likelihood of the DS parameters $\mathcal{L}(\Theta_{GMR}|x)$ or the uncertainty of the predicted velocities via $\text{var}\{p([\dot{x} \ \ddot{x}]^T | [x \ \dot{x}]^T)\}$, the remaining blocks of GMM parameters [equation (3.57)] must be estimated from the set of DS system parameters Θ_f.

Exercise 3.13 *Derive the equations that lead to the change of variables presented in equation (3.59) for* $\mathbb{E}\{p([\dot{x}\ \ddot{x}]^T\,|[x\ \dot{x}]^T)\}$ *in equation (3.58).*

Exercise 3.14 *Can this change of variable technique be applied to the first-order LPV-DS formulation? Derive the equation to recover the GMM parameters of the joint density* $p(x, \dot{x})$.

3.5.2 Stable Estimator of Second-Order DSs

Given the set of reference trajectories $\{\mathbf{X}, \dot{\mathbf{X}}, \ddot{\mathbf{X}}\} = \{X^m, \dot{X}^m, \ddot{X}^m\}_{m=1}^M = \{\{x^{t,m}, \dot{x}^{t,m}, \ddot{x}^{t,m}\}_{t=1}^{T_m}\}_{m=1}^M$, and assuming the attractor to be at the origin $x^* = 0$ (i.e., the desired target), we begin by estimating the GMM parameters for the mixing function, $\gamma_k(x, \dot{x})$ of the second-order LPV-DS formulation [equation (3.52)]—that is, $\Theta_\gamma = \{\pi_k, \boldsymbol{\mu}^k, \Sigma^k\}_{k=1}^K$. This can be achieved by using any of the GMM fitting methods described in the previous section. Then, we estimate the set of linear DS system matrices $\Theta_f = \{A_1^k, A_2^k\}_{k=1}^K$ by minimizing the MSE as in the SEDS [MSE] variant [equation (3.33)] and the first-order LPV-DS [equation (3.46)], and yet on the acceleration level,

$$\min_{\Theta_f} J(\Theta_f) = \frac{1}{2L} \sum_{m=1}^M \sum_{t=0}^{T_m} \left\| f(x^{t,m}, \dot{x}^{t,m}) - \ddot{x}^{t,m} \right\|^2. \tag{3.60}$$

This is subject to

$$\begin{cases} \begin{bmatrix} 0 & I \\ A_1^k & A_2^k \end{bmatrix}^T P + P \begin{bmatrix} 0 & I \\ A_1^k & A_2^k \end{bmatrix} \prec 0 \\ 0 \prec P, \qquad P^T = P \\ b^k = 0 \end{cases} \quad \forall k = 1, \ldots, K, \tag{3.61}$$

where $L = \sum_{m=1}^M T_m$ is the total number of training data-points and $f(\cdot, \cdot)$ is computed by equation (3.54). As in the SE-LPVDS learning algorithm, the optimization problem defined in equation (3.60) with constraints [equation (3.61)] can be solved via nonlinear SDP solvers. In this case, we use PENLAB [40], together with the YALMIP MATLAB toolbox [93]. One could also estimate jointly the parameters of Θ_γ and Θ_f by using the constrained GMR interpretation described in section 3.5.1.1. For this, we would need to add the constraints on the GMM parameters as in the SEDS optimization [equation (3.32)], which converts it to an NLP problem. Although it can be solved, we advocate for the LPV-DS formulation and SPD solver solution.

3.5.3 Learning Algorithm Evaluation

In figure 3.31, we show that the SE-SODS learning approach is capable of training a Second Order-Linear Parameter Varying DSs (SO-LPVDS) (3.52) to reproduce a self-intersecting motion. We then perform a qualitative and quantitative evaluation of SE-SODS on a subset of the LASA Handwriting Dataset. Figure 3.32 shows the qualitative results of the estimated motions. In all the experiments, we ran the EM algorithm several times to compute different

Figure 3.31
SO-LPVDS learned on a set of self-intersecting reference trajectories in red. The reproductions from the DS are the trajectories in blue.

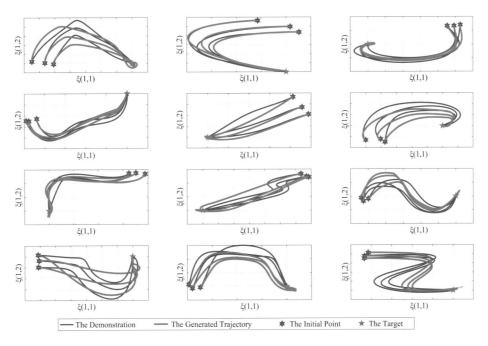

Figure 3.32
The qualitative performance evaluation of our algorithm. The red trajectories are the learned DS, and the blue lines are the training data points.

Table 3.1
Performance of SE-SODS learning scheme for 11 LASA Handwriting Dataset
motions.

Average/Range of $\dot{e}$	Average/Range of Number of the Gaussian Components	Average/Range of Training Time (s)
15.94 (6.85 − 22.35)	2.8182 (1 − 6)	78.91 (7.46 − 244.11)

GMM fits, and we illustrate the result from the best trial. The quantitative results of the
proposed method are presented in table 3.1. Further quantitative evaluation of this method
can be found in [100]. Note that while SO-LPVDS might not be as accurate at reproducing
the reference trajectories, the motions that they generate are significantly smoother than
from first-order DSs.

Programming Exercise 3.5 *The aim of this instructional exercise is to familiarize the
readers with the SE-SODS learning algorithm.*

Instructions: *Open MATLAB and set the directory to the folder corresponding to chapter 3
exercise; within this folder, you will find the following scripts:*

```
ı   ch3_ex5_soDS.m
```

*In this script, you will find instructions in block comments that will enable you to do the
following:*

- *Draw 2D trajectories on a GUI.*
- *Load the LASA Handwriting Dataset shown in figure 3.7.*
- *Learn second-order nonlinear DSs with SE-SODS.*

Note: *The MATLAB script provides options to load/use different types of data sets and to select
different options for the learning algorithm. Please read the comments in the script and run
each block separately.*

*We recommend readers to test the following with either self-drawn self-intersecting, trajecto-
ries from figure 3.31, or motions from the LASA Handwriting Dataset:*

*1. Compare qualitatively (RMSE) and quantitatively (phase plot) results with different GMM
fitting approaches as follows:*
   (i) Manually select the number of K Gaussian functions.
   (ii) EM with Model Selection via BIC

*2. Compare results in terms of RMSE and computation time when using $M = 1$ versus $M \gg 1$
trajectories.*

3.5.4   Robotic Implementation

Figure 3.33 illustrates an application of second-order control to allow the iCub to perform
a loop motion with its right hand. The motion lies in a vertical plane and thus contains a
self-intersection point. Here, the task is shown to the robot five times through kinesthetic
teaching and the number of the Gaussian components is seven. The motion of the robot

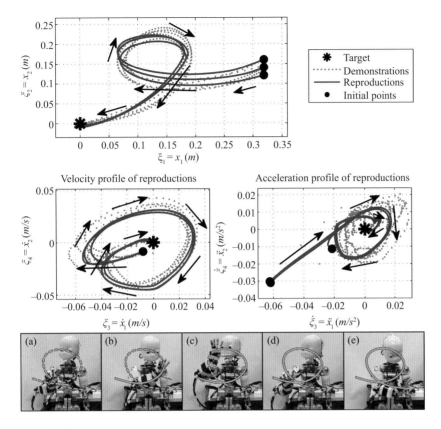

Figure 3.33
Executing a self-intersecting motion with second-order dynamics. [72]

is generated in the task level and converted to the joint space level by solving a velocity inverse kinematic solver (i.e., see figure 1.18b in chapter 1).

Other examples of applications of second order DS are presented in chapter 7. The second-order system comes in handy to allow one to reach moving targets and control for force at impact.

3.6 Conclusion

The learning approaches presented in this chapter all guarantee that the control law is stable at the desired attractor. They can enable control of velocity and acceleration of the robot. As the approach is agnostic to the structure of the robot, they can be applied to different robotic platforms. We showed a number of examples of applications for manipulating objects with a robot arm manipulator and for navigating a wheelchair or humanoid robot.

When learning robot control laws, two key factors play a role. The choice of data and the choice of hyperparameters for the learning algorithm. The methods presented here require a small amount of data. In each implementation, only a handful of demonstrations were sufficient to obtain a good model. While the first learning method, SEDS, presented in section 3.3, is sensitive to the choice of hyperparameter (K number of Gauss functions), the Bayesian treatment and the LPVDS method described in section 3.4 remove the need to determine the hyperparameter.

The approaches covered in this chapter, however, are restricted to learning one control law at a time and accepting only one attractor. These approaches are, hence, not suitable to learn motions that require (1) switching or transition across targets and dynamics and (2) following a sequence of distinct targets or dynamics.

In the next two chapters, we present DS formulations and learning approaches that tackle these problems. Namely, in chapter 4, we introduce learning approaches to encode switching strategies for multiple attractor or multiple behavior dynamics. Further, in chapter 5, we present methods to learn sequences of dynamics and how/when to transit across these dynamics.

4 Learning Multiple Control Laws

In chapter 3, we saw methods to learn a control law by embedding it in a single DS globally asymptotically stable at a single attractor. In this chapter, we explore how we can learn multiple control laws with distinct dynamics and distinct attractors. We show that we can embed these multiple control laws in a single continuous function, which makes it easy to switch at run time across control laws. While this provides more flexibility in replanning, this also increases the complexity of the underlying learning problem.

We start, in section 4.1, with an approach that learns a partitioning of the state space, whereby each partition entails one dynamical system (DS) with its own attractor. The method is called the *Augmented-SVM* framework (A-SVM). It inherits the region-partitioning ability of the well-known Support Vector Machine (SVM) classifier augmented with novel constraints for learning the demonstrated dynamics and ensuring local stability at each attractor. This is a supervised learning problem in which we know both the number of dynamics and to which dynamics each data point belongs.

While the previous approach assumes that each DS applies in a distinct part of the state space, in section 4.2 we relax this assumption and show how we can learn a DS valid through the entire state, whose characteristics can be modified through an external parameter. This is akin to the principle of bifurcation in DS theory. We show that we can embed in a single DS fixed point attractor DS and limit cycles. Furthermore, we show that we can learn complex limit cycles through diffeomorphism.

4.1 Combining Control Laws through State-Space Partitioning

We seek to automatically build a partitioning of the space and to learn a dynamics that is local to each partition that sends each DS to its own attractor. An example of such a system for eight regions of attraction with associated DSs and attractors is given in figure 4.1 (left). Before proceeding to describe how to achieve this, we start with a naive example in which one would construct each partition separately and combine them. To build each partition, we use SVM; readers who are not familiar with SVM should refer to section B.4 of the appendix B.[1]

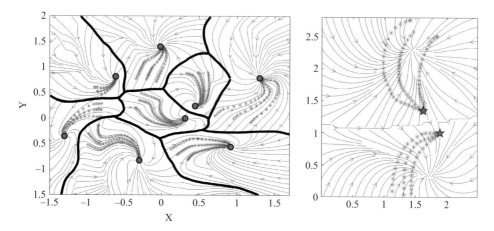

Figure 4.1
Left: Eight attractor DSs. *Right*: Modulated trajectories.

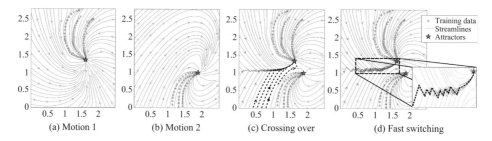

Figure 4.2
Combining motions using naive SVM classification-based switching.

4.1.1 Naive Approach

A naive approach to building a multiattractor DS would be to first partition the space and then learn a DS in each partition separately. Unfortunately, this rarely results in the desired compound system. Consider, for instance, two DSs with distinct attractors, as shown in figures 4.2a and 4.2b. First, we build an SVM classifier to separate the data points of the first DS, labeled $+1$, from data points of the other DS, labeled -1. We then estimate each DS separately using one of the techniques reviewed in the previous section. Let $h : \mathbb{R}^N \mapsto \mathbb{R}$ denote the classifier function that separates the state space $x \in \mathbb{R}^N$ into two regions, with the labels $y_i \in \{+1, -1\}$. Also, let the two DSs be $\dot{x} = f_{y_i}(x)$ with stable attractors at $x^*_{y_i}$. The combined DSs is then given by $\dot{x} = f_{\mathrm{sgn}(h(x))}(x)$.

Figure 4.2c shows the trajectories resulting from this approach. Due to the nonlinearity of the dynamics, trajectories initialized in one region cross the boundary and converge to the attractor located in the opposite region. In other words, each region partitioned by the SVM hyperplane is not a region of attraction for its attractor. In a real-world scenario in which the attractors represent grasping points on an object and the trajectories are to be followed by robots, crossing over may take the trajectories toward kinematically unreachable regions. Also, as shown in figure 4.2d, trajectories that encounter the boundary may switch rapidly between the different dynamics, leading to jittery motion.

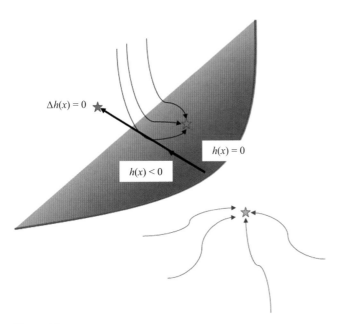

Figure 4.3
The partition function of SVM provides a negative gradient in the direction of the negative class, which is akin to a Lyapunov function. Its zero, however, is not necessarily at the desired attractor.

To ensure that the trajectories do not cross the boundary, and remain within the region of attraction of their respective attractors, one could adopt a more informed approach, in which each of the original DSs is modulated such that the generated trajectories always move away from the classifier boundary. Recall that by construction, the absolute value of the classifier function $h(x)$ increases as one moves away from the classification hyperplane. Therefore, the gradient $\nabla h(x)$ is positive, respectively negative, as one moves inside the region of the positive, respectively negative, class, as illustrated in figure 4.3.

We can use this observation to deflect selective components of the velocity signal from the original DS along, respectively opposite to, the direction $\nabla h(x)$. Concretely, if $\dot{x}_O = f_{\text{sgn}(h(x))}(x)$ denotes the velocity obtained from the original DS and

$$\lambda(x) = \begin{cases} \max\left(\epsilon, \nabla h(x)^T \dot{x}_O\right) & \text{if } h(x) > 0 \\ \min\left(-\epsilon, \nabla h(x)^T \dot{x}_O\right) & \text{if } h(x) < 0 \end{cases}, \tag{4.1}$$

the modulated DS is given by

$$\dot{x} = \tilde{f}(x) = \lambda(x)\nabla h(x) + \dot{x}_\perp. \tag{4.2}$$

Here, ε is a small positive scalar, and $\dot{x}_\perp = \dot{x}_O - \left(\frac{\nabla h(x)^T \dot{x}_O}{\|\nabla h(x)\|^2}\right)\nabla h(x)$ is the component of the original velocity perpendicular to ∇h. This results in a vector field that flows along increasing values of the classifier function in the regions of space where $h(x) > 0$, and along decreasing values for $h(x) < 0$. As a result, the trajectories move away from the classification hyperplane and converge to a point located in the region where they were initialized. Such modulated systems have been used extensively for estimating stability regions of interconnected power networks [91] and are known as *quasi-gradient systems* [29]. If $h(x)$ is upper bounded,[2] all trajectories converge to one of the stationary points $\{x : \nabla h(x) = 0\}$ and $h(x)$ is a Lyapunov function of the overall system (for more, refer to [29], proposition 1).

Figure 4.1 shows the result of applying this modulation to our pair of DSs. As expected, it forces the trajectories to flow along the gradient of function $h(x)$. Although this solves the problem of crossing-over the boundary, the trajectories obtained are deficient in two major ways: they depart heavily from the original dynamics, and they do not terminate at the desired attractors. This is because the function $h(x)$ used to modulate the DS was designed solely for classification and contained no information about the dynamics of the two original DSs. In other words, the vector field given by $\nabla h(x)$ was not aligned with the flow of the training trajectories, and the stationary points of the modulation function did not coincide with the desired attractors.

Exercise 4.1   *Consider four two-dimensional (2D) points with coordinates $[0, -0.1]^T$; $[0, 0.1]^T$ for class 1 and $[-0.1, 0.2]^T, [-0.2, 0.0]^T$:*

1. Build a linear classifier $y = \text{sgn}(h(x))$ and $h(x) = (\mathbf{w}^T x + b)$, such that points of class 1 have the label $y = 1$ and points of label 0 have class label $y = -1$.

2. Draw the boundary line.

3. Compute the gradient of the $h(x)$ and determine that it is zero. Is it located on one of the data points?

In subsequent sections, we show how we can learn a new modulation function that takes into account the three issues highlighted in this preliminary discussion. We will seek a system that (a) ensures strict classification across regions of attraction (ROAs) for each DS, (b) follows closely the dynamics of each DS in each ROA, and (c) ensures that all trajectories in each ROA reach the desired attractor. Satisfying requirements (a) and (b) is equivalent to performing classification and regression simultaneously. Here, we take advantage of the fact that the optimization in support vector classification and support vector regression takes the same form, in order to phrase this problem in a single constrained optimization framework. In the next sections, we show that in addition to the usual SVM support vectors (SVs), the resulting modulation function is composed of an additional class of SVs. We geometrically analyze the effect of these new SVs on the resulting dynamics. While this preliminary discussion considered only binary classification, we will now extend the problem to multiclass classifications.

Programming Exercise 4.1   *The aim of this exercise is to familiarize the reader with the naive classification approach presented in figure 4.2. This programming exercise can be done in MATLAB, or any other programming language. If you use MATLAB, go to directory* `chapter 4-multi-attractors/A-SVM` *and launch* `ch4_ex1_ simple_DSclassification.m.`

1. Plot the set of four data points of exercise 4.1.

2. Plot the boundary of the classifier using the line function.

3. *Generate two DSs of the form* $\dot{x} = \nabla h(x)$ *for* x *on the negative side of the boundary, and* $\dot{x} = -\nabla h(x)$ *for* x *on the positive side of the boundary. Draw the streamlines. Where do they converge?*

4. *Generate two DSs of the form* $\dot{x} = A_i x + b_i$ *with* $i = 1, 2$ *and setting nonzero entries on the offdiagonal elements of matrices* A_1 *and* A_2, *such that their attractors are located at the coordinates* $[-0.15\ 0.15]^T$ *and* $[0.15\ 0]^T$, *respectively. Using* $\dot{x} = f_{\mathrm{sgn}(h(x))}(x)$, *plot the streamline closest to the boundary. Modify the two DSs to have two nonlinear systems. Plot the streamlines closest to the boundary.*

4.1.2 Problem Formulation

The N-dimensional state space of the system represented by $x \in \mathbb{R}^N$ is partitioned into M different classes, one for each of the M motions to be combined. We collect trajectories in the state space, yielding a set of P data points $\{x^i; \dot{x}^i; l_i\}_{i=1\ldots P}$, where $l_i \in \{1, 2, \ldots, M\}$ refers to the class label of each point.[3] To learn the set of modulation functions $\{h_m(x)\}_{m=1\ldots M}$, we proceed recursively. We learn each modulation function in a one-versus-all classifier scheme and then compute the final modulation function, $\tilde{h}(x) = \max\limits_{m=1\cdots M} h_m(x)$. In the multiclass setting, the behavior of avoiding boundaries is obtained if the trajectories move along *increasing* values of function $\tilde{h}(x)$. To this effect, the deflection term $\lambda(x)$ presented in the binary case [equation (4.1)] becomes $\lambda(x) = \max\left(\varepsilon, \nabla\tilde{h}(x)^T \dot{x}_O\right); \forall x \in \mathbb{R}^N$. Next, we describe the procedure for learning a single $h_m(x)$ function.

We follow classical SVM formulation and lift the data into a higher-dimensional feature space through the mapping $\phi : \mathbb{R}^N \mapsto \mathbb{R}^F$, where F denotes the dimension of the feature space. We also assume that each function $h_m(x)$ is linear in feature space [i.e., $h_m(x) = \mathbf{w}^T\phi(x) + b$, where $\mathbf{w} \in \mathbb{R}^F, b \in \mathbb{R}$]. We label the current (mth) motion class as positive and all others negative, such that the set of labels for the current subproblem is given as

$$y_i = \begin{cases} +1 & \text{if } l_i = m \\ -1 & \text{if } l_i \neq m \end{cases}; \qquad i = 1 \ldots P.$$

Also, the set indexing the positive class is then defined as $\mathscr{I}_+ = \{i : i \in [1, P]; l_i = m\}$. With this, we formalize the three constraints explained in section 4.1.1 as follows:

Region separation: The constraint that each point must be classified correctly yields P constraints:

$$y_i \left(\mathbf{w}^T\phi(x^i) + b\right) \geq 1 \quad \forall i = 1 \ldots P. \tag{4.3}$$

Lyapunov constraint: To ensure that the modulated flow is aligned with the training trajectories, the gradient of the modulation function must have a positive component along the velocities at the data points. That is,

$$\nabla h_m(x^i)^T \hat{\dot{x}}^i = \mathbf{w}^T J(x^i)\hat{\dot{x}}^i \geq 0 \quad \forall i \in \mathscr{I}_+, \tag{4.4}$$

where $J \in \mathbb{R}^{F \times N}$ is the Jacobian matrix given by $J = \left[\ \nabla\phi_1(x)\nabla\phi_2(x)\cdots\nabla\phi_F(x)\ \right]^T$, and $\hat{\dot{x}}^i = \dot{x}^i/\|\dot{x}^i\|$ is the normalized velocity at the ith data point.

Stability: Finally, the gradient of the modulation function must vanish at the attractor of the positive class x^*. This constraint can be expressed as

$$\nabla h_m(x^*)^T e^i = \mathbf{w}^T J(x^*) e^i = 0 \quad \forall i = 1 \ldots N, \tag{4.5}$$

where the set of vectors $\{e^i\}_{i=1\ldots N}$ is the canonical basis of $\mathbb{R}^N$.

4.1.2.1   Primal and dual forms

As in the standard SVM [128], we optimize for maximal margin between the positive and negative class, subject to the constraints expressed in equations (4.3) through (4.5). This can be formulated as follows:

$$\min_{\mathbf{w},\xi_i} \frac{1}{2}\|\mathbf{w}\|^2 + C \sum_{i\in\mathscr{I}_+} \xi_i \quad \text{subject to} \quad \left.\begin{array}{rcll} y_i\left(\mathbf{w}^T\phi(x^i)+b\right) & \geq 1 & \forall i = 1\cdots P \\ \mathbf{w}^T J(x^i)\hat{x}^i + \xi_i & > 0 & \forall i \in \mathscr{I}_+ \\ \xi_i & > 0 & \forall i \in \mathscr{I}_+ \\ \mathbf{w}^T J(x^*)e^i & = 0 & \forall i = 1\cdots N \end{array}\right\}. \tag{4.6}$$

Here, $\xi_i \in \mathbb{R}$ are slack variables that relax the Lyapunov constraint in equation (4.4). We retain these in our formulation to accommodate noise in the data representing the dynamics. $C \in \mathbb{R}_+$ is a penalty parameter for these slack variables. The Lagrangian for this problem can be written as

$$\mathcal{L}(\mathbf{w},b,\alpha,\beta,\gamma) = \frac{1}{2}\|\mathbf{w}\|^2 + C\sum_{i\in\mathscr{I}_+}\xi_i - \sum_{i\in\mathscr{I}_+}\mu_i\xi_i - \sum_{i=1}^{P}\alpha_i\left(y_i(\mathbf{w}^T\phi(x^i)+b)-1\right)$$

$$- \sum_{i\in\mathscr{I}_+}\beta_i\left(\mathbf{w}^T J(x^i)\hat{x}^i + \xi_i\right) + \sum_{i=1}^{N}\gamma_i\mathbf{w}^T J(x^*)e^i, \tag{4.7}$$

where $\alpha_i, \beta_i, \mu_i, \gamma_i$ are the Lagrange multipliers, with $\alpha_i, \beta_i, \mu_i \in \mathbb{R}_+$ and $\gamma_i \in \mathbb{R}$.

Exercise 4.2   *Compute the dual of equation (4.7) and show that it has the form of the following constrained problem:*

$$\min_{\alpha,\beta,\gamma} \frac{1}{2}\begin{bmatrix}\alpha^T & \beta^T & \gamma^T\end{bmatrix}\begin{bmatrix} K & G & -G_* \\ G^T & H & -H_* \\ -G_*^T & -H_*^T & H_{**}\end{bmatrix}\begin{bmatrix}\alpha \\ \beta \\ \gamma\end{bmatrix} - \alpha^T\bar{\mathbf{1}}$$

subject to $\ 0 \leq \alpha_i \qquad \forall i = 1\ldots P$

$\qquad\qquad 0 \leq \beta_i \leq C \qquad \forall i \in \mathscr{I}_+$

$$\sum_{i=1}^{P}\alpha_i y_i = 0$$

Note: Since the matrices K, H, and H_{**} are symmetric, the overall Hessian matrix for the resulting quadratic program is also symmetric. However, unlike the standard SVM dual, it may not be positive definite, resulting in multiple solutions to this problem. In the implementation provided in the MATLAB code, we use the interior point solver IPOPT [153] to find a local optimum. We initialize the iterations using the α found by running a standard SVM classification problem. All entries of β and γ are set to 0[4].

The solution to this problem yields the following modulation function:

$$h_m(x) = \sum_{i=1}^{P} \alpha_i y_i k(x, x^i) + \sum_{i \in \mathscr{I}_+} \beta_i \hat{x}_i^T \frac{\partial k(x, x^i)}{\partial x^i} - \sum_{i=1}^{N} \gamma_i (e^i)^T \frac{\partial k(x, x^*)}{\partial x^*} + b, \qquad (4.8)$$

which can be further expanded depending on the choice of kernel.

The modulation function [equation (4.8)] learned using the A-SVM has the following noticeable similarities with the standard SVM classifier function:

• The first summation term is composed of the α support vectors (α-SV), which act as support for the classification hyperplane.

• The second term entails a new class of support vectors, which perform a linear combination of the normalized velocity $\hat{x}^i$ at the training data points x^i. These β support vectors (β-SVs) collectively contribute to the fulfillment of the Lyapunov constraint by introducing a positive slope in the modulation function value along the directions $\hat{x}^i$.

• The third summation term is a nonlinear bias, which does not depend on the chosen support vectors and performs a local modification around the desired attractor x^* to ensure that the modulation function has a local maximum at that point.

• b is the constant bias, which normalizes the classification margins as -1 and $+1$, also found in the original SVM derivation. As in SVM, we calculate its value by using the fact that for all the α-SV x^i values, we must have $y_i h_m(x^i) = 1$. We use the average of the values obtained from the various support vectors.

To better appreciate the effect of the new set of support vectors, the β-SV, we illustrate their effect locally in figure 4.4 using the RBF kernel $k(x^i, x^j) = e^{1/2\sigma^2 \|x^i - x^j\|^2}$, with x^i at the origin and $\hat{x}^i = [\frac{1}{\sqrt{2}} \frac{1}{\sqrt{2}}]^T$. These SVs generate a velocity direction. We contrast two values of kernel width σ. The smaller σ is, the larger the velocity vector is. This can be seen visible by the increase of the isoline values.

The effect of all three terms of the modulation function [equation (4.8)] is illustrated in figure 4.5 by progressively adding them and overlaying the resulting DS flow in each case.

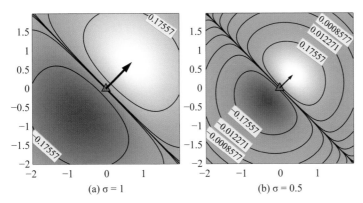

Figure 4.4
Isocurves of $f(x) = \hat{x}_i^T \frac{\partial k(x, x^i)}{\partial x^i}$ at $x^i = [0\ 0]^T$, $\hat{x}^i = [\frac{1}{\sqrt{2}} \frac{1}{\sqrt{2}}]^T$ for the RBF kernel.

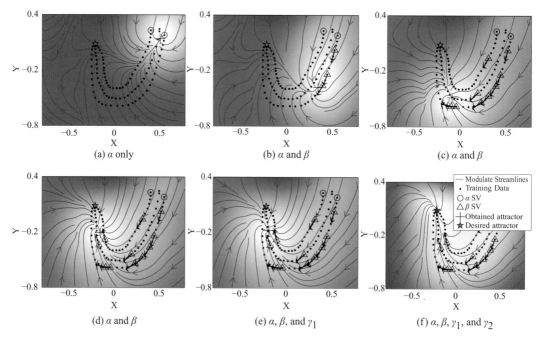

Figure 4.5
Progressively adding support vectors to highlight their effect on shaping the dynamics of the motion. (a) α-SVs largely affect classification. (b) through (d) β-SVs guide the flow of trajectories along their respective associated directions $\hat{x}^i$ shown by arrows. (e) and (f) The two γ terms force the local maximum of the modulation function to coincide with the desired attractor along the x- and y-axes, respectively.

The value of the modulation function $h_m(x)$ is shown by the color plot (white indicates high values). As the β-SV values are added in figures 4.5b through 4.5d, they *push* the flow of trajectories along their associated directions. In figures 4.5e and 4.5f, adding the two γ terms shifts the location of the maximum of the modulation function to coincide with the desired attractor. Once all the SVs have been taken into account, the streamlines of the resulting DS achieve the desired criteria (i.e., they follow the training trajectories and terminate at the desired attractor).

4.1.3   Scaling and Stability

The approach is not limited in the number of dynamics that it can learn. The various dynamics, however, must lie in distinct regions of the state space. It is not possible to have overlapping trajectories because the approach creates a hard partitioning of the space. The more dynamics, however, the more trajectories, and the more computational intensive the problem becomes. Therefore, there may be an intrinsic limit due to the costs of gathering data and running the optimization.

The partitioning of space created by A-SVM results in M ROAs for the M attractors. Given the complexity of the system, we cannot derive a theoretical bound on the ROA, nor can we asses that there are no spurious attractors in the system. To assess the size of the ROA and the existence of spurious attractors, one must proceed to a numerical estimate. For each class, we compute the isosurfaces of the corresponding modulation function $h_m(x)$ in

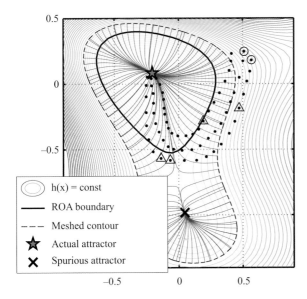

Figure 4.6
Test trajectories generated from several points on an isocurve (dotted line) to determine spurious attractors.

the range $[0, h_m(x^*)]$. These hypersurfaces incrementally span the volume of the mth region around its attractor. We mesh each of these test surfaces and compute trajectories starting from the obtained mesh points, looking for spurious attractors. Here, h_{ROA} is the isosurface of maximal value that encloses no spurious attractor and marks the ROA of the corresponding motion dynamics. We use the example in figure 4.5 to illustrate this process. Figure 4.6 shows a case where one spurious attractor is detected using a larger test surface (dotted line), whereas the actual ROA (solid line) is smaller. Once h_{ROA} is calculated, we define the size of the ROA as $r_{ROA} = (h(x^*) - h_{ROA})/h(x^*)$. Here, $r_{ROA} = 0$ when no trajectory except those originating at the attractor itself lead to the attractor; and $r_{ROA} = 1$ when the ROA is bounded by the isosurface $h(x) = 0$.

4.1.4 Precision of the Reconstruction

As in standard SVM, we have allowed a relaxation of the constraints by introducing slack variables. We then penalized this relaxation by introducing a cost on the slack variable in the objective function. In practice, this means that the system is allowed to deviate from the original dynamics. The error incurred in the reconstruction can be assessed by computing the average percentage error between the original velocity (read off of the data) and the modulated velocity [calculated using equation (4.2)] for the mth class as $e_m = \left\langle \frac{\|\dot{x}^i - \tilde{f}(x^i)\|}{\|\dot{x}^i\|} \times 100 \right\rangle_{i:l_i=m}$, where $< . >$ denotes the average over the indicated range.

Figure 4.7 shows the cross-validation error (mean and standard deviation over the five folds; 2:3 training-to-testing ratio) for a range of kernel width values computed on the four attractor problem (see figure 4.7a). Each motion class was generated from different closed-form dynamics and containing 160 data points. The color plot indicates the value of the combined modulation function $\tilde{h}(x) = \max_{m=1\cdots M} h_m(x)$, where each of the functions $h_m(x)$

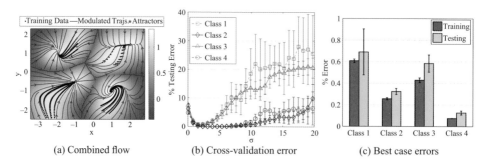

(a) Combined flow   (b) Cross-validation error   (c) Best case errors

Figure 4.7
Synthetic 2D case with four attractors.

are learned using the presented A-SVM technique. A total of nine support vectors were obtained, which is < 10 percent of the number of training data points. A total of ten trajectories per motion class was created. The trajectories obtained after modulating the original DSs flow along increasing values of the modulation function, thereby bifurcating toward different attractors at the region boundaries. For each class of motion, we can see a band of optimum values of the kernel width for which the testing error is the smallest. The region covered by this band of optimal values may vary depending on the relative location of the attractors and other data points. In figure 4.7a, motion classes 2 (upper left) and 4 (upper right) are better fitted and show less sensitivity to the choice of kernel width than classes 1 (lower left) and 3 (lower right). A comparison of testing and training errors for the least error case is shown in figure 4.7c. We see that the testing errors for all the classes in the best case scenario are less than 1 percent. This shows that, at least, in this test, the reconstruction is fairly good.

Programming Exercise 4.2   *The aim of this exercise is to familiarize the reader to the A-SVM algorithm and its sensitivity to choice of hyperparameters. Open MATLAB and set the directory to chapter 4-multi-attractors/A-SVM. Run app in ex4_2_ASVM.*

This app allows you to test A-SVM with hand-drawn data. For each class, make sure to provide a couple of trajectories.

1. Generate a three-attractor system by drawing trajectories for each of the three classes.

2. Explore the sensitivity of the algorithm to the choice of hyperparameters by running the algorithm on a range of parameters.

3. Report the performance in terms of number of SVs and reconstruction errors in a table.

4.1.5 Robotic Implementation

The multiattractor DS presented in this section finds several robotic applications. It is particularly useful when one needs to switch at run time across different attractors. This happens when, for instance, the robot is tasked to grasp an object that falls. The grasping points may be different, forming different attractors in the robot's joint space (i.e., different hand aperture and finger positionings on the object). As the object falls, the robot has little time to switch across the different attractors to catch the object before it hits the floor.

Figure 4.8a shows an application of the approach using a KUKA-LWR arm with 7 degrees of freedom (DOF) to reach a pitcher. We use the 7-DOF KUKA-LWR arm mounted on the 3-DOF KUKA-Omnirob base for executing the modulated Cartesian trajectories in simulation. We control all 10 DOFs of the robot using the damped least-square inverse kinematics. The attractors represent manually labeled grasping points on a pitcher. The three-dimensional (3D) model of the object was taken from the Robot Operating System (ROS) IKEA object library. Training data for this implementation were obtained by recording the end-effector positions $x^i \in \mathbb{R}^3$ from kinesthetic demonstrations of reach-to-grasp motions directed toward these grasping points, yielding a three-class problem. Each class was represented by 75 data points, all of which were used for training the model. Figure 4.8b shows the isosurfaces $h_m(x) = 0; m \in \{1, 2, 3\}$ learned using the presented method. Figure 4.8c shows on the robot executing two trajectories when started from two locations and converging on a different attractor (grasping point). Figure 4.8e shows the flow of motion around the object. Importantly, the time required to generate each trajectory point is linear in the number of support vectors, $O(S)$, where S denotes the number of support vectors. In the robotic implementation, with a total of 18 SVs, the trajectory points were generated at 1,000 Hz which is well suited for real-time control.

This DS embedding of the multiattractors offers an ideal means to generate new trajectories very rapidly. To demonstrate how rapid the approach can be, we implemented the approach to allow the robot to switch on the fly between two end-effector poses (attractors in joint space), so as to catch a falling bottle. For this purpose, a KUKA robot arm was mounted with a three-finger Barrett hand. To track the falling bottle, we use a set of 12 infrared (IR) based OptiTrack cameras running at 250 frames per second. As soon as the bottle is released, we compute a linear approximation of its trajectory, assuming a constant rotation and translation. The robot is embedded with two linear DSs to catch the bottle either at the neck or at the bottom. The attractors are generated manually as specific aperture for the robot's gripper. We choose two sizes of aperture of the fingers, given the different widths of the neck and the bottom. At run time, the robot switches between the two attractors. This was demonstrated live to allow a KUKA robot to catch a bottle falling in 0.2 seconds (see the video on the book's website). Snapshots of this implementation are shown in figure 4.9.

4.2 Learning of DSs with Bifurcations

One interesting property of DSs is the fact that they can drastically change dynamics depending on external parameters. Such sharp change has been known as *bifurcation*. A bifurcation may typically correspond to a sudden change from a stable dynamics to an unstable one. It may also correspond to a change across two stable dynamics, but with different attractor types. For instance, the system may change from a single, fixed-point attractor dynamics to a limit cycle. Such bifurcation is known as a *Hopf bifurcation*. In this section, we present an algorithm to learn a DS that embeds such Hopf bifurcations, so as to encode different types of dynamics and to offer a smooth transition across these dynamics in a single DS. We learn the bifurcation parameters, which offer an explicit way to control

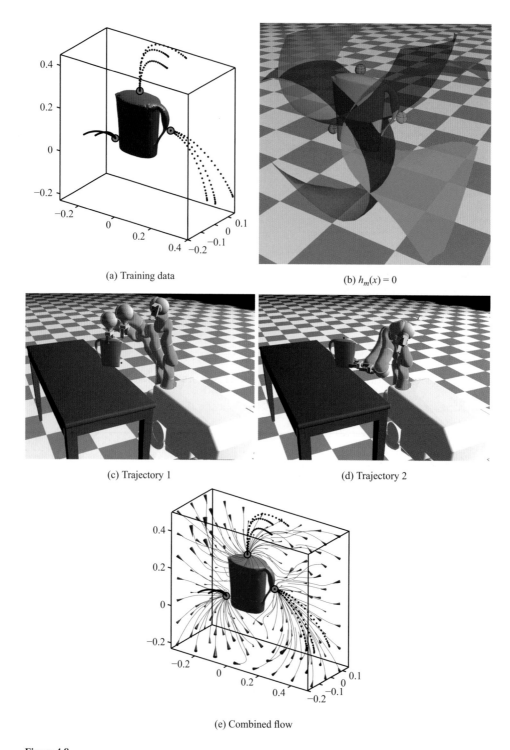

(a) Training data

(b) $h_m(x) = 0$

(c) Trajectory 1

(d) Trajectory 2

(e) Combined flow

Figure 4.8
A 3D experiment. (a) training trajectories for three manually chosen grasping points. (b) the isosurfaces $h_m(x) = 0$; $m = 1, 2, 3$, along with the locations of the corresponding attractors. In (c) and (d), the robot executes the generated trajectories starting from different positions, and hence converging on different grasping points. (e) the complete flow of motion [134].

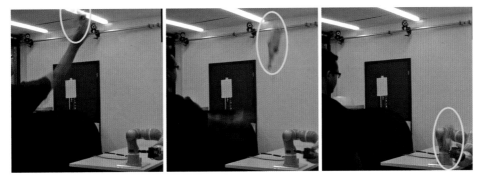

Figure 4.9
Snapshots of a video showing the robot catching a falling bottle in flight.

for the change of dynamics at run time. We show that we can use this method to control for the speed of the motion, the location of the attractor, and the shape of the limit cycle, and we can switch at run time through these various dynamics.[5]

4.2.1 DSs with Hopf Bifurcation

We reduce our model to a first-order DS, with one bifurcation parameter μ. Here, we consider *local bifurcations*, whereby the change of dynamics at the bifurcation can be studied in the neighborhood of an equilibrium point or a closed orbit (see the appendixes):

$$f: \ \mathbb{R}^N \to \mathbb{R}^N, \dot{x} = f(x, \mu), \quad x \in \mathbb{R}^N, \quad \mu \in \mathbb{R}. \tag{4.9}$$

We assume that the system in equation (4.9) presents a bifurcation at $\mu^* \in \mathbb{R}$, such that, for small changes around μ^*, the system incurs a topological change, leading in turn to a change in the type of equilibria from stable fixed point to limit cycle. Such bifurcations may incur in systems with at least two dimensions and are characterized by the appearance of a limit cycle bifurcating from an equilibrium point, thus changing its stability. They have been extensively studied within the framework of Hopf [56] and Poincaré-Andronov-Hopf bifurcations [116]. Here, we are interested in reproducing the *supercritical* Hopf bifurcation case, with a stable limit cycle appearing from a stable equilibrium when crossing the bifurcation value so as to transit smoothly across periodic and discrete movement.

Exercise 4.3   *Consider the following two-dimensional (2D) DS:*

$$\dot{x}_1 = x_2 - x_1 \mu,$$
$$\dot{x}_2 = -x_1 - x_2 \mu, \tag{4.10}$$

parameterized by $\mu \in \mathbb{R}$.

* *For which values of μ, do you observe changes in the attractor and dynamics toward the attractor?*
* *Plot the vector field for different values of μ.*

Programming Exercise 4.3   *The aim of this exercise is to familiarize the reader with DS entailing bifurcations. Open MATLAB and set the directory to folder* `ch4-multi-attractors/bifurcation` *and run* `ch4_ex3_bifurcation`:

1. Program the following DS:

$$\dot{x}_1 = -x_2 + x_1(\mu + \sigma(x_1^2 + x_2^2)),$$

$$\dot{x}_2 = x_1 + x_2(\mu + \sigma(x_1^2 + x_2^2))\mu, \tag{4.11}$$

parameterized by $\mu, \sigma \in \mathbb{R}$.

2. Test the change induced in the phase plot of the DS for different values of the two bifurcation parameters μ *and* σ. *For which values of the bifurcation parameters do you observe a change in the stability of the DS?*

4.2.2 Desired Shape for the DS

We start with the normal form of a DS, with $x \in \mathbb{R}^2$, where $x = [x_1 \ x_2]^T$, presenting a supercritical Hopf bifurcation at $\mu = 0$ (see [52, 53]). We simplify the system by converting the task coordinates $x \in \mathbb{R}^2$ to polar coordinates $r \in \mathbb{R}^2$, with polar radius $\rho = \sqrt{x_1^2 + x_2^2}$ and polar angle $\theta = \tan(\frac{x_2}{x_1})$:

$$f(\rho, \theta, \mu) = \begin{cases} \dot{\rho} = \rho(\rho^2 - \mu) \\ \dot{\theta} = \omega, \end{cases} \tag{4.12}$$

with μ the bifurcation parameter, and $\omega \in \mathbb{R}, \omega \neq 0$ an open parameter determining the frequency.

As shown in [53], for $\mu > 0$, the bifurcation parameter is equal to the squared radius of the stable limit cycle.[6] We can identify the bifurcation parameter μ as being a measure of the squared radius of the stable limit cycle (for $\mu > 0$) as $\mu = \frac{a}{d}(x_1^2 + x_2^2)$, for $x_1, x_2 \in \Omega$, where Ω is the ω-limit set of the DS (i.e., an invariant set of points such that for time $t_i \to \infty$, $f(x, \mu, t_i) \in \Omega$, as explained in [52]), and it includes either the attractor or the limit cycle, depending on the value of μ.

We retain this dependency when reformulating the 2D DS for our approach and, as simplification, we rename the bifurcation parameter as *target radius* ρ_0, defined as

$$\rho_0 = \sqrt{x_1^2 + x_2^2}. \tag{4.13}$$

The formulation is inspired by the dynamics underlying the motions of satellites around planets. The streamlines of this system generate parabolic motions that stabilize either an attractor point (here, the origin of the system) or transform into ellipses at limit cycles, around the origin. The energy E of the system in polar coordinates is given as

$$E(\dot{\rho}, \dot{\theta}, \rho) = \frac{N}{2}(\dot{\rho}^2 + \rho^2\dot{\theta}^2) + U(\rho, \rho_0), \tag{4.14}$$

where $\dot{\rho}$ and $\dot{\theta}$ are the radial and angular velocities and N is the constant inertia of the system. U is the potential function of the system, which is parametrized by the radius of the orbit, and it is given by

$$U(\rho) = \frac{K_\rho}{2}(\rho - \rho_0)^2. \tag{4.15}$$

> **Exercise 4.4** *Show that the potential function* U *is given by equation (4.15). HINT: Start by writing the complete energy of the system (i.e., the composition of kinetic and potential energy).*

Taking the derivative of the potential function $dU(\rho) = K_U(\rho - \rho_0)$, we compute the dynamics in polar coordinates with respect to ρ_0 as follows:

$$f(\rho, \theta, \rho_0) = \begin{cases} \dot{\rho} = -\sqrt{M}(\rho - \rho_0) \\ \dot{\theta} = Re^{-M^2(\rho - \rho_0)^2}. \end{cases} \tag{4.16}$$

$M = \frac{1}{2N} > 0$ is a parameter that modulates the strength of the attraction and can be used to modulate the speed at which the robot moves. We can see from equation (4.16) that the differential equation of the polar radius vanishes at $\rho = \rho_0$. The differential equation of the polar angle θ is numerically different from 0 only in a neighborhood of ρ_0, which is defined by M. Therefore, parameter M determines both the speed of convergence to the limit cycle and the distance at which the dynamics starts to rotate toward the stable ellipse. Finally, $R \in \mathbb{R}$ is an additional parameter that permits one to easily change the angular speed (e.g., by doubling R, the speed will double), as well as to invert the rotation of dynamics around the limit cycle. Thus, through modulation of M and R, one can control the convergence rate and maximum velocity of the designed DS. Also when learning M and R from demonstration, the acquired DS captures the velocity limits of the demonstrated trajectories. Figure 4.10a shows these three parameters in an example of the introduced DS in 2D, and figure 4.11a shows examples of the DS on the robot used for simulations.

By constraining $\rho_0 \geq 0$, we have two behaviors:

- For $\rho_0 = 0$, the origin is a stable focus,
- For $\rho_0 > 0$, a stable limit cycle with radius ρ_0 appears.

> **Exercise 4.5** *Extend the system of equation (4.16) to 3D. [HINT:] Use spherical coordinates in the place of the polar coordinates.*

4.2.3 Two Steps Optimization

Assuming that we have collected a data set of sampled trajectories as examples of the DS to be learned, in order to identify the parameters of the DS from equation (4.16), we use two least-squares minimizations:

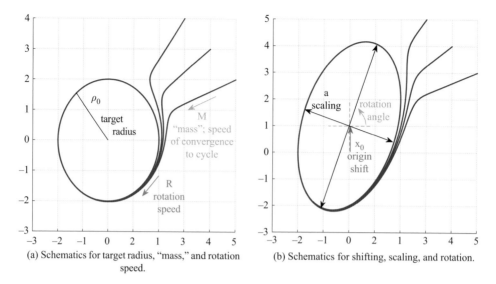

(a) Schematics for target radius, "mass," and rotation speed.

(b) Schematics for shifting, scaling, and rotation.

Figure 4.10
Examples of trajectories and parameters to be learned [69].

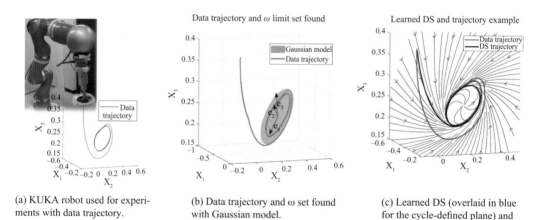

(a) KUKA robot used for experiments with data trajectory.

(b) Data trajectory and ω set found with Gaussian model.

(c) Learned DS (overlaid in blue for the cycle-defined plane) and reproduction of trajectory.

Figure 4.11
Example of Gaussian model for rotation and optimization results [69].

1. $\min \| -\sqrt{2M}(\rho_{data} - \rho_0) - \dot{\rho}_{data} \|^2$

2. $\min \| R\, e^{-4M^2(\rho_{data}-\rho_0)^2} - \dot{\theta}_{data} \|^2$ $\qquad\qquad$ (4.17)

under constraints $\rho_0 \geq 0, \quad M > 0.$

$\rho_{data}, \dot{\rho}_{data}$, and $\dot{\theta}_{data}$ are the radius, the radial velocity, and the (azimuth) angular velocity of the data x, respectively, in polar coordinates $r = [\rho, \theta]$ for the 2D case, and in spherical coordinates $r = [\rho, \phi, \theta]$ for 3D. The optimization is the same in 3D (i.e., $x \in \mathbb{R}^3$, where $x = [x_1\ x_2\ x_3]^T$); the radius is then $\rho = \sqrt{x_1^2 + x_2^2 + x_3^2}$ and the radial speed $\dot{\rho}$ already accounts for changes in x_3.

Original data with ω limit set found

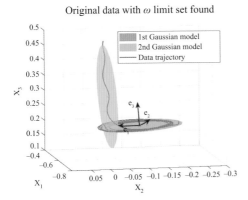

Kinesthetic data trajectory and learned DS

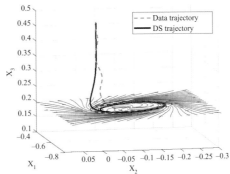

(a) Kinesthetic data trajectory and ω set found considering the first Gaussian of the GMM.

(b) Data trajectory (red), learned DS trajectory (black), and vector field of DS shown for the cycle-defined plane (blue arrows).

Figure 4.12
Learning the DS from kinesthetic demonstrations [69].

By repeating the two steps until the error is below a fixed threshold, we find the parameters ρ_0, M, and R.

When working with real data, it is often important to scale and rotate the demonstration to flatten the space in 2D. This can be done using a Gaussian mixture model (GMM) and principal component analysis (PCA), as illustrated in figure 4.12. Pseudocode of the complete algorithm, from scaling, shifting, and rotation to optimization, is given in algorithm 4.1.

Algorithm 4.1 Steps of the optimization

Input: $\{\mathbf{x}_{data}, \dot{\mathbf{x}}_{data}\} = \{x_{data}^i, \dot{x}_{data}^i\}_{i=1}^M$ for M data points.   ▷ Data
Output: $\{\rho_0, M, R, \mathbf{a}, x_0, \theta_0\}$ ▷ Estimated Parameters
  1: Find the Gaussian model for the ω-limit set;
  2: **if** the limit the cycle is rotated in space, **then**
       Perform PCA to find the frame rotation matrix
       Find Euler angles θ_0 from the rotation matrix
       Multiply x_{data} by the rotation matrix
  3: **else**
       Set Euler angles θ_0 to $[0\ 0\ 0]$
  4: **end if**
  5: **procedure** Convert to spherical coordinates$(\mathbf{a}, \mathbf{x}_{data}, x_0, \dot{\mathbf{x}}_{data})$
       $\mathbf{r}_{data} = 2\mathtt{Spherical}(\mathbf{a}(\mathbf{x}_{data} + x_0))$
       $\dot{\mathbf{r}}_{data} = 2\mathtt{SphericalVelocity}(\dot{\mathbf{x}}_{data}, \mathbf{a}(\mathbf{x}_{data} + x_0))$
  6: **end procedure**
  7: **while** error $>$ threshold **do**
       $\min \| -\sqrt{2M}(\rho_{data} - \rho_0) - \dot{\rho}_{data} \|^2,$
       $\min \| Re^{-4M^2(\rho_{data}-\rho_0)^2} - \dot{\theta}_{data} \|^2,$
       $u.c. \quad \rho_0 \geq 0, \quad M > 0$
  8: **end while**

4.2.4   Extension to Nonlinear Limit Cycles

The system presented previously is linear. To extend the approach to learning DSs, that entail nonlinear limit cycles, we can use the concept of diffeomorphism. The principle is to learn a bidirectional map from the original to another space in which the dynamics is linear. Here, we seek to learn mapping function $s(\cdot)$ to send the nonlinear DS into a space where the DS appears linear. We start from the DS in equation (4.16). This is our base linear DS. We search a diffeomorphism from this base oscillator to the desired nonlinear limit cycle, and back again.

For limit cycles and phase oscillators, scaling the polar radius depending on the phase angle is sufficient to change a limit cycle into a nonlinear one [2]. Therefore, we assume that there is a phase-based radial mapping function $s(\theta)$ that scales the acquired linear DS in equation (4.16) to the desired nonlinear limit cycle:

$$\rho_n = s(\theta)\rho_b, \tag{4.18}$$

where ρ_b and ρ_n are the radius of base and nonlinear limit cycles, respectively. For better readability, we drop the subscript n in ρ_n, and refer to $s(\theta)$ as s. Given this, if $\forall \theta : s \neq 0$ and s is differentiable at least once, then s will define the Jacobian determinant of the desired diffeomorphism [2]. To form the nonlinear DS, we can take the derivative of equation (4.18) as follow:

$$\dot{\rho}(\rho, \theta) = s\dot{\rho}_b\left(\frac{\rho}{s}\right) + \frac{\rho}{s}\frac{\partial s}{\partial \theta}\dot{\theta}\left(\frac{\rho}{s}\right). \tag{4.19}$$

Using equation (4.16) in equation (4.19), we obtain the following expression for the nonlinear limit cycle:

$$\begin{cases} \dot{\rho} = -\sqrt{M}\left(\rho - s\rho_0\right) + \frac{R\rho}{s}\frac{\partial s}{\partial \theta}e^{-M^2\left(\frac{\rho}{s}-\rho_0\right)^2} \\ \dot{\theta} = Re^{-M^2\left(\frac{\rho}{s}-\rho_0\right)^2}. \end{cases} \tag{4.20}$$

Extracting s and real-time computation of $\partial s/\partial \theta$ are the added complexities of equation (4.20) compared to equation (4.16).

Learning of the diffeomorphism can be done using a GMM, which estimates a joint density of velocities in the original and projected spaces. Mapping from one to the other can be obtained by conditioning the model. Figure 4.13 illustrates the steps followed to learn a nonlinear DS through diffeomorphism.

Implementation of the method with both synthetic data and real data is available in the supplementary code (see programming exercise 4.4 and figure 4.14).

4.2.5   Robotic Implementation

We show in this section one example of use of learned DS with bifurcation to control a robot. The robot is tasked to switch from performing a point-to-point motion toward an attractor to a limit cycle and back. The task is to wipe in turn two objects moving from one object to the other. As the objects have different sizes, this requires changing the radius of the associated limit cycle, which can be controlled with ρ_0. To wipe the correct object can be done by changing the orientation θ_0 of the DS and origin x_0, placing the latter in the middle of the region that needs to be wiped.

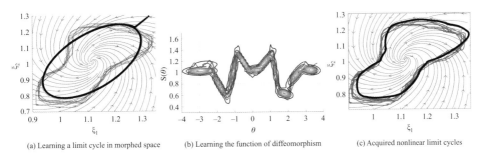

(a) Learning a limit cycle in morphed space (b) Learning the function of diffeomorphism (c) Acquired nonlinear limit cycles

Figure 4.13
The process of learning a nonlinear limit cycle from a set of demonstrations: (a) the result of learning the closest linear limit cycle to the demonstrated data; (b) using GMR to learn the mapping function of the diffeomorphism, and then to morph the obtained linear polar DS to the desired limit cycle; (c) the acquired nonlinear limit cycle determined by morphing the learned linear one [69].

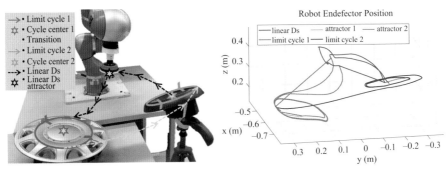

(a) Hard and smooth transitions across a DS to wipe two objects.

(b) Position trajectories of a robot over each task segment.

Figure 4.14
An example of a robotic task with hard switches between linear DSs (black) and bifurcation DSs (red/blue lines) for wiping the surface of two objects with different orientations and diameters. Switching occurs within limit cycles and attractors by changing ρ_0, x_0, and θ_0 [69].

For this task, we used a 7-DOF KUKA LWR 4+ mounted with a wiping tool (as seen in figure 4.14a), and then modified the controller at run time by switching to each task segment for the desired object and behavior. In the first segment, a linear DS is used to generate desired velocity trajectories rather than learned DSs in equation (4.16). In the second segment, we switch from the linear DS to the learned DS set in limit cycle mode.

Using bifurcations in place of switching across limit cycles and linear DSs is advantageous, as it ensures the continuity of the control function.

Programming Exercise 4.4   *This exercise allows the reader to use the learning approach for DS with bifurcations and test its sensitivity to choice of hyperparameters. Open MATLAB and set the directory to the ch4-multi-attractors folder.*

1. Select option 3 to load real-life data and test the sensitivity of the algorithm to the number of Gauss functions. For each dataset, determine what is the optimal number.

2. Select option 4 to draw data and generate examples of a DS with a single fixed-point attractor or a limit cycle. Is the algorithm sensitive the location and orientation of your drawing?

3. Optional: Compare the performance with an alternative technique (namely, DMP) by uncommenting the last lines of code.

5 Learning Sequences of Control Laws

One of the main advantages of using control laws based on dynamical systems (DSs) as motion generators is their inherent property of *robustness to perturbations*. As shown in previous chapters, no matter how many times a robot (or a target) is perturbed, the controller will lead the robot to ultimately reach the target. However, when a robot is perturbed during task execution, it might fail to achieve some of the high-level requirements/objectives of the task at hand, such as: (i) Following precisely (or tracking) the reference trajectories used to learn the DS; and (ii) Reaching via-points or subgoals that might be crucial to accomplish a task.

In this chapter we offer approaches that endow the controller with the capabilities of accomplishing task objectives (i) and (ii), while preserving state dependency and convergence guarantees. We posit that (i) and (ii) can be achieved via the DS motion generation paradigm by modeling the robots' task as a *sequence of DS*. The principle is detailed next.

Tracking a reference trajectory: As illustrated in figure 5.1, nonlinear DSs, such as those presented in chapter 3, learn a multitude of trajectories in space, all of which lead to the attractor. While they follow the demonstrated trajectories accurately, upon incurring a disturbance, the algorithm will not return the system to the demonstrated trajectory, but rather simply send the robot to move through an alternative trajectory bound to reach the attractor. Hence, if precisely following a reference trajectory is the objective of the task, such an approach would fail.

In section 5.1, we introduce a DS formulation that is capable of trajectory-tracking behavior around a reference demonstration, while preserving global asymptotic stability (GAS) at a final attractor. This DS formulation, referred to as locally active globally stable DS (LAGS-DS), models a complex nonlinear motion as a *sequence of local virtual attractor dynamics*. These *local dynamics* encode different trajectory tracking behaviors in linear regions around the nonlinear reference trajectory. Further, they are *virtual*, as the DS does not stop at the local attractors but rather smoothly transits through them if the robot is within the locally active regions.

Reaching via-points: If reaching and stopping at intermediary targets constitute the objective of the task, as illustrated in figure 5.2, then the task could be modeled as a *sequence*

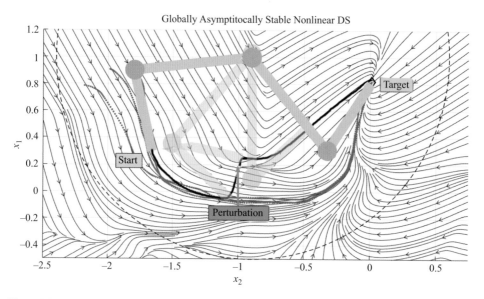

Figure 5.1
Illustrative example of a robot being guided by a first-order, nonlinear DS such as those introduced in chapter 3. Although the robot reaches the target after being perturbed, it fails to follow the reference trajectories (red) used to learn the DS. The **black trajectory** indicates the trace of the end effector of the simulated robot.

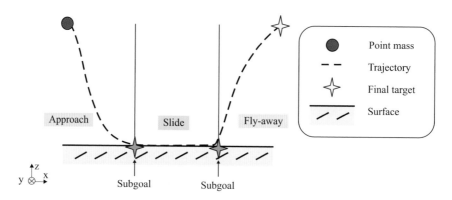

Figure 5.2
Illustrative example of a task involving a robot (represented by a point-mass) reaching a sequence of subgoals (i.e., subtasks) before converging on a final target.

of single-attractor nonlinear DSs, such as those introduced in chapter 3, where each nonlinear DS's attractor is located at the intermediary target. Such intermediary targets are often referred to as *via-points*. The challenge resides, then, in determining a mechanism to transition across each nonlinear DS.

In section 5.2, we introduce a formulation based on hidden Markov models (HMMs) that sequences nonlinear DS in a probabilistic manner. We refer to this approach as the *HMM-LPV*, as it uses the LPV expression for DS introduced in chapter 3. We show that we can ensure that the stability guarantees of each individual DS at its respective attractor are preserved throughout the sequencing of the task. The overall dynamics is, hence, guaranteed to transition through each via-point.

Note: Reviewing of chapter 3 is compulsory in order to follow this chapter. Furthermore, for an optimal understanding of section 5.2, we recommend reading [17] and [117] if you are not familiar with HMM and the Baum-Welch algorithm for HMM parameter estimation.

5.1 Learning Locally Active Globally Stable Dynamical Systems

The problem of tracking a reference trajectory with a DS motion generator emerges if the robot is compliant. In figure 5.1, we illustrate a DS learned by a set of demonstrations (the red trajectories) that guide the motion of the end effector of a two-dimensional (2D) robot controlled via a DS-based impedance control law (see chapter 10). In the absence of any external perturbation, and if the robot initiates its motion near the initial points of the reference trajectories, it can precisely follow the motion delineated by the reference trajectories. However, if the robot is perturbed along the way, it will follow the next integral curve of $f(x)$, initiated from the perturbed state. The robot will reach the target, regardless of the region in the state space where it is perturbed. Yet it will no longer faithfully follow the reference trajectories.

To provide a "trajectory-tracking" like behavior while preserving compliance at the control level, this behavior should be encoded in the DS itself. To achieve this, we require a DS, $f(x)$, to symmetrically converge to the reference trajectory (or trajectories), as shown in figure 5.3. As can be seen, this DS exhibits a behavior inside the shaded regions that is qualitatively similar to a stiffness attraction around a reference trajectory. When the robot

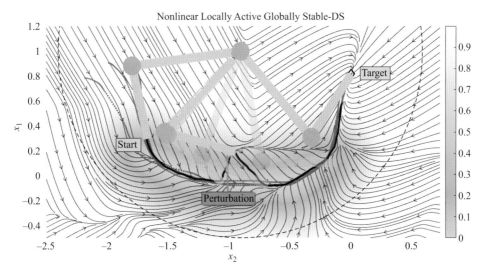

Figure 5.3
Illustrative example of a 2 DOFs robot-arm being guided by a LAGS-DS that not only converges symmetrically to the reference trajectories (in red), but also reaches the target.

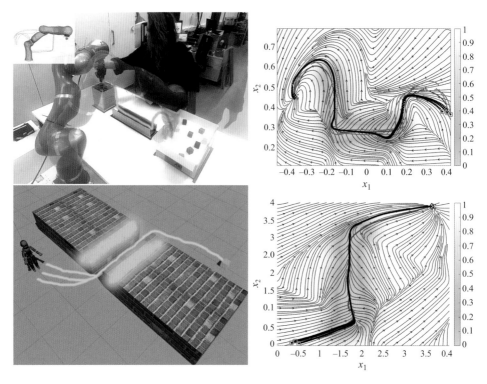

Figure 5.4
Scenarios where we would like a LAGS-DS to encode trajectory-tracking behaviors in shaded regions of the state space *(top)* and manipulation *(bottom)* locomotion.

is perturbed and still inside the shaded region, the integral curve of $f(x)$ will pull the robot back to the reference trajectory. On the other hand, if the robot exhibits a large perturbation, the integral curve of $f(x)$ will guide the robot toward the target. Such locally attractive behavior can be useful in applications that go beyond point-to-point motions, such as (1) *compliant manipulation*, where the DS can exhibit different reference trajectory attraction behaviors in local regions of the state-space; and (2) even to teach a floating-base robot to navigate in constrained environments (see figure 5.4).

To improve readability, in this section, we introduce the following variables and notation:

Notation

$x_g^* \in \mathbb{R}^N$	Global attractor
$x_l^* \in \mathbb{R}^N$	Single local virtual attractor
$x_k^* \in \mathbb{R}^N$	kth local virtual attractor
$\tilde{x}_g = (x - x_g^*)$	Error vector between the current state of the robot, x, and $x_g^* \in \mathbb{R}^N$
$\tilde{x}_l = (x - x_l^*)$	Error vector between the current state of the robot, x, and $x_l^* \in \mathbb{R}^N$
$\tilde{x}_k = (x - x_k^*)$	Error vector between the current state of the robot, x, and $x_k^* \in \mathbb{R}^N$
$f_g(\cdot) : \mathbb{R}^N \to \mathbb{R}^N$	Global DS with attractor $x_g^* \in \mathbb{R}^N$
$f_l(\cdot) : \mathbb{R}^N \to \mathbb{R}^N$	Local DS with attractor $x_l^* \in \mathbb{R}^N$
$f_l^k(\cdot) : \mathbb{R}^N \to \mathbb{R}^N$	kth local DS with attractor $x_k^* \in \mathbb{R}^N$

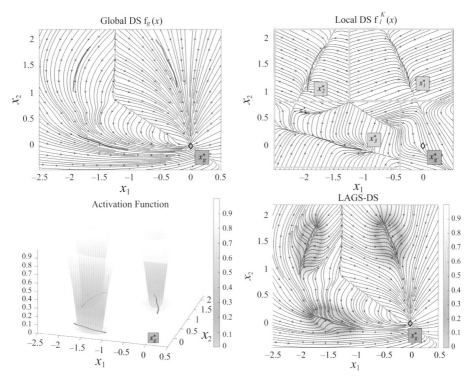

Figure 5.5
Illustration for constructing the LAGS-DS. The *top-left* plot shows a globally stable DS. The *top-right* plot shows the set of local DSs corresponding to each linear reference trajectory. The *bottom-left* plot shows the activation regions of the local DS, and the *bottom-right* plot shows the resulting LAGS-DS.

The main idea behind LAGS-DS is defined here (see also figure 5.5):

Let $f_g(x)$ be a global/nominal DS that should strictly converge to the global attractor x_g^*, as shown in figure 5.5 (*top left*). We also have a set of local DSs $f_l^k(x)$ for $k = 1, \ldots, K$, as the ones shown in figure 5.5 (*right top and bottom images*), that exhibit a specific trajectory-tracking behavior around a local virtual attractor $x_k^* \neq x_g^*$. Finally, we have a set of local activation regions, indicating where the local DS $f_l^k(x)$ should be active. The goal of a LAGS-DS is to evolve according to $f_g(x)$, where the local activation regions are inactive, whereas in the regions where they are active, the state evolves according to the locally active DS $f_l^k(x)$. In the absence of perturbations, if the state is in a locally active region, it will transit through the local virtual attractor x_k^* and then to $f_g(x)$, or to another local DS, ultimately reaching the global attractor x_g^*.

We begin in section 5.1.1 by describing a linear LAGS-DS formulation with a single locally active region. We use the theoretical findings from this section to introduce a nonlinear LAGS-DS formulation with multiple locally active regions (section 5.1.2).[1]

5.1.1 Linear LAGS-DS with a Single Locally Active Region

Given a set of linear reference trajectories $\{\mathbf{X}, \dot{\mathbf{X}}\} = \{X^m, \dot{X}^m\}_{m=1}^M = \{\{x^{t,m}, \dot{x}^{t,m}\}_{t=1}^{T_m}\}_{m=1}^M$, we seek to design a DS whose global dynamical behavior is shaped by the overall motion pattern delineated by $\{\mathbf{X}, \dot{\mathbf{X}}\}$ and converges on x_g^* [i.e., $f_g(x)$], while also symmetrically

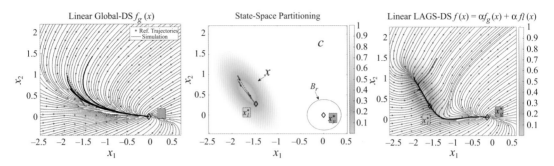

Figure 5.6
Illustrative 2D example for Linear LAGS-DS Exposition. *Left*: Linear Global-DS, $f_g(x)$ [equation (5.2)], shaped with the reference trajectory. *Center*: State-space partitioning via activation function $\alpha(x)$ (5.3). *Right*: Linear LAGS-DS [equation (5.1)] with a symmetrically converging local dynamics.

converging to a linear reference trajectory in a locally activated region of the state space. In order to achieve this desiderata, we propose the following LAGS-DS formulation:

$$\dot{x} = \underbrace{\alpha(x)f_g(x)}_{\text{Global dynamics}} + \underbrace{\overline{\alpha}(x)f_l(h(x),x)}_{\text{Local dynamics}} \tag{5.1}$$

Here, $f_l(\cdot)$ denotes the locally active dynamics parametrized by a partition function $0 \leq h(x) \leq 1$. Further, $\alpha(x) : \mathbb{R}^M \to \Re$ is a continuous activation function in the range of $\alpha(x) \in [0, 1]$, indicating the regions in state space where the local dynamics are active, $\overline{\alpha}(x) = (1 - \alpha(x))$. When $\alpha(x) = 1$, the global DS $f_g(x)$ is activated, while when $\alpha(x) = 0$, the local DS $f_l(\cdot)$ is active (see figure 5.6). When $\alpha(x) \in (0, 1)$, this results in a mixture of $f_g(x)$ and $f_l(\cdot)$. Next, we describe the properties and roles of each term in equation (5.1), as well as the derivation of sufficient conditions for GAS.

Global dynamics: $f_g(x)$ is a linear DS of the form:

$$f_g(x) = A_g x + b_g, \tag{5.2}$$

where $A_g \in \mathbb{R}^{M \times M}$ and $b_g = -A_g x_g^*$. The role of this DS is to impose the global converging behavior toward x_g^*. Techniques to ensure such convergence are covered in chapter 3.

Activation function: The activation function $\alpha(x)$ activates the local dynamics when $\alpha(x) < 1$—that is, solely in the compact set $\chi \subset \mathbb{R}^M$ (see figure 5.1). On the other hand, in both the converging region $\mathscr{C} \subset \mathbb{R}^M$ and the ball $\mathscr{B}_r$ of radius r centered at x_g^*, the local dynamics are inactive and the global DS guides the motion [i.e., when $\alpha(x) = 1$]. We parameterize this activation function as follows:

$$\alpha(x) = \left(1 - r(x)\right)\left(1 - \frac{1}{Z}\mathcal{N}(x|\mu, \tilde{\Sigma})\right) + r(x), \tag{5.3}$$

where $\mu, \tilde{\Sigma}$ are the parameters of a Gaussian distribution computed as the sample mean and the rescaled covariance matrix of the reference position trajectories, $\{\mathbf{X}\}$. By tuning the eigenvalues of $\tilde{\Sigma}$, the region of activation can be enlarged or contracted. $Z = \mathcal{N}(\mu|\mu, \tilde{\Sigma})$ is a normalization factor that flattens the probability values on the reference trajectory such that regions where $\alpha(x) \approx 0$ are wider within the compact set χ. Finally, $r(x)$ is

$$r(x) = \exp\left(-c||x - x_g^*||\right), \tag{5.4}$$

which is a a radial exponential function centered at x_g^*, with c proportional to the radius r of $\mathscr{B}_r$, enforcing $\alpha(x) = 1$ in the region within $\mathscr{B}_r$.

Lyapunov candidate function: To determine sufficient conditions for the GAS of equation (5.1), we propose the following Lyapunov candidate function:

$$V(x) = \tilde{x}_g^T P_g \tilde{x}_g + \beta(x)\left(\tilde{x}_g^T P_l \tilde{x}_l\right)^2, \tag{5.5}$$

with,

$$\beta(x) = \begin{cases} 1 & \forall x : \tilde{x}_g^T P_l \tilde{x}_l \geq 0 \\ 0 & \forall x : \tilde{x}_g^T P_l \tilde{x}_l < 0. \end{cases} \tag{5.6}$$

Equation (5.5) is of class $\mathscr{C}^1$ and is radially unbounded, and both $P_g = P_g^T, P_l = P_l^T \succ 0$ are symmetric positive definite (SPD) matrices. The first term in equation (5.5) is a standard P-QLF centered at the global attractor x_g^*, while the second term is an asymmetric local P-QLF shaped by the local virtual attractor x_l^* and activated via $\beta(x)$. Equation (5.5) follows the form of the weighted sum of asymmetric quadratic functions (WSAQF) proposed in [75]. Hence, we refer to equation (5.5) as a dynamics-driven WSAQF (DD-WSAQF), as it is parameterized by the global, x_g^*, and local virtual attractor, x_l^*, from the desired dynamics [equation (5.1)].

Locally active dynamics: The $f_l(\cdot)$ component in equation (5.1) denotes the dynamics that will induce the desired attractive behavior in the locally active regions. We design $f_l(\cdot)$ such that there is no equilibrium point within the compact set χ (i.e., the local virtual attractor x_l^* vanishes). To achieve this, we propose $f_l(\cdot)$ as a weighted combination of a *locally active* and a *locally deflective* DS. Both are centered on the x_l^* and modulated via a hyperplane partitioning function $h(x)$ as follows (with $\overline{h}(x) = 1 - h(x)$):

$$f_l(h(x), x) = h(x)f_{l,a}(x) + \overline{h}(x)f_{l,d}(x) - \lambda(x)\nabla_x h(x). \tag{5.7}$$

The *locally active* DS, $f_{l,a}(x)$, is defined as

$$f_{l,a}(x) = A_{l,a}x + b_{l,a}. \tag{5.8}$$

To induce a locally attractive behavior toward the reference trajectory $\{\mathbf{X}^l, \dot{\mathbf{X}}^l\}$, the parameters of equation (5.8) are constructed as follows:

$$A_{l,a} = U_{l,a}\Lambda_{l,a}U_{l,a}^T, \; b_{l,a} = -A_{l,a}x_l^* \tag{5.9}$$

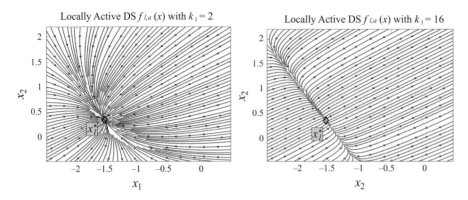

Figure 5.7
Effect of κ_i stiffness values on the locally active DS $f_{l,a}(x)$ from equation (5.8).

for

$$U_{l,a} = \begin{bmatrix} \bar{x}_0 & \bar{x}_{0\perp}^1, \ldots, \bar{x}_{0\perp}^{M-1} \end{bmatrix}, \quad \Lambda_{l,a} = \mathrm{diag}([\lambda_{l,a}^1, \ldots, \lambda_{l,a}^M]), \tag{5.10}$$

where $\bar{x}_0 \in \mathbb{R}^{M\times 1}$ is the direction of motion of the reference trajectory toward the local virtual attractor x_l^*; and $\bar{x}_{0\perp}^i \in \mathbb{R}^{M\times 1}$ are orthonormal vectors indicating the direction of convergence toward the reference trajectory. By imposing $\lambda_{l,a}^i < 0$, we can achieve $A_{l,a} \prec 0$. To ensure symmetric convergence to the reference trajectory, the eigenvalues must comply with the condition $|\kappa_i \lambda_{l,a}^i| > |\lambda_{l,a}^1|$, where $\kappa_i \in \Re_+$ indicates the stiffness of the DS around the reference trajectory in each ith orthogonal direction. Hence, under a perturbation, this DS will pull the robot toward the reference trajectory in a springlike manner. See figure 5.7 for the effect of κ_1 on a 2D system.

Local linear basis and virtual attractor placement: The orthonormal basis vectors that form $U_{l,a}$ are estimated by computing the covariance matrix of the reference position trajectories (i.e., $\Sigma_X = \mathbb{E}\{XX^T\}$), and extracting its eigenvalue decomposition, $\Sigma_X = VLV^T$. Assuming that the eigenvectors $V = [v_1, \ldots, v_M]$ are sorted in descending order with regard to the eigenvalues $L = \mathrm{diag}([l_1, \ldots, l_M])$ for $l_1 > \cdots > l_M$, we state that v_1 is the direction of motion of the reference trajectory. The remaining eigenvectors are directions pointing toward this reference trajectory. Hence, we define $\bar{x}_0 = v_1, \bar{x}_{0\perp} = v_2$ for $M = 2$. This can be generalized to $M > 2$, with $U_{l,a} = [v_1, v_2, \ldots, v_M]$.

Given the local direction of motion, $\bar{x}_0$, and the Gaussian distribution, $\mathcal{N}(x|\mu, \tilde{\Sigma})$ used in $\alpha(x)$ [equation (5.3)], we find the optimal placement of the local virtual attractor x_l^* by searching for the point along $\bar{x}_0$ where $\mathcal{N}(\mu, \tilde{\Sigma}) \approx \epsilon_l$. Here, ϵ_l is set empirically to a small value such that the local attractor lies on the outer tail of the distribution. To find this point, we perform a local line search in the direction of $\bar{x}_0$ by evaluating the following update step:

$$x_l^* \leftarrow x_l^* + \rho \bar{x}_0 \tag{5.11}$$

until $\mathcal{N}(x_l^*|\mu, \tilde{\Sigma}) \approx \epsilon_l$, and $\rho \in \Re_+$ is the update rate, which can be set to a very small number. Notice that, by construction, the locally active DS, $f_{l,a}(x)$, as defined in equation (5.9), is GAS at the local virtual attractor x_l^* as $\Re(\lambda_i(A_{l,a})) < 0 \ \forall i = 1, \ldots, M$. To preserve the

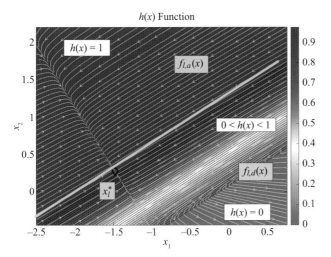

Figure 5.8
Hyperplane partition function $\tilde{h}(x)$ color scaling corresponds to value of $h(x)$ with locally active component $f_l(x) = h(x)f_{l,a}(x) + \bar{h}(x)f_{l,d}(x)$ (vector field) [see equation(5.7)].

desired locally attractive behavior of $f_{l,a}(x)$ while vanishing the attractor in the composed system, we introduce the *locally deflective* term, $f_{l,d}(x)$, defined as

$$f_{l,d}(x) = A_{l,d}x + b_{l,d}, \tag{5.12}$$

with

$$A_{l,d} = \lambda_d \mathbb{I}_M, \ b_{l,d} = -A_{l,d}x_l^* \tag{5.13}$$

for $\lambda_d > 0$ and $|\lambda_d| \le |\lambda_{l,1}^1|$; that is, it is a repulsive DS centered on the local virtual attractor x_l^*, bounded by the rate of convergence of the main direction of the locally active DS. To combine equations (5.8) and (5.12), we propose a nonnegative partition function $h(x)$ to indicate which region of the state-space each DS belongs to, parameterized as follows:

$$h(x) = \min\left(1, \frac{1}{2}(\mathbf{w}^T x + b + |\mathbf{w}^T x + b|)\right), \tag{5.14}$$

with $\mathbf{w} = -\bar{x}_0$ and $b = 1 - \mathbf{w}^T x_l^*$.

Equation (5.14) is a simple linear classifier, where $h(x) = 1$ denotes the region of the + class, corresponding to the *locally active* DS; and $h(x) < 1$ denotes the region of the − class, corresponding to the *locally deflective* DS, as shown in figure 5.8. Via the upper bound of 1, equation (5.14) becomes an activation function of the locally active and deflected linear DS bounded by $0 \le h(x) \le 1$. To ensure that there are no induced equilibrium points around the local virtual attractor x_l^*, the term $-\lambda(x)\nabla_x h(x)$ adds a velocity component in the direction of the negative gradient of the partition function $-\nabla_x h(x)$, modulated with

$$\lambda(x) = \begin{cases} \exp\left(-c_l\|x - x_l^*\|\right) & \forall x : \dfrac{\nabla_x \tilde{h}(x)^T \nabla_x V(x)}{\|\nabla_x \tilde{h}(x)\|\|\nabla_x V(x)\|} \ge 0 \\ 0 & \forall x : \dfrac{\nabla_x \tilde{h}(x)^T \nabla_x V(x)}{\|\nabla_x \tilde{h}(x)\|\|\nabla_x V(x)\|} < 0, \end{cases} \tag{5.15}$$

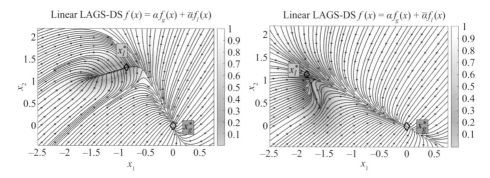

Figure 5.9
Examples of linear LAGS-DS with conditions from theorem 5.1 ensured.

where c_l is proportional to the radius r_l of the $\mathcal{B}_{r_l}$. The addition of this term is necessary when, in the composed system [equation (5.1)], the direction of motion of the global DS $f_g(\cdot)$ and the local DS $f_l(\cdot)$ are either perpendicular or opposite to each other. Furthermore, we constrain the addition of this velocity vector with the gradient of the Lyapunov function $\nabla_x V(x)$. This ensures that $-\lambda(x)\nabla_x h(x)$ does not induce instabilities in equation (5.1); see theorem 5.1.

Global asymptotic stability: Up to now, we have constructed a local DS [equation (5.7)] whose equilibrium point vanishes inside the compact set χ. This, however, does not ensure that the composed system defined in equation (5.1) has a single equilibrium point at x_g^*. Next, we propose the necessary conditions for the GAS of equation (5.1). In figure 5.9, we illustrate DSs that ensure such conditions for different behaviors.

Theorem 5.1   *The DS in equation (5.1) is GAS at the attractor* $x_g^* \in \mathbb{R}^N$ *if, for the Lyapunov candidate function [equation (5.5)], the following conditions hold:*

1. For the global DS parameters:

$$\begin{cases} b_g = -A_g x_g^*, \; A_g^T P_g + P_g A_g = Q_g, \; A_g^T P_l = Q_g^l. \\ Q_g = Q_g^T \prec 0, \;\; Q_g^l = (Q_g^l)^T \prec 0 \end{cases} \tag{5.16}$$

2. For the activation/partition/modulation functions:

$$\begin{cases} 0 < \alpha(x) \le 1, 0 \le h(x) \le 1, \; \lambda(x) \ge 0. \end{cases} \tag{5.17}$$

3. For the local DS with $\mathbf{A_l} = h(x)A_{l,a} + \bar{h}(x)A_{l,d}$:

$$\begin{cases} b_{l,a} = -A_{l,a} x_l^*, \; b_{l,d} = -A_{l,d} x_l^* \\ \mathbf{A}_l^T + \mathbf{A}_l \prec \epsilon \mathbb{I}, \; \mathbf{A}_l^T P_l = Q_l, 2\mathbf{A}_l^T P_g = Q_l^g \\ Q_{l+} \not\succ 0, Q_{l+}^g \not\succ 0, \end{cases} \tag{5.18}$$

where $\epsilon \in \mathfrak{R}_+$ *is a small, positive, real number. The positive subscript* $\mathbf{Q}_+$ *indicates the symmetric part of the matrix and* $\not\succ 0$ *indicates nonpositive definiteness (i.e., it can be negative-definite or indefinite, but never positive). Finally, the combined system must be bounded* $\forall x \in \chi \subseteq \mathbb{R}^N$, *where* χ *is the locally active region as follows:*

$$\left\{ \lambda_{\max}(Q_G(x))||\tilde{x}_g||^2 + \lambda_{\max}(Q_{L+}(x))||\tilde{x}_l||^2 < -\tilde{x}_l^T Q_{LG}(x)\tilde{x}_g, \right. \tag{5.19}$$

where $Q_G(x) = \alpha(Q_g + \beta_l^2 Q_G^l)$, $Q_L(x) = \overline{\alpha}\beta_l^2 Q_l$, $Q_{LG}(x) = \alpha\beta_l^2 Q_G^l + \overline{\alpha}(Q_l^g + \beta_l^2 Q_l)$ *and* $\beta_l^2(x) = 2\beta(x)\tilde{x}_g^T P_l\tilde{x}_l$. *Recall that* $P_g = P_g^T, P_l = P_l^T \succ 0$.

Proof See section D.3.2 appendix D. $\qquad\square$

Exercise 5.1 *Prove that the Lyapunov candidate function defined in equation (5.5) is radially unbounded.*

Exercise 5.2 *Derive the gradient of the Lyapunov candidate function defined in equation (5.5) [i.e., $\nabla_x V(x)$].*

Exercise 5.3 *Derive the gradient of the hyperplane partition function defined in equation (5.14) [i.e., $\nabla_x h(x)$].*

Exercise 5.4 *Can the locally active DS defined in equation (5.8) be a deflective behavior (i.e., $A_{l,a} \not\prec 0$)? Derive the sufficient stability conditions for LAGS-DS [equation (5.1)] to remain GAS at x_g^*.*

5.1.2 Nonlinear LAGS-DS with Multiple Locally Active Regions

The nonlinear LAGS-DS formulation, which can be used to encode nonlinear reference trajectories,$\{X, \dot{X}\}$ with multiple locally active regions, takes the following form:

$$\dot{x} = \underbrace{\alpha(x) \overbrace{\sum_{k=1}^{K} \gamma_k(x) f_g^k(x)}^{f_g(x)}}_{\text{Nonlinear global DS}} + \underbrace{\overline{\alpha}(x) \overbrace{\sum_{k=1}^{K} \gamma_k(x) f_l^k(h_k(x), x)}^{f_l(\cdot)}}_{\text{Set of } K \text{ local dynamics}}. \tag{5.20}$$

As in the linear case, $f_g(x)$ is a GAS DS that converges on a global attractor x_g^*, $f_l^k(x)$ represent the kth local DS with local virtual attractors x_k^* defined by equation (5.7), and $\alpha(x)$ is the activation function that selects which local DS should be activated.

Global dynamics: The nonlinear global DS, $f_g(x)$, is an LPV-DS (as in chapter 3) that converges on the global attractor x_g^*, formulated as follows:

$$f_g(x) = \sum_{k=1}^{K} \gamma_k(x)(A_g^k x + b_g^k), \tag{5.21}$$

where $\gamma_k(x)$ is a state-dependent mixing function, which must be $0 < \gamma_k(x) \le 1$; $\sum_{k=1}^{K} \gamma_k(x) = 1$; and $b_g^k = -A_g^k x_g^*$. See chapter 3 for further information on the convergence condition.

Remark 5.1.1   *Each kth linear DS of* $f_g(x)$ *from equation (5.22) is tied to a locally active DS,* $f_l^k(\cdot)$*, via the mixing function* $\gamma^k(x)$*. Hence, equation (5.20) becomes a weighted, nonlinear combination of linear LAGS-DS (5.1), as follows:*

$$\dot{x} = \alpha(x) \sum_{k=1}^{K} \gamma_k(x)(A_g^k x + b_g^k) + \overline{\alpha}(x) \sum_{k=1}^{K} \gamma_k(x) f_l^k(h_k(x), x)$$

$$= \sum_{k=1}^{K} \gamma_k(x) \Big(\alpha(x)(A_g^k x + b_g^k) + \overline{\alpha}(x) f_l^k(h_k(x), x) \Big).$$

(5.22)

Locally linear state-space partitioning: To parameterize each kth linear LAGS-DS of equation (5.22) from a data set of reference position trajectories, $\{\mathbf{X}\}$, we partition the state space into compact sets, $\chi = \chi^1 \cup \cdots \cup \chi^K \subset \mathbb{R}^N$. Each compact set $\chi^k \subset \mathbb{R}^M$ encapsulates a locally linear region of the reference trajectories. To achieve such locally linear partitioning, we employ the physically consistent Gaussian mixture model (PC-GMM) introduced in chapter 3. The PC-GMM ensures that the data points assigned to each kth component correspond to a linear DS. Hence, the state space can be represented with the GMM density, $p(x|\Theta_\gamma) = \sum_{k=1}^{K} \pi_k \mathcal{N}(x|\mu^k, \Sigma^k)$, with parameters $\Theta_\gamma = \{\pi_k, \mu^k, \Sigma^k\}_{k=1}^{K}$. Figure 5.10 top left shows the GMM density of the reference trajectories of a nonlinear motion fitted with the PC-GMM approach. Further, given this locally linear state-space partitioning, for each kth compact set χ^k, we take the corresponding Gaussian parameters $\mathcal{N}(x|\mu^k, \Sigma^k)$ and can automatically estimate the local basis, $U_{l,a}^k$, and local virtual attractors, x_k^*, with the approach introduced in section 5.1.1, as shown in figure 5.10 (bottom left).

As in chapter 3, we parametrize $\gamma_k(x)$ in (5.22) with the a posteriori probability for the kth Gaussian component [i.e., $\gamma_k(x) = p(k|x, \Theta_\gamma)$], defining the contribution of each linear LAGS-DS [equation (5.22)]. Note that, when $\alpha(x) = 1$, equation (5.22) reduces to equation (5.21), which is the LPV-DS introduced in chapter 3. This allows us to have $\leq K$ locally active DSs, while preserving the overall motion pattern of the reference trajectories mimicked by the global DS.

Activation function: To define in which of the compact sets $\chi^k \subset \mathbb{R}^N$ a local DS component, $f_l^k(\cdot)$, should be activated, we propose the following alternatives for $\alpha(x)$, which, as in the linear LAGS-DS case, should have the following bounds: $0 < \alpha(x) \leq 1$.
GMM-based: Following the Gaussian-based activation function for the linear LAGS-DS [equation (5.3)], we use the GMM density from $\gamma(x)$ to parameterize $\alpha(x)$ as follows:

$$\alpha(x) = \Big(1 - r(x)\Big)\Big(1 - \frac{1}{Z}p(x|\Theta_\alpha)\Big) + r(x),$$

(5.23)

where $Z = \min_{\mu^k} p(\mu^k|\Theta_\alpha)$ is a normalization factor used for peak truncation. This modeling approach has the following two properties:

• When all kth compact sets are to be activated, the parameters of the GMM used to parameterize equation (5.23) are equivalent to those of the mixing function $\gamma(x)$ (i.e., $\Theta_\alpha = \Theta_\gamma$).

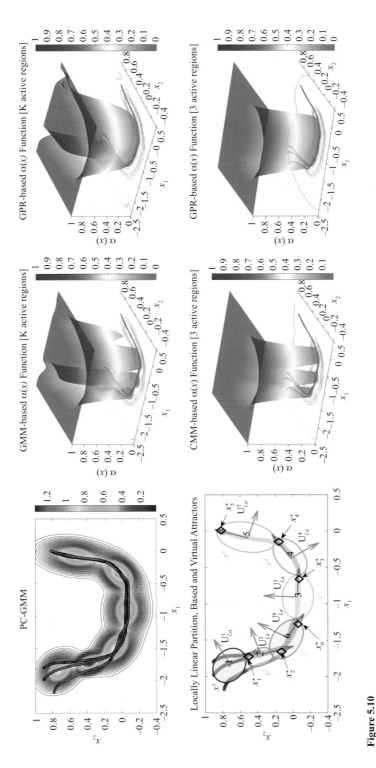

Figure 5.10

(*top-left*) *PC*-GMM probability density function. (*bottom-left*) Locally linear partition of state space *via PC*-GMM with local basis, $U_{l,a}^k$, and local virtual attractors, x_k^*. (*center-column*) GMM-based activation function [equation (5.23)] for (*top*) all K local regions to be locally active and for (*bottom*) solely three local regions being locally active (i.e., $k = 3, 4, 6$). (*right-column*) The GPR-based activation function defined in equation (5.23).

- In the case when only a subset of the K compact sets are to be activated, $\Theta_\alpha = \{\pi_k, \mu^k, \Sigma^k\}_{k=1}^{K_+}$, where $K_+ < K$. The parameters of the Gaussians μ^k, Σ^k are preserved, and the priors π_k are reestimated to ensure $\sum_{k=1}^{K_+} \pi_k = 1$.

Finally, $r(x)$ is parameterized by equation (5.4). Figure 5.10 (center column) we shows illustrations of equation (5.23) with all compact sets and a subset of compact sets activated.

GPR-based: An alternative is to use a probabilistic regression approach where the reference position trajectories $\{\mathbf{X}\}$ are considered the input and $\kappa \in \mathbb{R}^{1 \times L}$, a vector of 1s corresponding to each input, is the output for L data points. $\alpha(x)$ is be formulated as

$$\alpha(x) = \Big(1 - r(x)\Big)\Big(1 - \mathbb{E}\{p(\kappa|x, \mathbf{X}, \kappa)\}\Big) + r(x), \tag{5.24}$$

where $\mathbb{E}\{p(\kappa|x, \mathbf{x}, \kappa)\}$ is the Gaussian Process Regression prediction function (see section B.5.2 in appendix B). We use the squared exponential (SE) kernel function discussed in section 3.1.1.3 in chapter 3. Equation (5.24) yields a smooth, continuous local activation along the trajectories, as illustrated in figure 5.10 (right column). Learning proceeds as follows:

- When all kth compact sets are to be activated, rather then using the entire data set, $\{\mathbf{X}\}$, we take samples of the data set to reduce the computational complexity of GPR.

- When only a subset of the $K_+ < K$ compact sets are activated, we use the GMM density from the mixing function, $\gamma_k(x)$, to select the data points that belong to the chosen activation regions via GMM clustering (see section 3.4.1 in appendix B).

kth locally active dynamics: Each $f_l^k(h_k(x), x)$ in equation (5.22) is parameterized by equation (5.7) with independent local virtual attractors x_k^* and partition functions $h_k(x)$ [equation (5.14)] corresponding to each compact set $\chi^k \; \forall k = 1, \ldots, K$. From section 5.1.1, we know that by construction, each kth locally active DS, $f_{l,a}^k(x)$, is GAS with regard to its corresponding kth local virtual attractor x_k^*, in the absence of the other $K - 1$ locally active DS. When multiple locally active DSs are present, they become locally asymptotically stable systems that converge to their respective local virtual attractors.

Proposition 5.1 *Let $\mathscr{B}_k$ be a ball encapsulating the kth local virtual attractor x_k^* and the compact set $\chi^k \subset \mathbb{R}^N$. If $\mathrm{x} \in \mathscr{B}_k$, the kth locally active DS defined as*

$$f_{l,a}^k(x) = A_{l,a}^k x + b_l^k \tag{5.25}$$

is locally asymptotically stable within $\mathscr{B}_k$ if the following conditions hold:

$$\Big\{ b_{l,a}^k = -A_{l,a}^k x_k^*, \; (A_{l,a}^k)^T + A_{l,a}^k \prec -\epsilon \mathbb{I} \tag{5.26}$$

Exercise 5.5 *Prove that the sufficient conditions expressed in equation (5.26) for local asymptotic stability of equation (5.25) hold.*

Lyapunov candidate function: The nonlinear LAGS-DS variant inherits the same stability issues as the linear case. Following the same procedure as in section 5.1.1, we extend the dynamics-driven WSAQF Lyapunov function [equation (5.5)] to multiple AQF's corresponding to each locally active region of equation (5.22) as follows:

$$V(x) = \tilde{x}_g^T P_g \tilde{x}_g + \sum_{k=1}^{K} \beta_k(x) \left(\tilde{x}_g^T P_l^k \tilde{x}_k \right)^2, \tag{5.27}$$

where

$$\beta_k(x) = \begin{cases} 1 & \forall x : \tilde{x}_g^T P_l^k \tilde{x}_k \geq 0 \\ 0 & \forall x : \tilde{x}_g^T P_l^k \tilde{x}_k < 0. \end{cases} \tag{5.28}$$

As before, $P_g = P_g^T \succ 0$ and $P_l^k = (P_l^k)^T \succ 0 \ \forall k = 1, \ldots, K$. As in the linear case, the first term in equation (5.27) corresponds to the global DS [equation (5.21)], which is a P-QLF centered on the global attractor x_g^*. The second term, on the other hand, corresponds to a set of K locally active, asymmetric QLFs shaped by their corresponding local attractors x_k^*. Further, equation (5.27) is of class $\mathscr{C}^1$ and is radially unbounded [75].

Exercise 5.6 *Prove that the Lyapunov candidate function defined in equation (5.27) is radially unbounded.*

Exercise 5.7 *Derive the gradient of the Lyapunov candidate function defined in equation (5.27) [i.e., $\nabla_x V(x)$].*

Global asymptotic stability: Given the proposed Lyapunov candidate function [equation (5.27)], we state the sufficient conditions for the GAS of equation (5.22) as described in theorem 7.

Theorem 5.2 *The DS in equation (5.22) with activation function [equation (5.23) or (5.24)] is GAS at the attractor $x_g^* \in \mathbb{R}^N$ if, for the Lyapunov candidate function defined in equation (5.27), the following conditions hold:*

1. For each $k = 1, \ldots, K$ *component of the global DS:*

$$\begin{cases} b_g^k = -A_g^k x_g^*, \ (A_g^k)^T P_g + P_g A_g^k = Q_g^k \\ (A_g^k)^T + A_g^k \prec -\epsilon \mathbb{I}, \ (A_g^k)^T \mathbf{P}_l(x) = Q_G^{l,k} \\ Q_g^k = (Q_g^k)^T \prec 0, (Q_g^{l,k})_+ \prec 0, \end{cases} \tag{5.29}$$

where $\mathbf{P}_l(x) = \mathbf{P}_l(x)^T = \sum_{j=1}^{K} \beta_j^2(x) P_l^j \succ 0$ *is the weighted sum of local symmetric positive definite matrices.*

2. For the activation/partition/modulation functions:

$$\begin{cases} 0 < \alpha(x) \leq 1, \\ 0 \leq h_k(x) \leq 1, \ \lambda_k(x) \geq 0 \qquad \forall k = 1, \ldots, K, \\ 0 < \gamma_k(x) \leq 1, \sum_{k=1}^{K} \gamma_k(x) = 1 \end{cases} \tag{5.30}$$

where $\epsilon \in \Re_+$ *is a small, positive real number.*

3. *For each* $k = 1, \ldots, K$ *locally active DS with* $\mathbf{A}_l^k = h_k(x)\mathbf{A}_{l,a}^k + \bar{h}_k(x)\mathbf{A}_{l,d}^k$:

$$
\begin{cases}
b_{l,a}^k = -A_{l,a}^k x_k^*, \; b_{l,d}^k = -A_{l,d}^k x_k^* \\
(\mathbf{A}_l^k)^T + \mathbf{A}_l^k \prec -\epsilon \mathbb{I}, \; (\mathbf{A}_l^k)^T \mathbf{P}_l(x) = Q_l^k, \\
2(\mathbf{A}_l^k)^T P_g = Q_l^{g,k}, \; (Q_l^k)_+ \not\succ 0, \; (Q_l^{g,k})_+ \not\succ 0
\end{cases}
\tag{5.31}
$$

with the following bounds $\forall x \in \chi^k \subseteq \mathbb{R}^N$:

$$
\begin{cases}
\lambda_{\max}(Q_{G+}^k(x))\|\tilde{x}_g\|^2 < -\tilde{x}_k^T Q_{LG}^k(x)\tilde{x}_g \\
-\sum_{j=1}^{K} 2\beta_j \left(\alpha\lambda_{\max}((A_g^k)_+)(\tilde{x}_g^T P_l^j \tilde{x}_j)^2 + \bar{\alpha}\lambda_{\max}(\mathbf{A}_l^k)(\tilde{x}_k^T P_l^j \tilde{x}_j)^2 \right),
\end{cases}
\tag{5.32}
$$

with $Q_{LG}^k(x) = \bar{\alpha}(Q_l^{g,k} + Q_l^k)$, $Q_G^k(x) = \alpha(Q_g^k + Q_G^{l,k})$ *and* $\beta_j^2(x) = 2\beta_j(x)\tilde{x}_g^T P_l^k \tilde{x}_j$.

Proof See section D.3.3 in appendix D.   □

Remark 5.1.2   *The global DS in equation (5.20) can also be a linear DS, which would yield a LAGS-DS of the following form:*

$$
\dot{x} = \alpha(x)f_g(x) + \bar{\alpha}(x)\sum_{k=1}^{K}\gamma_k(x)f_l^k(h_k(x), x)
$$

$$
\tag{5.33}
$$

$$
= \alpha(x)(A_g x + b_g) + \bar{\alpha}(x)\sum_{k=1}^{K}\gamma_k(x)f_l^k(h_k(x), x).
$$

> **Exercise 5.8**   *Derive the sufficient conditions to prove the GAS of equation (5.33) with Lyapunov candidate function [equation (5.27)].*

5.1.3   Learning Nonlinear LAGS-DS

Given a set of reference trajectories $\{\mathbf{X}, \dot{\mathbf{X}}\}$, we perform the following learning steps:

1. Learn the locally linear state-space partitioning with the PC-GMM approach. Once this step is performed, we have the set of parameters for the mixing function, Θ_γ, the number of K linear DSs and their local virtual attractors x_k^* and basis $U_{l,a}^k$.

2. Learn the set of SPD matrices, $\Theta_P = \{P_g, \{P_l^k\}_{k=1}^K\}$, that parameterize the Lyapunov candidate function [equation (5.27)].

3. Learn the set of linear DS system matrices, $\Theta_{f_g} = \{A_g^k\}_{k=1}^K$, for the global DS-component, $f_g(x)$ [equation (5.21)].

4. Given the user-defined, locally active regions, the activation function $\alpha(x)$ can be parameterized via GMM [equation (5.23)] or GPR [equation (5.24)].

5. Given Θ_γ, $\alpha(x)$, $\{x_k^*, U_{l,a}^k\}_{k=1}^K$, and Θ_{f_g}, we estimate the system parameters of the locally active components $f_l^k(\cdot, x)$.

Steps 1 and 4 are described in sections 5.1.3.1-4, respectively. Next, we provide further details into performing steps 2, 3, and 5.

5.1.3.1 Learning a Candidate Lyapunov Function for LAGS-DS

Given the set of reference trajectories $\{\mathbf{X}, \dot{\mathbf{X}}\}$, the global attractor x_g^*, and the set of local virtual attractors $\{x_k^*\}_{k=1}^K$, we can learn $\Theta_P = \{P_g, \{P_l^k\}_{k=1}^K\}$ that parameterizes [equation (5.27)]. This is achieved by solving a constrained optimization problem such as the one introduced in chapter 3 for learning the P matrix for a P-QLF. While we use the same objective function as in equation (3.47), the optimization problem is subject to the following constraints:

$$\min_{\Theta_P} J(\Theta_P)$$

subject to

$$(5.34)$$

$$\begin{cases} P_g \succ 0, \ P_g = P_g^T \\ P_l^k \succ 0, \ P_l^k = (P_l^k)^T \\ \operatorname{Tr}(P_l^k) \leq \operatorname{Tr}(P_g) \ \forall \ k = 1, \ldots K. \end{cases}$$

The first and second constraints impose the structure of the P matrices (namely, that they should be SPD). The last constraint is introduced to control the amplitude of the eigenvalues of both P_g and P_l^k, while ensuring that the rate of decrease of energy in the global component is higher than in the local ones. As in equation (3.47) in chapter 3, in order to find a locally optimal solution for Θ_P, we use the IPOPT [20] with the fmincon solver in MATLAB, as in [75].

5.1.3.2 Global DS Estimation for LAGS-DS

Given the set of reference trajectories $\{\mathbf{X}, \dot{\mathbf{X}}\}$, the global attractor x_g^*, the mixing function $\gamma_k(x)$, and the set of SPD matrices $\Theta_P = \{P_g, \{P_l^k\}_{k=1}^K\}$, we estimate the parameters for the global DS $\Theta_{f_g} = \{A_g^k, b_g^k\}_{k=1}^K$ by minimizing the mean squared error (MSE) between the approximated and observed velocity as in the stable estimator of dynamical systems (SEDS) and stable estimator of LPV dynamical systems (SE-LPVDS) approaches (see chapter 3). While we use the same objective function as in equations (3.33) and (3.46), the optimization problem is subject to the following constraints:

$$\min_{\theta_{f_g}} J(\theta_{f_g})$$

subject to ($\forall k = 1, \ldots, K$)

$$\begin{cases} b_g^k = -A_g^k x_g^* \\ (A_g^k)^T P_g + P_g A_g^k = Q_g^k, \ \ Q_g^k \prec 0 \\ (A_g^k)^T \sum_{k=1}^K P_l^k = Q_G^{l,k}, (Q_g^{l,k})_+ \prec 0 \\ (A_g^k)^T + A_g^k \prec -\epsilon \mathbb{I} \ \text{if} \ x_k^* = x_g^*. \end{cases} \quad (5.35)$$

The constraints in equation (5.35) are derived from the necessary stability conditions expressed in equation (5.29). While the first two lines of constraints are the same as in equation (5.29), the third constraint is a conservative, nonstate-dependent version of the condition defined in equation (5.29). Due to this conservative constraint, we can relax the fourth constraint to solely the kth matrix A_g^k, whose corresponding local virtual attractor x_k^* is equivalent to the global attractor x_g^*. Equation (5.35) is a nonconvex semidefinite program, which we solve with PENLAB [40], interfaced through the YALMIP MATLAB toolbox [93].

5.1.3.3 Local DS Estimation for LAGS-DS

Given the mixing function $\gamma(x)$, the activation function $\alpha(x)$, the set of SPD matrices Θ_P, the set of global DS parameters Θ_{f_g}, and the set of local basis and virtual attractors $\{x_k^*, U_{l,a}^k\}_{k=1}^K$, we can now estimate the set of eigenvalues of the system matrices $\theta_{f_l^k} = \{\Lambda_{l,a}^k, \Lambda_{l,d}^k\}$ for each $k \in K_+$ locally active dynamics $f_l^k(\cdot)$. Via equation (5.10), the ratio of the eigenvalues $\Lambda_{l,a}^k$ of each locally active dynamics represents the stiffness κ_i of the locally active DS around the reference trajectory for each ith orthogonal direction of motion. To achieve this, we formulate an optimization problem in which the stiffness factors κ_i are maximized, while minimizing the velocity error between the desired velocity given by $f_l^k(\cdot)$ [equation (5.7)] and the observed velocity from the reference trajectories $\dot{x}^{\text{ref}}$. Given the reference trajectories that belong to each kth local compact set $\{\mathbf{X}, \dot{\mathbf{X}}\}_k$, the following optimization problem for each kth locally active region is solved as follows:

$$\min_{\theta_{f_l^k}} J(\theta_{f_l^k}) = \frac{1}{M-1} \sum_{i=1}^{M-1} \frac{1}{\kappa_i} + \bar{v} \sum_{m=1}^{M} \sum_{t=1}^{T^m} ||\dot{x}^{t,m} - f_l^k(x^{t,m})||^2,$$

subject to

$$\begin{cases} A_{l,a}^k = U_{l,a}^k \Lambda_{l,a}^k (U_{l,a}^k)^T, \;\; b_{l,a}^k = -A_{l,a}^k x_k^*, \;\; \Lambda_{l,a}^k \prec 0 \\ A_{l,d}^k = \lambda_{l,d} \mathbb{I}, \;\; b_{l,d}^k = -A_{l,d}^k x_k^*, \;\; 0 < \lambda_{l,d} < |(\lambda_{l,a}^k)^1| \\ (\lambda_{l,a}^k)^1 \leq R(A_g^k, U_{l,a}^k[:,1]) < 0 \\ (\lambda_{l,a}^k)^i \leq \kappa_i (\lambda_{l,a}^k)^1, \;\; 1 \leq \kappa_i \leq \kappa_{\max} \quad \forall i = 2, \ldots, M. \end{cases} \tag{5.36}$$

Given $\mathbf{A}_l^k = h_k(x) A_{l,a}^k + \bar{h}_k(x) A_{l,d}^k$, $(\mathbf{A}_l^k)^T \mathbf{P}_l(x) = Q_l^k$, $2(\mathbf{A}_l^k)^T P_g = Q_l^{g,k}$

$Q_{LG}^k(x) = \bar{\alpha}(Q_l^{g,k} + Q_l^k), Q_G^k(x) = \alpha(Q_g^k + Q_G^{l,k})$:

$$\begin{cases} \lambda_{\max}(Q_{G+}^k(x))||\tilde{x}_g||^2 < -\tilde{x}_k^T Q_{LG}^k(x)\tilde{x}_g \hspace{2cm} \forall x \in \chi^k \subseteq \mathbb{R}^N \\ -\sum_{j=1}^{K} 2\beta_j \left(\alpha \lambda_{\max}((A_g^k)_+)(\tilde{x}_g^T P^j \tilde{x}_j)^2 + \bar{\alpha}\lambda_{\max}(\mathbf{A}_l^k)(\tilde{x}_k^T P^j \tilde{x}_j)^2 \right), \end{cases}$$

where $\bar{v} > 0$ is a small, positive scalar (i.e., $\bar{v} << 1$). The first two lines of constraints in equation (5.36) define the structure of the parameters for the locally active [equation (5.8)] and locally deflective [equation (5.12)] DSs. By imposing these strict constraints, the stability conditions [equation (5.31)] from theorem 5.2 are directly ensured. The third line defines an upper bound on the first eigenvalue of $A_{l,a}^k$, where $R(A_g^k, U_{l,a}^k[:,1]) = \frac{\bar{x}_0^T(A_g^k) + \bar{x}_0}{\bar{x}_0^T \bar{x}_0}$ is the Rayleigh quotient of the global DS matrix A_g^k with regard to the main direction of motion of the locally active DS. This will ensure that the norm of the velocities produced by $f_l^k(\cdot)$ is equivalent or within the same range as the corresponding global component $f_g^k(\cdot)$. The fourth line defines the stiffness ratio κ_i for each ith orthogonal direction and sets upper/lower bounds for the desired κ_i. Finally, the fifth constraint imposes an upper bound on the dissipation rate of the local component $f_l^k(\cdot)$ with regard to the global component $f_g^k(\cdot)$ as defined in theorem 5.2. Note that this constraint is state-dependent, and yet it should be enforced only within the compact set χ^k. Hence, when solving equation (5.36), we enforce this constraint on the reference trajectories and a set of discrete sample points that are drawn from inside the continuous region $\chi^k \subset \mathbb{R}^N$ (i.e., $\{x\}_k \cup \{x_{\chi^k}\}$). Equation (5.36) is a nonlinear, nonconvex optimization problem. We find a locally optimal solution for $\theta_{f_l^k}$ by using the IPOPT [20] with the fmincon solver in MATLAB.

It is straightforward to see that ensuring theorem 5.2 relies on ensuring the upper bound [equation (5.32)] on the continuous compact set χ^k represented by the set of sample points $\{x_{\chi^k}\}$. Because each compact set χ^k has a continuous representation via a corresponding Gaussian distribution, we draw points from $\mathcal{N}(x|\mu^k, \Sigma^k)$ to collect our samples $\{x_{\chi^k}\}$.

Efficient sampling and relearning for χ^k To reliably integrate the continuous constraints within the compact set χ^k, we follow the approach presented in [107], in which an interplay between learning and verification is proposed. At the first iteration $i = 0$, we uniformly sample points, $\{x_{\chi^k}\} = x_{\chi^k}^0$, from the outermost isocontour of the Gaussian distribution parameterized by $\mathcal{N}(x|\mu^k, \Sigma^k)$ and use these points to estimate the initial parameters $\theta_{f_l^k}^0$ by solving equation (5.36). We then verify equation (5.32) on M_{ver} random samples of the Gaussian distribution; in our experiments, we set $M_{ver} = 10^5$. We then collect the samples that violate the constraints and add them to our evaluation set $\{x_{\chi^k}\} = x_{\chi^k}^0 \cup x_{\chi^k}^i$ and relearn of the parameters $\theta_{f_l^k}^i$ by solving equation (5.36) with the new sample points and initialized with $\theta_{f_l^k}^{i-1}$. This procedure is iterated until no violations are obtained in the compact set or a maximum number of iterations N_{iter} is reached. Figure 5.12 shows an example of a run of this sampling/relearning scheme for a single locally active region.

5.1.4 Learning Algorithm Evaluation

The LAGS-DS illustrations provided in figures 5.5 and 5.11 have been learned by using the scheme presented in this section from traced-out demonstrations with a MATLAB graphical user interface (GUI). Figure 5.13 shows global nonlinear DS versus LAGS-DS, learned on exemplars from the LASA Handwriting Dataset. In these examples, we selected locally linear regions to be activated. As can be seen, our learning scheme finds the maximum allowable stiffness κ_i^k on each kth locally active region in order for GAS to be ensured.

The approach presented here is capable of generating a GAS vector field that tightly converges on linear regions of a nonlinear reference trajectory. However, due to this linearization, depending on the reference trajectory, we might lose some of the smoothness in the overall motion pattern and exhibit "corner cutting" (as in the C-shape in figure 5.13) or "sharper edges" (as in the Khamesh-shape in figure 5.13).

Programming Exercise 5.1   *The aim of this instructional exercise is to familiarize readers with the LAGS-DS learning algorithm.*

Instructions: *Open MATLAB and set the directory to the folder corresponding to chapter 5 exercises; in this folder, you will find the following scripts:*

```
1  ch5_ex1_linear_lagsDS.m
2  ch5_ex1_nonlinear_lagsDS.m
```

In these scripts, you will find instructions in block comments that will enable you to do the following:

- *Draw 2D trajectories on a GUI with a 2D simulated robot.*
- *Load the LASA Handwriting Dataset shown in figure 3.7 in chapter 3.*
- *Learn linear LAGS-DS (ch5_ex1) or a nonlinear (ch5_ex1_nonlinear_ *.m) LAGS-DS.*

Note: *The MATLAB scripts provide options to load/use different types of data sets and to select different options for the learning algorithm. Please read the comments in the script and run each block separately.*

After following each of these instructions, one can simulate the robot's behavior as follows:

1. *Draw arbitrary linear trajectories and define the global attractor, and then do the following (ch5_ex1_linear_lagsDS.m):*

    - *Write a function to estimate the linear LAGS-DS parameters from a desired κ_i and a set of linear trajectories while ensuring the conditions in theorem 5.1.*
    - *Test the optimization above for a range of values of κ_i. Can you find the maximum allowable value of κ_i that will remain globally asymptotically stable?*
    - *Tune the hyperparameters for radial exponential functions $r(x)$ and $\lambda(x)$ (i.e., c_g and c_l), How do these parameters affect the smoothness resulting vector field in terms of DS shape and stability?*
    - *Compare the allowable stiffness and computation time achieved by your optimization vs. one provided in code. How do the resulting DS differ qualitatively?*
    - *Tune the eigenvalues of the covariance matrix Σ for the activation function $\alpha(x)$, and then estimate the allowable stiffness κ_i with the provided optimization. What is the effect of the width of $\alpha(x)$ around a reference trajectory?*

2. *Draw nonlinear trajectories or load motions from the LASA Handwriting Dataset, then:*

    - *Fit PC-GMM and select the locally active regions on the nonlinear trajectory. Implement a function that finds the local attractors and basis functions following equation (5.11)*
    - *Find the allowable stiffnesses κ_i^k for each kth region (independently) using the optimization you implemented above and the one provided in the code (5.36). Does the mixture of linear LAGS-DS ensure a globally stable nonlinear DS?.*
    - *Run the constrained optimization solver for nonlinear LAGS-DS provided in the code.*
    - *Tune the hyper-parameters for radial exponential functions $r_k(x)$ and $\lambda_k(x)$ (i.e., c_g^k and c_l^l) as the width of the $\alpha(x)$ function. Analyze the effect of these parameters on the resulting nonlinear motion.*

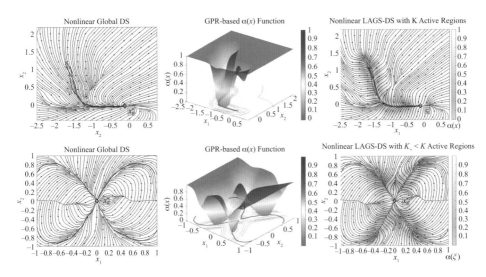

Figure 5.11
Examples of global DS. (*left column*) $f_g(x)$, equation (5.22), (*middle column*) activation functions $\alpha(x)$ [equation (5.23) or (5.24)], and (*right column*) LAGS-DS $f(x) = \alpha f_g(x) + \overline{\alpha} f_l^k(\cdot)$, equation (5.20).

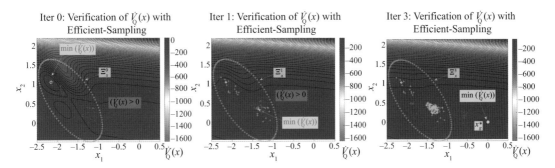

Figure 5.12
Illustration of efficient sampling and relearning scheme.

5.1.5 Robotic Implementation

We present here two implementations of the LAGS-DS method to learn stiff control laws for robotic systems. We showcase that the approach does not depend on the type of robot used. Rather, we consider an implementation on two very different platforms and different tasks: (1) a KUKA LWR 4+ robotic arm with 7 degrees of freedom (DOF) for handwriting tasks; and (2) a single or set of iCub humanoid robots for navigation and comanipulation tasks.

5.1.5.1 Handwriting task

The handwriting task involves learning a complex 2D motion corresponding to a letter or symbol that should be written on a whiteboard. The letters/symbols used are Angle-shape and Khamesh-shape from the LASA Handwriting Dataset used to evaluate the approaches in chapter 3. A marker is attached to the end effector of a 7-DoF KUKA LWR4+ robot arm. The goal of this task is to showcase the capabilities of LAGS-DS to recover from a perturbation by following the reference trajectories as faithfully as possible.

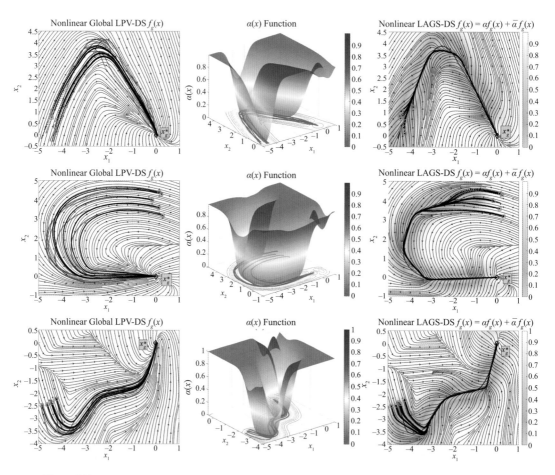

Figure 5.13
Global DS $f_g(x)$ **(left column)**, activation functions $\alpha(x)$ **(middle column)**, and LAGS-DS **(right column)** learned on exemplars from the LASA Handwriting Dataset.

Data collection A description of how the LASA Handwriting Dataset was generated was provided in section 3.3 in chapter 3. For each shape, we use five of the ten trajectories provided. We also scale the positions and velocities such that they are large enough to be written on the whiteboard and visualize the recovery from perturbations.

Learning After applying preprocessing steps as in section 3.4.5 for the LPV-DS experiments, we perform the sequence of optimizations listed and described in detail in section 5.1.3 to learn the LAGS-DS parameters.

Robot control and execution We use the 7-DoF KUKA LWR4+ robot arm to perform the task learned with LAGS-DS and adopt the passive DS-based control strategy, such that the robot allows perturbations from the human while excuting the task. Figure 5.14 shows examples of a KUKA LWR robot manipulator writing the Angle-shape while being perturbed by a human. As shown, after the perturbations, the robot no longer follows the reference trajectory and instead follows the next integral curve of the global DS, leading it to the final target. On the other hand, when using LAGS-DS, the robot goes back to following the reference trajectory after being perturbed. Figure 5.14 also shows the trajectories of the

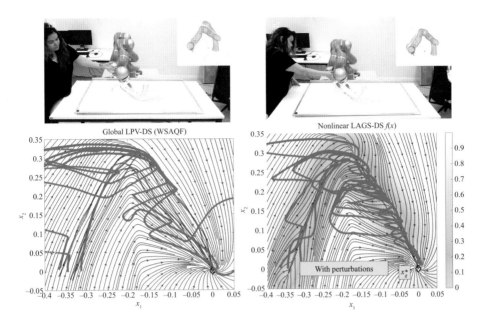

Figure 5.14
Trajectories of the end effector executed for the Angle-shape. Red trajectories are demonstrations, and blue and green trajectories are executions.

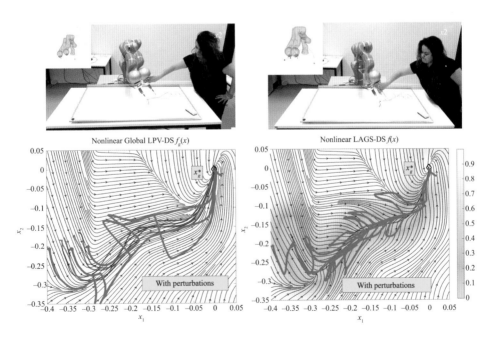

Figure 5.15
Trajectories of the end effector executed for the Khamesh-shape. Red trajectories are demonstrations, and blue and green trajectories are executions.

end effector being perturbed continuously using both DS variants, starting from the same initial states. Notice how in the unperturbed executions, LAGS-DS follows precisely the mean reference trajectory, while a global LPV-DS generalizes the motion pattern. Similar behavior is achieved with the Khamesh-shape, as shown in figure 5.15.

5.1.5.2   iCub navigation/comanipulation task

Similar to the LPV-DS experimental validation in section 3.4.5, we further validated the use of LAGS-DS for motion planning in two scenarios: (1) constrained locomotion with one iCub humanoid robot; and (2) a comanipulation task between the two iCub robots. Illustrations of the two scenarios are provided in figures 5.16 and 5.17.

Data collection For these scenarios, we collected data by teleoperating the robots within the Gazebo physics simulator as described in section 3.4.5 in chapter 3 for the LPV-DS robotic experiments.

Learning After applying preprocessing steps as in section 3.4.5 for the LPV-DS experiments, we perform the sequence of optimizations listed and described in detail in section 5.1.3 to learn the LAGS-DS parameters.

Robot control and execution The learned LAGS-DS generates the desired velocities for the object (in the two robot scenarios), which are then transformed to velocities in the frame of each robot's base and fed to a compliant, low-level walking controller [38] [37] or fed directly as the desired reference velocity for the base of the robot (in the single robot scenario). Further details of the control architecture used for the iCub humanoid robots are detailed in [44].

Scenario 1 (Constrained Locomotion): The desired task of the iCub robot is to navigate from a starting point to a target by following a particular path and orientation that will allow it to walk through a narrow passage. Figure 5.16 illustrates the executed path and plots of the robot's trajectory during multiple executions of the task, superimposed on the global-DS or LAGS-DS. In this scenario, we abstain from using any collision avoidance strategy to showcase how LAGS-DS can implicitly solve for this problem by providing stiffness around the desired reference trajectory. The global-DS is more prone to get stuck in a collided configuration and fail to reach the target. On the other hand, when using LAGS-DS for motion planning, regardless of how many external perturbations the robot is subject to, it will follow the reference trajectory and reach the target.

Scenario 2 (Object Comanipulation): The desired task of the two iCub robots is to pick up an object from one conveyor belt and slide it onto a second conveyor belt until the desired target is reached. Figure 5.17 illustrates snapshots of the executed path of the object's trajectory during multiple executions of the task, superimposed on the global-DS and LAGS-DS. The global-DS is prone to failing to achieve the task due to a slight perturbation along the path. On the other hand, LAGS-DS can handle multiple perturbations along the path, while always managing to slide the object on the second conveyor belt.

5.2   Learning Sequences of LPV-DS with Hidden Markov Models

In order to reach subgoals or via-points within a complex task (as illustrated in figure 5.2), we propose the following formulation, which sequences distinct LPV-DSs as follows:

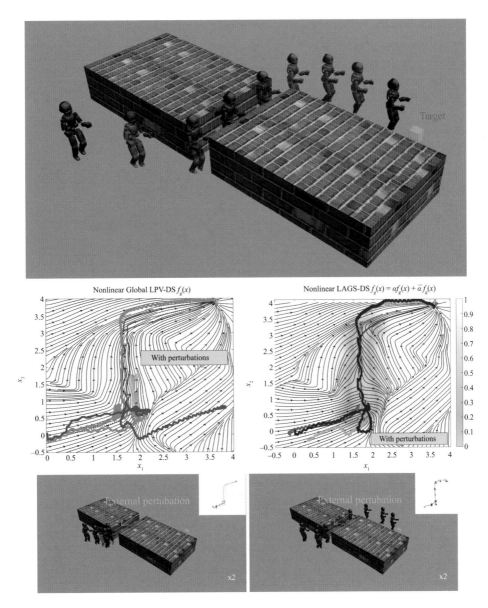

Figure 5.16
Scenario 1: Constrainted Navigation with one iCub. (top) Simulated scenario defining the task. *(center row)* Vector fields for Global-DS versus LAGS-DS learned from the demonstrated trajectories highlighted in red, **black** trajectories are open-loop trajectories generated by integrating the DS, and different-colored trajectories are comanipulations of the robot while simulating the task with different external perturbations. *(bottom row)* Snapshots of the simulated scenario with similar external perturbations. As can be seen, LAGS-DS is capable of recovering and finishing the task after an external perturbation, while a Global-DS gets stuck in the wall.

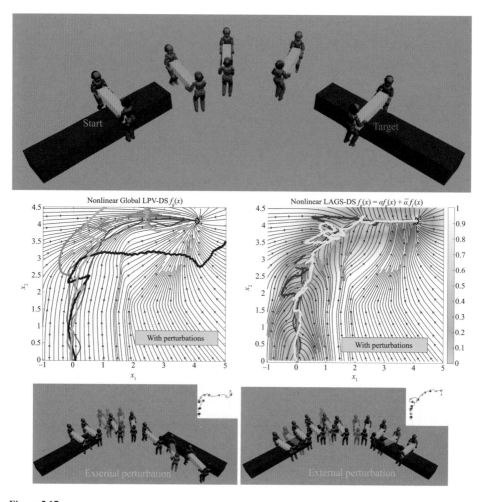

Figure 5.17
Scenario 2: Object Comanipulation with two iCubs. (top) Simulated scenario defining the task. *(center row)* Vector fields for Global-DS versus LAGS-DS learned from the demonstrated trajectories highlighted in red, **black** trajectories are open-loop trajectories generated by integrating the DS, and different-colored trajectories are comanipulations of the robot while simulating the task with different external perturbations. *(bottom row)* Snapshots of the simulated scenario with similar external perturbations. As can be seen, LAGS-DS is capable of recovering and finishing the task after an external perturbation, while a Global-DS tries to reach the target but follows a path that leads to task failure.

$$\dot{x} = \sum_{s=1}^{S} h_s(x) f_s(x), \tag{5.37}$$

where S is the total number of subtasks (i.e., subgoals or via-points) and $f_s(x)$ is the sth LPV-DS with attractor $x_s^* \in \mathbb{R}^N$ corresponding to the sth subtask; and $h_s(x)$ is a sequencing function that dictates when the robot should switch to the sth subtask. This switching occurs solely when the robot was at the sth subtask and reaches its corresponding attractor.

One of the assumptions made in chapter 3 is that the attractors were known for each DS learning approach. In this case, we seek to learn a sequence of LPV-DSs, with unknown attractors; hence, we introduce the inverse LPV-DS formulation and a learning algorithm to parameterize both the DS system matrices $\{A^k\}_{k=1}^K$ and the attractor x^* from data.

Given the inverse LPV-DS formulation, we seek to jointly learn the sequencing function, $h_s(x)$, and the set of S inverse LPV-DS that represent each of the subtasks. To achieve this, we parameterize equation (5.37) as an HMM where $h_s(x)$ is defined by the probability of the robot being in the sth state (i.e., the *active* subtask), and terminating in the sth attractor. Furthermore, because the inverse LPV-DS is learned as a joint probability distribution (as in the SEDS approach from chapter 3), it is considered the emission model of the sth state in the HMM. The resulting DS velocities are then computed from the expected dynamics of the HMM-LPV model. Hence, this section begins with describing and evaluating the inverse LPV-DS formulation and learning approach and ends with learning and evaluation of the HMM-LPV model.[2]

5.2.1 Inverse LPV-DS Formulation and Learning Approach

Let us restate the LPV-DS formulation with QLF stability conditions in a more generic form. Namely, a nonlinear DS can be represented as

$$\dot{x} = \sum_{k=1}^K \gamma_k(z)(A^k x + b^k),\tag{5.38}$$

where $z \in \mathbb{R}^N$ is a measurable external parameter or operating point that defines the mixing coefficient $\gamma_k(z)$, $\sum_{c=1}^C \gamma_k(z) = 1$, and $\gamma_k(z) > 0$. In this case, if

$$A^k + (A^k)^T \prec 0, \; b^k = -A^k x^* \; \forall k = 1 \cdots K,\tag{5.39}$$

the system globally asymptotically converges to the unique attractor x^* as shown in chapter 3. This is a specific instance of a stable LPV system [9], and its parameterization can be estimated from data *if the attractor is known a priori* using GMMs [60] or fuzzy models [39, 28, 48]. When the parameter is a function of the system's state (i.e., $z = x$), then equation (5.38), which becomes fully autonomous, is referred to in the literature as a *quasi-LPV* system [39], and it can also be learned under stability constraints, as shown in chapter 3.

Note: In chapter 3, we did not make the distinction between quasi-LPV and LPV systems as $z = x$ in all the presented formulations. In this section, however, we do make that distinction, as the presented LPV formulation is more generic.

When the attractor, x^*, is unknown, the estimation of the DS parameters $\Theta_f = \{A^k, b^k\}_{k=1}^K$ becomes a challenging problem due to the nonconvexity arising from the product of all A^k and b^k values for each k constraint from equation (5.39). Hence, instead of considering the standard forward dynamic LPV-DS model [equation (5.38)], the main idea behind the method presented in this section is to avoid this dependency by considering its inverse. From equation (5.39), the equation of an *unknown* attractor can be derived as follows:

$$b^k = -A^k x^* \implies x^* = -(A^k)^{-1} b^k \qquad \forall k = 1, \ldots, K,\tag{5.40}$$

thus inverting equation (5.38), and using the equation of the *unknown* attractor [equation (5.40)] leads to the mixture of local *inverse* linear DSs:

$$x = \sum_{k=1}^{K} \gamma_k(z) \left((A^k)^{-1}\dot{x} - (A^k)^{-1}b^k \right)$$

$$x = \sum_{k=1}^{K} \gamma_k(z) \left((A^k)^{-1}\dot{x} + x^* \right) \tag{5.41}$$

$$= x^* + \sum_{k=1}^{K} \gamma_k(z)(A^k)^{-1}\dot{x}.$$

This alternative representation simplifies the identification of LPV and quasi-LPV systems when the attractor x^* is unknown, as it becomes a bias in the model.

Problem formulation: Given a set of M independent and identically distributed samples from reference trajectories $\{\mathbf{X}, \dot{\mathbf{X}}, \mathbf{Z}\} = \{x^i, \dot{x}^i, z^i\}_{i=1}^{M}$, the problem considered in this section is the estimation of the dynamic behavior represented in the observations assuming that it is GAS and has a single unknown attractor x^*.

5.2.2   Learning Stable Inverse LPV-DS with GMMs

The model presented in this section assumes that the joint distribution of z and the observed dynamic behavior represented by $\dot{x}, x$ are given by a GMM. Specifically, observations are distributed as

$$p\left(\begin{bmatrix} z \\ \dot{x} \end{bmatrix} \middle| x, \Theta_{\text{GMM}} \right) = \sum_{k=1}^{K} p(k|\Theta_{\text{GMM}})p\left(\begin{bmatrix} z \\ \dot{x} \end{bmatrix} \middle| x, k, \Theta_{\text{GMM}} \right), \tag{5.42}$$

where $p(k|\Theta_{\text{GMM}}) = \pi_k$ represents the prior and the conditional probability density functions of each component is defined as follows:

$$p\left(\begin{bmatrix} z \\ \dot{x} \end{bmatrix} \middle| x, k, \Theta_{\text{GMM}} \right) = p(\dot{x}|x, k, \Theta_{\text{GMM}})p(z|k, \Theta_{\text{GMM}})$$

$$p(\dot{x}|x, k, \Theta_{\text{GMM}}) = \mathcal{N}\left(A^k(x - x^*), \Sigma_{\dot{\epsilon}}^k \right)$$

$$p(z|k, \Theta_{\text{GMM}}) = \mathcal{N}(\mu_z^k, \Sigma_z^k). \tag{5.43}$$

This has dynamics matrix A^k, and estimation noise $\Sigma_{\dot{\epsilon}}^k$ and μ_z^k, Σ_z^k denote the mean and variance of z, respectively. Note that all mixture components share the same attractor x^* by design.

The parameter set is given by $\Theta_{\text{GMM}} = \{x^*, \theta_1, \cdots, \theta_k\}$, where $\theta_k = \{\pi_k, \mu_z^k, \Sigma_z^k, A^k, \Sigma_{\dot{x}}^k\}$. For the sake of simplicity, in the following discussion, we will omit the dependency with regard to Θ_{GMM}.

For a standard LPV system (i.e., $z \neq x$), equation (5.43) accordingly assumes that z is independent of the system's state or its derivative. In the case of a quasi-LPV system (i.e., $z = x$), equation (5.43) reduces to the joint density $p\left(\begin{bmatrix} x \\ \dot{x} \end{bmatrix} \middle| k \right) = p(\dot{x}|x, k)p(x|k)$, as in the SEDS approach (section 3.3 in chapter 3). For later convenience, we also define the inverse linear dynamics of each component, which, under the asymptotic stability constraint, are

$$p(x|\dot{x}, k) = \mathcal{N}\left(x^* + (A^k)^{-1}\dot{x}, \Sigma_\epsilon^k\right), \tag{5.44}$$

and, again, all components share the same attractor x^*. Notice that in equation (5.44), the estimation noise covariance, $\hat{\Sigma}_\epsilon^k$, is now in terms of the inverse rather than the forward model $\hat{\Sigma}_\epsilon^k$.

5.2.2.1 Expected dynamics of the GMM

Given an observed $\{z, x\}$, the dynamic behavior specified by the conditional distribution of $\dot{x}$ from equation (5.42) is also Gaussian, with the expected mean

$$\mathbb{E}\left[p(\dot{x}|x, z)\right] = \sum_{k=1}^{K} \gamma_k(z)\left(A^k(x - x^*)\right), \tag{5.45}$$

where

$$\gamma_k(z) = \frac{\pi_k \mathcal{N}\left(z|\mu_z^k, \Sigma_z^k\right)}{\sum_{j=1}^{K} \pi_j \mathcal{N}\left(z|\mu_z^j, \Sigma_z^j\right)}.$$

Note that $\gamma_k(z) \geq 0$ and $\sum_{k=1}^{K} \gamma_k(z) = 1$. Sufficient conditions for convergence of equation (5.45) are given in proposition 5.2.

Proposition 5.2 *Dynamics [equation (5.45)] globally asymptotically converge to the attractor* x* *if*

$$A^k + (A^k)^T \preceq -\epsilon\mathbb{I} \qquad \forall k = 1 \cdots K, \tag{5.46}$$

with $\epsilon \in \mathbb{R}_+$ being a small positive number.

> **Exercise 5.9** *Prove that the conditions expressed in equation (5.46) are sufficient for the stability of the expected dynamics defined by equation (5.45).*

5.2.2.2 Learning asymptotically stable LPV systems with the EM algorithm

Given a set of observations, the LPV parameters Θ_{GMM} that maximize the likelihood are estimated by means of the EM algorithm [17] (see section B.3.1 in appendix B). To ensure convergence of equation (5.45), it suffices to constrain the maximization problem with the set of convex constraints [equation (5.46)], yielding

$$\underset{\Theta_{\text{GMM}}}{\arg\max} \sum_{i=1}^{M} \log p\left(\begin{bmatrix} z^i \\ \dot{x}^i \end{bmatrix} \middle| x^i, \Theta_{\text{GMM}}\right)$$

$$s.t. \quad A^k + (A^k)^T \preceq -\epsilon\mathbb{I} \qquad \forall k = 1 \cdots K. \tag{5.47}$$

The EM algorithm aims for a locally optimal solution of this problem by maximizing a lower bound of the log-likelihood in an iterative process. Let K_i be a latent variable representing the membership of the ith obsevation. The *expectation step (E-step)* consists of the maximization of the distribution of K_i—that is, the *responsibility* of each component, considering the current parameters Θ_{GMM}, which corresponds to

$$p(K_i = k) = \frac{p(k)p\left(\begin{bmatrix} z^i \\ \dot{x}^i \end{bmatrix} \Big| x^i, k\right)}{\sum\limits_{k=1}^{K} p(k)p\left(\begin{bmatrix} z^i \\ \dot{x}^i \end{bmatrix} \Big| x^i, k\right)}. \tag{5.48}$$

Leveraging this distribution, the *maximization step (M-step)* computes the optimal parameters solving the now-simplified optimization problem

$$\hat{\Theta}_{\text{GMM}} = \arg\max_{\Theta_{\text{GMM}}} \sum_{i=1}^{M} \sum_{k=1}^{K} p(K_i = k) \log p(k)p(\dot{x}|x, k)p(z|k)$$

$$s.t. \quad A^k + (A^k)^T \preceq -\epsilon \mathbb{I} \qquad \forall k = 1 \cdots K. \tag{5.49}$$

The optimal $\{\hat{\pi}_k, \hat{\mu}_z^k, \hat{\Sigma}_z^k\}$ are computed in closed form; see, for instance, [17]. To efficiently estimate the attractor $\hat{x}^*$ and avoid nonconvexity, we consider inverse dynamics [equation (5.44)] yielding the surrogate problem

$$\hat{\Theta}_{\text{GMM}} = \arg\max_{\Theta_{\text{GMM}}} \sum_{i=1}^{M} \sum_{k=1}^{K} p(K_i = k) \log p(k)p(x|\dot{x}, k)p(z|k)$$

$$s.t. \quad (A^k)^{-1} + (A^k)^{-T} \preceq -\epsilon_{\text{inv}} \mathbb{I} \qquad \forall k = 1 \cdots K, \tag{5.50}$$

with $\epsilon_{\text{inv}} \in \mathfrak{R}_+$ being a small, positive number and which, from applying Bayes's rule, is a lower bound of equation (5.49) if the underlying probabilities are $p(x|k) \leq p(\dot{x}|k)$. Note that the problem now constrains inverse dynamic matrices to be nonsymmetric negative definite to ensure convergence on the attractor, which at the same time guarantees that the matrices are invertible. The stationary point for the estimation noise covariance, $\hat{\Sigma}_\epsilon^k$, in terms of the constrained optimal inverse dynamic parameters $\hat{x}^*, \hat{A}_k^{-1}$, is given by

$$\hat{\Sigma}_\epsilon^k = \frac{\sum\limits_{i=1}^{M} p(K_i = k) \left(x^i - (\hat{x}^* + (\hat{A}^k)^{-1}\dot{x}^i)\right) \left(x^i - (\hat{x}^* + (\hat{A}^k)^{-1}\dot{x}^i)\right)^T}{\sum\limits_{i=1}^{M} \sum\limits_{k=1}^{K} p(K_i = k)}.$$

Substituting this expression into equation (5.50) and neglecting constant terms, the constrained maximization step for $\hat{x}^*, (\hat{A}^{-1})^{1..K}$ is

$$\hat{x}^*, (\hat{A}^{-1})^{1..K} = \arg\max_{x^*, (\hat{A}^{-1})^{1..K}} \sum_{k=1}^{K} -\frac{1}{2} \log |\hat{\Sigma}_\epsilon^k| \left(\sum_{i=1}^{M} p(K_i = K)\right)$$

$$s.t. \quad (A^k)^{-1} + (A^k)^{-T} \preceq -\epsilon_{\text{inv}} \mathbb{I} \qquad \forall k = 1 \cdots K, \tag{5.51}$$

which can be proved convex by adding a regularizing term in the covariance and an additional trace term [59, 80]. As an alternative, if the estimation noise covariance is assumed diagonal, the problem simplifies to the convex quadratic program

$$\hat{x}^*, (\hat{A}^{-1})^{1..K} = \underset{x^*, (\hat{A}^{-1})^{1..K}}{\arg\max} \sum_{k=1}^{K} -\frac{1}{2} \operatorname{tr}\left(\hat{\Sigma}_\epsilon^k\right) \left(\sum_{i=1}^{M} p(K_i = k)\right)$$

$$\text{s.t.} \quad (A^k)^{-1} + (A^k)^{-T} \preceq -\epsilon_{\text{inv}} \mathbb{I} \qquad \forall k = 1 \cdots K. \tag{5.52}$$

Although $(\hat{A}^{-1})^k$ is a valid estimate of the inverse of the forward dynamics matrix A^k of each component, a direct optimization of equation (5.49) (with the forward dynamics conditional probability density function), considering the resulting $\hat{x}^*$ after solving equation (5.52), yields more accurate results. The corresponding convex program is

$$\hat{A}^{1..K} = \underset{A^{1..K}}{\arg\max} \sum_{k=1}^{K} -\frac{1}{2} \operatorname{tr}(\hat{\Sigma}_\epsilon^k) \left(\sum_{i=1}^{M} p(K_i = k)\right)$$

$$\text{s.t.} \quad A^k + (A^k)^T \preceq -\epsilon \mathbb{I} \qquad \forall k = 1 \cdots K, \tag{5.53}$$

where

$$\hat{\Sigma}_\epsilon^k = \frac{\sum_{i=1}^{M} p(K_i = k) \left(\dot{x}^i - \hat{A}^k(x^i - \hat{x}^*)\right)\left(\dot{x}^i - \hat{A}^k(x^i - \hat{x}^*)\right)^T}{\sum_{i=1}^{M} \sum_{k=1}^{K} p(K_i = k)}.$$

Given an initial Θ_{GMM}, the EM algorithm iteratively applies the E-step [equation (5.48)], followed by the M-steps for $\{\hat{\pi}_k, \hat{\mu}_z^k, \hat{\Sigma}_z^k\}$ from equation (5.49), attractor $\hat{x}^*$ from equation (5.52), and the linear dynamics matrix $\hat{A}^{1..K}$ and estimation noise $\hat{\Sigma}_\epsilon^k$ from equation (5.53) until convergence.

5.2.2.3 Considering prior information about the attractor

Although the previous subsection assumes an unknown attractor, in many scenarios, prior information might be available, such as a specific area where we expect the system to converge, or a set of regions of interest. In those cases, a maximum a posteriori estimate takes this additional information into account by solving

$$\underset{\Theta_{\text{GMM}}}{\arg\max} \sum_{i=1}^{M} \log p\left(\begin{bmatrix} z_i \\ \dot{x}_i \end{bmatrix} \middle| x^i, \Theta_{\text{GMM}}\right) p(x^*)$$

$$\text{s.t.} \quad A^k + (A^k)^T \preceq -\epsilon \mathbb{I} \qquad \forall k = 1 \cdots K,$$

where $p(x^*)$ is the prior distribution of the attractor. The solution to this problem can also be solved by means of the EM algorithm explained in the previous subsection, except for equation (5.52), which, in this case, yields an additional term, $\log p(x^*)$. If this term is convex, such as with a Gaussian or a mixture of Gaussians, the solution remains convex and can be efficiently estimated. In addition, any linear constraints on the attractor (e.g., to avoid unreachable regions of a robot's workspace) can be added to this problem without losing convexity.

5.2.2.4   Learning algorithm evaluation

We implemented our method in MATLAB using the SEDUMI [142] solver and the YALMIP [93] interface to solve equations (5.52) and (5.53). We limit our evaluation to a quasi-LPV system where $z = x$. We compare our method LPV-EM to the SEDS method, minimizing the MSE using the LASA Handwriting Dataset. For this evaluation, we consider four conditions:

• *SEDS with known attractor x^**. The attractor was fixed to the end point of the demonstrations.

• *LPV-EM with known attractor x^**. The attractor was fixed to the end point of the demonstrations, and only equation (5.53) is solved.

• *SEDS with unknown attractor x^**. The attractor is added as an open parameter to the solver provided in [72].

• *LPV-EM with unknown attractor x^**. The attractor is an open parameter, and both equations (5.52) and (5.53) are solved.

All conditions are evaluated between one and twenty-five components. To apply the EM algorithm, we initialize x^* with the average of the observed states and the rest of parameters with the initial clustering given by k-means. The choice of ϵ_{inv} is especially important, as it determines the maximum eigenvalues of the system dynamic matrices and strongly influences the estimated attractors. A value of $\epsilon_{inv} \approx 0$ will produce unnatural results, while an $\epsilon_{inv} \gg 0$ yields attractors very far from the original data. From our experience, values in the range $10^{-1} \leq \epsilon_{inv} \leq 10^{1}$ produced plausible results in all the tested data sets. In this specific experiment, we set $\epsilon_{inv} = 0.5$ and $\epsilon = 10^{-6}$. We evaluate our approach in terms of the root-mean-square error (RMSE) of the estimated velocity, the RMSE of the estimation of the attractor x^*, and the computation time.

In terms of the prediction RMSE, and as shown in figure 5.18, when the attractor is known, SEDS outperforms all other conditions because it directly minimizes the MSE

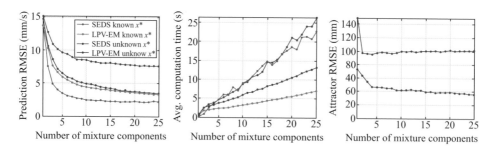

Figure 5.18
Results for the LASA Handwriting Dataset for SEDS and LPV-EM, both with and without knowing the attractor x^*. The left and the center plots show the prediction RMSE and the training time, respectively, for all four conditions as function of the number of mixture (mix.) components. The right plot shows the attractor estimation error for the conditions without knowing the attractor.

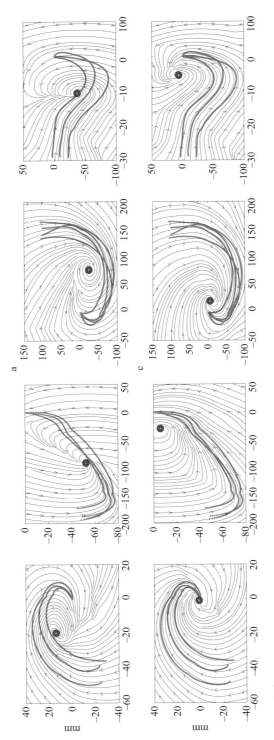

Figure 5.19
The resulting streamlines and attractor of four 2D recordings of the LASA Handwriting Dataset for SEDS (first row) and LPV-EM (second row), with unknown attractor x^* and $K = 7$ components. The red dots show the training samples, and the blue marker indicates the estimated attractor. The data set assumes that the attractors are at $[0, 0]$.

in its learning process. In contrast, LPV-EM maximizes the whole likelihood, including the distribution of z, not only the prediction error, resulting in a marginally less accurate prediction. However, this capability will be essential in the next section to formulate an HMM based on this model. When the attractor is unknown, LPV-EM clearly outperforms the SEDS variant. Interestingly, the prediction performance of LPV-EM with and without knowing the attractor is only marginally different. This does not happen in the case of SEDS, where, when the attractor is unknown, the RMSE remains almost constant after ten components, indicating unsatisfactory performance overall. In terms of computation time, both LPV-EM conditions yield a significantly faster computation time as the number of components increases. The condition with unknown attractors is approximately two times slower, as it must solve an additional optimization problem in each M-step. Concerning the RMSE of the estimation of the attractor, the LPV-EM condition yields significantly better results.

As illustrated in figure 5.19, when the attractor is unknown, the SEDS estimation might fall into local minima, yielding undesired attractors. In contrast, the LPV-EM variant always achieves consistent results due to the convexity of the maximization steps. Although the resulting attractors do not exactly fit the a priori assumed goals, the estimated solutions are plausible explanations of the data as goal-oriented motions. In fact, the human demonstrations of the data set are assumed to be converging at (0,0), but the data might not be exhibiting such behavior clearly. In summary, the LPV-EM model is a suitable alternative to SEDS when the attractor is known, and it becomes the only reliable option when the attractor is unknown.

Programming Exercise 5.2   *The aim of this instructional exercise is to familiarize readers with the LPV-EM algorithm for trajectories with unknown attractors.*

Instructions: *Open MATLAB and set the directory to the folder* `ch5-DS_Learning`; *in this folder, you will find the following script:*

```
ı   ch5_ex2_lpvEM.m
```

In this script, you will find instructions in block comments that will enable you to do the following:

- *Draw 2D trajectories on a GUI.*
- *Load the LASA Handwriting Dataset shown in figure 3.7.*
- *Learn inverse LPV-EM and compare with the SEDS and LPV-DS approaches.*

Note: *The MATLAB script provides options to load/use different types of data sets and to select different options for the learning algorithm. Please read the comments in the script and run each block separately.*

We recommend readers to test the following with either self-drawn trajectories or motions from the LASA Handwriting Dataset:

1. For a set of trajectories compare qualitatively (RMSE, convergence) and quantitatively (phase-plot) LPV-EM vs. inverse LPV-EM with unknown attractor.

2. Compare results when using M = 1 versus M >> 1 trajectories. How does the number of datapoints affect each method in terms of accuracy and computation time?

> 3. Add priors on the attractor locations using a single Gaussian or a GMM, does the estimation improve? Is it faster to converge?

5.2.3 Learning Sequences of LPV-DS with HMMs

In this section, to represent task sequences, we formulate an HMM representing multiple subtasks where each subtask is given by an asymptotically stable, quasi-LPV system. In contrast to previous work, the capability of the LPV-EM algorithm of robustly estimating the attractor from data enables learning HMMs with nonlinear stable dynamics as state emission probabilities. In addition, we incorporate the intuitive idea that a sub-task is finished when it reaches its goal in our model: following recent work [81, 34], transitions between latent states (subtasks) depend on observations by means of a termination policy. We then constrain the model such that terminations are most likely around the attractor (i.e., subtasks finish when their corresponding attractor is reached with a certain level of precision, which is also learned from data). We then study the dynamics of the *HMM-LPV* model and propose a learning and motion generation method that guarantees stability of the full sequence for left-to-right and periodic models.

Problem formulation: Given a set of demonstrated reference trajectories $\{\mathbf{X}, \dot{\mathbf{X}}, \mathbf{Z}\}$, the problem considered in this section is the acquisition of a dynamic model of the task assuming that it consists of a sequence of S goal-oriented GAS subtasks. Trajectories might represent either a part of the task, a part of a single subtask or the complete sequence of subtasks.

5.2.3.1 The HMM-LPV model

Let $s_i \in \{1, 2, \ldots, S\}$ be a discrete latent variable representing the active subtask at time step i, and $b_i \in \{0, 1\}$ a *termination* binary variable that represents the event of finishing ($b_i = 1$) or not ($b_i = 0$) subtask s_i. The HMM-LPV model assumes that observed *trajectories* are distributed as

$$p(x|\Lambda) = \left(\sum_{s_1=1}^{S} \sum_{b_{i-1}=0}^{1} p(s_1|\lambda_\pi) p\left(\begin{bmatrix} x^1 \\ \dot{x}^1 \end{bmatrix} \Big| s_1, \lambda_e \right) \right) \prod_{i=1}^{T} \sum_{s_i=1}^{S} \sum_{s_{i-1}=1}^{S} \sum_{b_{i-1}=0}^{1} p(s_i|s_{i-1}, b_{i-1}, \lambda_a)$$

$$\cdot p(b_{i-1}|s_{i-1}, x_{i-1}, \lambda_b) p\left(\begin{bmatrix} x^i \\ \dot{x}^i \end{bmatrix} \Big| s_i, \lambda_e \right), \tag{5.54}$$

with parameter set $\Lambda = \{\lambda_\pi, \lambda_a, \lambda_b, \lambda_e\}$, and where

- Initial subtask probabilities are $p(s|\lambda_\pi) = \lambda_\pi = \{\overline{\pi}_s\}$ for $1 \leq s \leq S$.

- Subtask transition parameters $\lambda_a = \{a_{jk}\}$ for $1 \leq j, k \leq S$, with $j \neq k$, are such that in the case of no termination, $p(j|j, b=0, \lambda_a) = 1$, and in the case of termination, $p(j|k, b=1, \lambda_a) = a_{jk}$ and $p(j|j, b=1, \lambda_a) = 0$. As we are only considering sequences, we limit our model to a left-to-right or periodic topology (i.e., if j precedes k, then $a_{jk} = 1$).

- Emission probabilities for each subtask are given by a quasi-LPV system; that is, $\lambda_e = \{\Theta_{\text{GMM}}^1 \cdots \Theta_{\text{GMM}}^S\}$ and $p\left(\begin{bmatrix} x \\ \dot{x} \end{bmatrix} \Big| s, \lambda_e \right) = p\left(\begin{bmatrix} x \\ \dot{x} \end{bmatrix} \Big| \Theta_{\text{GMM}}^s \right)$ as in equation (5.42). We denote the GMM parameters representing the quasi-LPV system of the sth subtask as $\Theta^s_{\text{GMM}} = \{x^*_s, \Theta^{s,1}_{\text{GMM}}, \cdots, \Theta^{s,K}_{\text{GMM}}\}$, with $\Theta^{s,k}_{\text{GMM}} = \{\pi_{s,c}, \mu^k_{s,z}, \Sigma^k_{s,z}, A^k_s, \Sigma^k_{s,\dot{x}}\}$.

• Termination probabilities $p(b|s, x, \lambda_b)$ are given by an x-dependent Bernoulli distribution parameterized by a radial basis function in the form

$$p(b = 1|s, x, \lambda_b) = \exp\{-(x - \mu^{b,s})^{\mathsf{T}} \Sigma^{-1}_{b,s}(x - \mu^{b,s})\}, \tag{5.55}$$

with parameters $\lambda_b = \{\mu^{b,1}, \Sigma^{b,1} \cdots \mu^{b,S}, \Sigma^{b,S}\}$. From the assumed distribution, the probability of no termination is $p(b = 0|s, x, \lambda_b) = 1 - p(b = 1|s, x, \lambda_b)$.

5.2.3.2 Expected dynamics of the HMM-LPV model

At time step i, and having observed the previous x-history since the first sample (i.e., $\{x^t\}^i_{t=1}$), the expected dynamics are

$$\dot{x}_{\text{hmm}} = \mathbb{E}[p(\dot{x}^i|\{x^t\}^i_{t=1}, \lambda)] \tag{5.56}$$

$$= \sum_{s=1}^S \underbrace{\tilde{h}_{s,i+1}(x^i)}_{h_s(x)} \underbrace{\mathbb{E}[p(\dot{x}^i|x^i, \Theta^s_{\text{GMM}})]}_{f_s(x)}, \tag{5.57}$$

where $\mathbb{E}[p(\dot{x}^i|x^i, \Theta^s_{\text{GMM}})]$ is the expected dynamics of the GMM given by equation (5.45), and $\tilde{h}_{s,i+1}(x_i) = p(s_i|\{x_t\}^i_{t=1}, \lambda)$ is the forward variable of the HMM considering only partial information—that is, x_t, which is computed recursively as

$$\tilde{h}_{s,i}(x^{i-1}) = \sum_{s_{i-1}=1}^S \sum_{b_{i-1}=0}^1 \tilde{h}_{s,i}(x^{i-1}) p(s_i|s_{i-1}, b_{i-1}, \lambda_a) \cdot$$

$$p(b_{i-1}|s_{i-1}, x^{i-1}, \lambda_b) p(x^{i-1}|s_i, \lambda_e), \tag{5.58}$$

and is normalized such that $\sum_{s=1}^S \tilde{h}_{s,i}(x^i) = 1$. Given the variables $h_s(x)$ and $f_s(x)$, we can see that the expected dynamics of the HMM [equation (5.56)] is equivalent to the proposed LPV-DS sequencing equation defined in equation (5.37). However, even if all subtasks fulfill proposition 5.2, the resulting dynamics are potentially unstable as they consider different attractors. In fact, we only preserve the stability properties of the sth subtask if $\tilde{h}_{s,i}(x^i) = 1$, when we recover dynamics [equation (5.45)]. To ensure the convergence of the whole sequence, we define s_{curr} as the *current* subtask and s_{next} as the *next* subtask with dynamics:

$$s_{\text{curr},i+1} = \begin{cases} s_{\text{next},i} & \text{if } \tilde{h}_{s_{\text{next}}}(x^i) = 1 \\ s_{\text{curr},i} & \text{otherwise} \end{cases}$$

$$s_{\text{next},i+1} = \begin{cases} s_{\text{next},i} + 1 & \text{if } \tilde{h}_{s_{\text{next}}}(x^i) = 1 \\ s_{\text{next},i} & \text{otherwise} \end{cases}. \tag{5.59}$$

We then propose a sufficient condition that guarantees that transitions between s_{curr} and s_{next} always evolve from $\tilde{h}_{s_{\text{curr}}}(x^i) = 1$ to $\tilde{h}_{s_{\text{next}}}(x^i) = 1$ as the system reaches the attractor of s_{curr}.

Proposition 5.3 *Let* s_{curr}, s_{next} *be the current and the next subtask respectively with dynamics [equation (5.59)] and let the conditional forward probability be computed considering these two states only—that is,* $S = \{s_{curr}, s_{next}\}$ *in equation (5.58). Any trajectory that reaches the attractor* $x^*_{s_{curr}}$ *will converge to the subtask distribution* $\tilde{h}_{s_{curr},i}(x_i) = 0$, $\tilde{h}_{s_{next},i}(x_i) = 1$ *if*

$$p(b = 1 | s_{curr}, x^*_{s_{curr}}, \lambda_b) = 1. \tag{5.60}$$

> **Exercise 5.10** *Prove that the conditions expressed in equation (5.60) are sufficient for the stability of transitions [equation (5.59)].*

To guarantee that $x^*_{s_{curr}}$ is reached, a straightforward approach generates motion considering only the current subtask, which is guaranteed to converge to its attractor from proposition 5.2. An alternative approach that continuously transitions between tasks results from modifying the original expected dynamics [equation (5.56)] by adding a stabilizing input [75], such that

$$\dot{x} = \dot{x}_{hmm} + \dot{x}_{corr} \tag{5.61}$$

$$\dot{x}_{corr} = \begin{cases} 0 & \text{if } l^\mathsf{T}\dot{x}_{hmm} > 0 \\ -\frac{l^\mathsf{T}\dot{x}_{hmm}}{\|l\|}l + \epsilon_{corr}l & \text{otherwise,} \end{cases}$$

where $l = (x^*_{s_{curr}} - z)$ and $\epsilon_{corr} > 0$. This way, the resulting dynamics is as close as possible to the model dynamics [equation (5.56)], which are corrected only in case of potential divergence.

If proposition 5.3 is fulfilled by every subtask, a left-to-right model representing a full sequence is guaranteed to converge to the attractor of the last subtask by following equation (5.61) in every transition. Similarly, for a periodic topology, the latent state evolution is guaranteed to exhibit a stable discrete limit cycle dynamics. Note that proposition 5.3 provides very conservative conditions; in practice, transitions usually converge without intervention of the corrective input.

5.2.3.3 Learning an HMM-LPV model with the Baum-Welch algorithm

To obtain the optimal parameters from demonstrations, we maximize the model likelihood by means of the Baum-Welch algorithm [117] and the EM algorithm for HMMs, and by constraining the problem with conditions from propositions 5.2 and 5.3 as

$$\arg\max_{\Lambda} \sum_{d=1}^{D} \log p(x_d | \Lambda) \tag{5.62}$$

$$s.t. \ p(b = 1 | s, x^*_s, \lambda_b) = 1 \tag{5.63}$$

$$A^k_s + (A^k_s)^\mathsf{T} \prec 0 \qquad \forall s = 1 \cdots S, \forall k = 1 \cdots K.$$

The E-step and the M-steps for λ_π, λ_a are detailed in [117] and [34]. The M-step for the emission probabilities λ_e (i.e., the LPV systems) is similar to equation (5.52), differing only in the responsibilities computed in the E-step. The M-step for the termination probabilities λ_b with the RBF function [equation (5.55)] yields a similar problem structure to logistic regression [34], but with an ellipsoid boundary function and therefore a nonconvex objective. However, equation (5.63) in this specific setting implies that for each state, $x_s^* = \mu^{b,s}$ and couples these two problems together. As a result, the maximization of λ_e drives solutions for λ_b toward regions, where the dynamics converge and where, at the same time, transitions are more likely to happen. Given an initial guess for Λ, the optimal parameters are computed by applying the E-step and the M-step iteratively until negligible improvement occurs.

MATLAB Exercise: In this exercise, we provide code to learn HMM-LPV (and visualize it) from 2D-drawn data sets.

Programming Exercise 5.3   *The aim of this instructional exercise is to familiarize readers with the HMM-LPV algorithm for trajectories of sequences of attractor dynamics.*

Instructions: *Open MATLAB and set the directory to the folder corresponding to chapter 5 exercise; in this folder, you will find the following script:*

```
1   ch5_ex3_hmmLPV.m
```

Within this script, you will find instructions in block comments that will enable you to do the following:

- *Draw complex 2D trajectories on a GUI.*
- *Learn HMM-LPV.*

Note: The MATLAB script provides options to load/use different types of data sets and to select different options for the learning algorithm. Please read the comments in the script and run each block separately.

We recommend the readers to test the following with self-drawn trajectories:

1. *Manually select the number of K Gaussian functions and the number of S subtasks.*
2. *Use model selection via BIC to estimate the number of S subtasks and tune the number of K Gaussians per sth LPV-DS.*
3. *Compare results when using M = 1 versus M >> 1 trajectories.*
4. *Add priors to the attractor locations using a single Gaussian or a GMM.*

5.2.4   Simulated and robotic implementation

We implemented our approach in MATLAB using the FMINSDP solver [147] to solve the joint maximization of λ_b and λ_e from equation (5.62). We initialize the model parameters with k-means. With this initial clustering, we apply the M-step of the LPV-EM algorithm explained in the previous section to initialize the parameters for λ_e including the attractors, while the covariance of the RBF termination function is initially set to the variance of the corresponding cluster. In our experiments, we set positive constants to $\epsilon_{corr} = 1$ in equation (5.61), and $\epsilon_{inv} = 0.5$, $\epsilon_{inv} = 10^{-6}$ in equations (5.52) and (5.53) respectively.

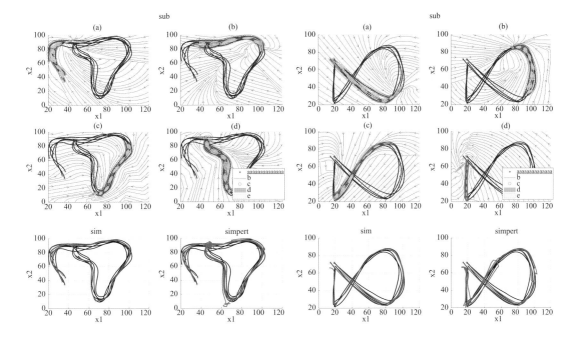

represent training samples, and the pink asterisk indicates the initial positions. The first two rows depict the LPV parameters of each subtask, their termination policy and their responsibility (the training samples that are most probably belonging to the subtask). The last row shows several simulated trajectories depicted by the green solid lines, generated following equation (5.61) and starting at every initial point. The left-to-right sequence of subtasks for both models corresponds to plots (a)-(b)-(c)-(d). Plots (a) and (c) show simulated trajectories without perturbations, while plots (b) and (d) show trajectories generated while the system is subjected to a perturbation every second. Small deviations from the main trajectories are observed each time a perturbation occurs.

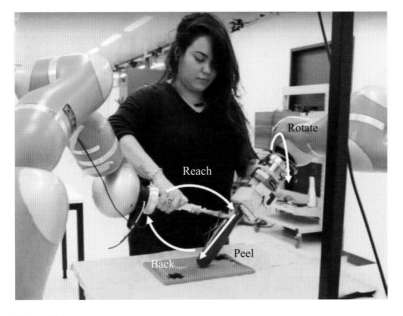

Figure 5.21
Illustration of a task involving a sequence of subtasks and robotic experiments of performing a peeling sequence for a vegetable.

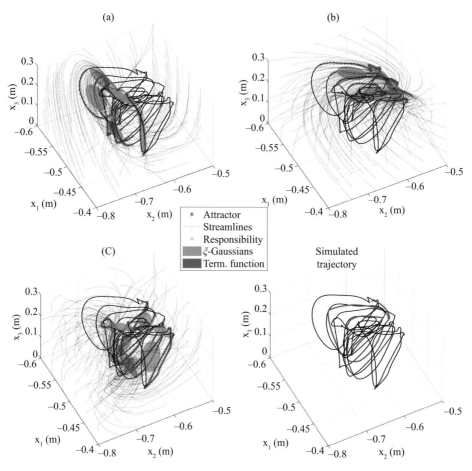

Figure 5.22
The resulting subtasks and simulated trajectories for the HMM-LPV model with three subtasks with five GMM components each for the zucchini-peeling task. The zucchini is placed on the right side in the demonstration. Figures (a), (b) and (c) depict the LPV parameters of each subtask, their termination policy and their responsibility (the training samples that are most probably belonging to the subtask). The bottom right plot shows simulated trajectories generated following equation (5.61). The **black** trajectory represent the training data, and the pink asterisk indicates the initial position. The underlying periodic structure yields cyclic subtask sequence (b)-(c)-(a), which correspond to the reach-peel-back phases depicted in figure 5.21.

We first illustrate the capability of our model in two exemplary sets of 2D human motions captured with a mouse. As shown in figure 5.20, the HMM-LPV model is able to extract meaningful subtask dynamics and termination policies. The attractors, and therefore the centers of the termination policies, are typically at the end of the direction of motion, but in some cases, such as example 1 (a) or example 2(a), they are placed farther away as the corresponding data does not exhibit convergence. As a result, the trajectories generated following equation (5.61) are as smooth as the training samples during the transitions; and due to proposition 5.3, they also converge on the last attractor. It is remarkable that in both examples, the generated trajectories without perturbations yielded no corrections from the stabilizing input. In simulations with disturbances, the time-independent and state-feedback nature of the model allows immediate replanning, thereby recovering from deviations. In addition, due to the stability properties of the model, trajectories converge on the last attractor.

To validate our approach in a more realistic setting, we test it on trajectories obtained during a kinesthetic teaching session where a human teacher drives a compliant passive robot to peel a zucchini [41] (see figure 5.21). The task consists of a repetitive process with three phases: reach, peel, and retract, while a second arm rotates the zucchini after each peel. We only modeled the motions of the peeling arm, and to capture the expected behavior, we train our model with a periodic structure and three states. The resulting subtasks are depicted in figure 5.22. The periodic topology of the model successfully captures the observed motion structure, and the reach-peel-back subtasks are represented by subtasks figure 5.22b, c, and a, respectively. Also, the simulated trajectory shows the state-dependent limit cycle behavior captured by the model, which matches the demonstrations.

In summary, the HMM-LPV model is a suitable method to represent complex dynamic behavior with multiple attractors from data. The resulting dynamic policies are time-independent, and therefore insensitive to perturbations. In addition, thanks to the stability properties and smooth transitions, dynamics are guaranteed to converge without discontinuities during transitions. All videos of the experiments presented in this chapter are available on the book's website.

III COUPLING AND MODULATING CONTROLLERS

6 Coupling and Synchronizing Controllers

In previous chapters, we always assumed that we were controlling a single agent. However, it is often useful to control several agents simultaneously. The agent may be several robots or simply several limbs of one robot. Simultaneous control allows the systems to act in synch. It also allows one to generate precedences across the movement of the various systems.

To generate such dependency across different systems, we use mathematical *coupling* across the dynamical systems (DSs) at the basis of the controllers of each agent. The notion of coupling is core to DS theory and appears when there is an explicit dependency across two or more dynamics. This chapter presents examples in which explicit couplings across dynamical systems-based control laws can simplify robot control while preserving the natural robustness against perturbation.

Section 6.1 introduces the type of dependencies considered in this book. Section 6.2 shows how coupling across DSs can be used to enforce that the DS follows a precise trajectory in space. This is used to control a robot to cut along a specific line in space. As the robot is controlled through dynamical systems, it retains robustness against external disturbances and hence can be moved away from the cutting line at all times by a human operator who verifies the quality of the cut. Section 6.3 presents another application of coupling across DSs for robot control. Coupling is used to control the arm and hand of a robot. One DS controls the arm movement in space, while a second DS controls the fingers of the robot's hand. The hand is coupled to the arm, as its DS depends on the state of the arm. The dependency creates a master-slave system in which the arm drives the movement of the hand. This allows one to synchronize the arm and hand movements to ensure that the hand closes on the object once the arm reaches the object. This coupled arm-hand DS offers a natural robustness to disturbances. For instance, if the object moves just when the robot is about to grasp it, the hand and arm will modify their movement in synch to adapt to the new position and orientation of the object. In section 6.4, the principle is extended to couple the eyes, arm, and hand of the robot through a hierarchical coupling across three DSs. The DS of the eyes drives the movement of the arm, while the DS of the arm drives the DS of the hand. This system enables the smooth visual pursuit and grasping of a moving object. It can also be used to track a visually moving obstacle and steer the arm and hand away from the obstacle.

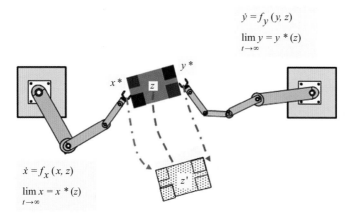

$$\dot{y} = f_y(y, z)$$
$$\lim_{t \to \infty} y = y^*(z)$$

$$\dot{x} = f_x(x, z)$$
$$\lim_{t \to \infty} x = x^*(z)$$

Figure 6.1
The DSs of two robot arms are coupled through an external variable z, the location of the object. When the object moves, the two robot arms move in synch. The coupling implicitly enforces synchrony.

Coupling across DSs is discussed further in other chapters of the book. Chapter 7 uses the coupling of DSs to control multiple robots in coordination. The dependency across the DS controlling the agents can be used to synchronize or sequence the movement of various robotic systems. In chapter 7, we show an application of this principle to catch a flying object with two robotic arms. The principle is based on three-way coupling. We couple the two robotic arms with one another and with the dynamics of the external object. This principle is extended in chapter 11 to ensure coupling in position and force, to enable a pair of robotic arms to reach and move jointly objects while balancing the forces involved.

6.1   Preliminaries

Consider two DSs of the form $\dot{x} = f_x(x)$ and $\dot{y} = f_y(y)$, each acting on two separate variables, x and y. We consider the two DSs to be coupled when they share an explicit dependency, which may be a shared external variable. For instance, if we write $\dot{x} = f_x(x, z)$ and $\dot{y} = f_y(y, z)$, both systems depend on variable z. Assume that the two variables x and y control for the state of two robotic arms, and z is the location of an object in space. Let $x^*(z)$ and $y^*(z)$ be the attractors of f_x and f_y, respectively, the two grasping locations on either side of the object. When the object moves, z moves to a new location, z'. Both attractors also move to new locations, $x^*(z')$ and $y^*(z')$. The two arms adapt to the new location of their respective attractors. As a result, they appear to move in synch in space. Such implicit coupling is used in chapter 7, in which we explain how we can learn and ensure the stability of the coupled system.

Another type of coupling is to let one DS depend on the state of the other system. For instance, if we write $\dot{x} = f_x(x)$ and $\dot{y} = f_y(y, x)$, y is explicitly dependent on x. Assume, as before, that both systems are asymptotically stable at x^* and y^*. Consider further a dependency of the form $\dot{y} = f_y(y)\|\dot{x}\|$. Such a dependency would lead the variable y to stop each time x stops. Furthermore, it would let y accelerate when x accelerates and vice versa.

In section 6.2, we develop an example of such coupling. The reader is encouraged to do exercise 6.1 to practice these concepts.

Exercise 6.1 *Consider two one-dimensional (1D) systems $\dot{x} = x$ and $\dot{y} = y + \alpha x$, $\alpha \in \mathbb{R}$.*

1. Consider $\alpha = 1$, for which values of x and y do the two systems stabilize?

2. Characterize the stability of the attractors.

3. Draw the phase plot of the two systems.

Programming Exercise 6.1 *Open MATLAB and set the directory to the following folder:*

```
1          ch6-coupled-DS
```

This code launches a MATLAB simulation of a two-dimensional (2D) coupled system of the form $\dot{x} = f_x(x, z)$ and $\dot{y} = f_y(y, z)$ with two linear DSs for f_x and f_y both stable at the origin, and z a 2D vector point. Do the following:

1. Make the attractor of both systems depend on z such that f_x stabilizes at point 1 and f_y at point -1. Where should z be?

2. Generate a linear displacement for z and observe the resulting movement of x and y.

3. The stability of the system depends on the speed at which z moves. Create a dynamics for z twice faster than the integration step of the two DSs. What happens?

6.2 Coupling Two Linear DSs

Assume a first linear DS of the form $\dot{x} = Ax + a$, $\dot{x}, x, a \in \mathbb{R}^N$, A negative definite matrix. Now assume a second *virtual* DS of the form $\dot{z} = Bz + b$, $\dot{z}, z, b \in \mathbb{R}^N$, which follows a virtual trajectory in space and stops at z^*.

We couple the two systems such that the dynamics of x is attracted to the state of the z, with $a(z) = Az$, similar to the example given in section 6.1. This let x move in synch with z. Furthermore, we can create a second dependency, this time from z to x, $\dot{z} = (Bz + b)\delta(x, z)$, to enforce that z moves when both systems are in the same state (i.e., $x = z$):

The complete system is given by

$$\dot{x} = (Ax + a(z)),$$

$$x^* = z,$$

$$a(z) = Az,$$

$$\dot{z} = (B\dot{z} + b)\delta(x, z), \tag{6.1}$$

$$z^* = B^{-1}b,$$

$$\delta(x, z) = \begin{cases} 1 \text{ if } x = z \\ 0 \text{ if } x \neq z. \end{cases}$$

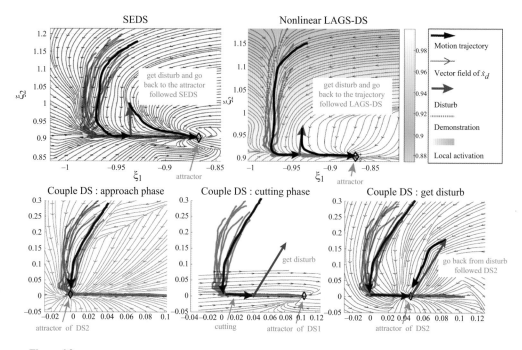

Figure 6.2
Comparison of the use of CDSs to ensure that the system follows a particular trajectory, compared to using SEDS and LAGS-DS. In contrast to SEDS and LAGS-DS, the CDS can return to the disturbance point and continue along the trajectory from there onward.

The coupling term $a(z) = Az$ changes the attractor of the first DS, continuously sending it to the state of the second DS. The coupling through $\delta(x, z)$ forces the second system to wait until the first system has reached its state. It then starts to increment its own state to move to its own attractor. That is, z remains static so long as $x \neq z$. Once both states are superimposed, the two systems move in synch along their common trajectory and toward the attractor z^*. The parameters of the system must be chosen carefully to ensure the stability of the entire system (see exercise 6.2).

This principle is illustrated in figure 6.2. Two coupled linear DSs are used to ensure that the system follows a particular trajectory. The dynamics is compared to using stable estimator of dynamical systems (SEDS) and locally active globally stable dynamical system (LAGS-DS), introduced in chapters 3 and 5. In contrast to SEDS and LAGS-DS, if the system is disturbed, the dynamics will return to the disturbance point once the disturbance is stopped, whereas SEDS will move to the final attractor following a different trajectory and LAGS-DS will return to the main trajectory, but farther down from the point where the disturbance took place.

6.2.1   Robot Cutting

Here, we illustrate an application of this for controlling a robot tasked to follow a line in space in simulation (see figure 6.4, and also programming exercise 6.2). This work was presented in detail in [156].

We apply a virtual disturbance while the robot moves along the line. As soon as the robot departs from the virtual trajectory, the second system stops incrementing. This enables the

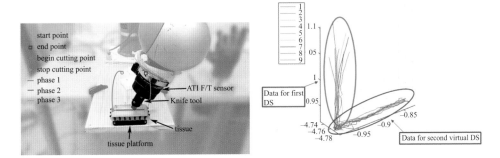

Figure 6.3
Application of CDSs to control the cutting movements of a robot. The robot is tasked to cut a piece of silicone on a flat surface (*left*). Data used from human demonstrations to train the robot's controller are shown at right [156].

Figure 6.4
Example of the use of CDSs to drive a robot to cut a piece of tissue. A first DS brings the robot's end point to the state of a secondary virtual DS, which is the transient attractor of the robot's DS. Once the robot has reached the state of the virtual system, the system starts moving and the attractor of the robot moves in synch with the system. If the robot is perturbed and moved away from the state of the virtual system, its attractor stays at the last state of the system and the system stops moving. Only once the robot has reached the state of the virtual system again does the attractor move again. From left to right, the six photos show: 1. robot approaches the beginning cutting point; 2. robot is disturbed during the approach phase; 3. robot returns to the disturbance point under control of DS2 and begins cutting under the control of DS1; 4. robot is disturbed during cutting. 5. DS2 is activated again and returns the robot to the disturbance point; 6. robot continues cutting [156].

system to keep a memory of where the robot was at the disturbance time and to move back to this point once the disturbance disappears.

In figure 6.4, we show an example of this approach to control a robot tasked to cut a piece of material along a specific line. To generate this controller, we used data gathered from human demonstrations following methods presented in chapter 2, as shown in figure 6.3. To generate the appropriate force for cutting the material, we combine the coupled DSs (CDSs) with impedance control, which is described in chapter 11.

Controlling the robot with the CDSs offered in equation (6.1) is advantageous, as it allows the robot to be robust to two types of disturbances when performing the task:

1. Assume that a human wishes to verify the goodness of the cut, the human would pick up the robot's arm and move it away from the cutting trajectory. As soon as the robot's state is no longer at the virtual state z, the second DS will stop incrementing z. Once the human releases the robot, the first DS drives the robot back to the position where the robot was cutting before the human intervened. Once the disturbance has disappeared and the robot has returned to the cutting position, the second DS starts incrementing again, which leads the robot to continue cutting as both x and z states move in synchrony.

2. If we fix the frame of reference in such a way that x and z are aligned with the plane on which the robot is tasked to cut into, the system can adapt to changes in the orientation of the cutting plane.

Exercise 6.2   *Assume that the system is three-dimensional (3D) (i.e., N = 3). Determine values for A, a, B, and b such that the CDS of 6.1 follows a line along the first axis and stops at $z^* = [1, 0, 0]^T$.*

Programming Exercise 6.2   *Open MATLAB and set the directory to the following folder:*

```
ch6-coupled-DS/ch6_ex_2
```

This code launches a robot simulator and generates the data shown in figure 6.4, using the CDS described in equation (6.1). Perform the following tests:

1. Generate a disturbance that sends the robot to a position ahead of the virtual trajectory. What happens then?

2. Change the parameters of the virtual trajectory to draw a circle in space instead of a line.

3. Modify the virtual trajectory for an ellipse.

6.3 Coupling Arm-Hand Movement[1]

In this section, we show another example of the use of coupling function across DS to control for fluent reach and grasp movements in robots. Indeed, when controlling for reach and grasp movements with DSs, we must account for the fact that the arm and hand movements depend on one another.

The reader may, however, be unconvinced by our previous statement that we need to couple the arm and hand. So let us consider the opposite—that we use two separare, not synchronized DSs to control for the movement of the arm and the closing of the fingers of the hand, respectively. Let us further assume that the first DS controls for the reaching movement and is asymptotically stable to the object's location, whereas the second DS is asymptotically stable at a desired joint configuration to achieve the desired grasp. If the object starts moving just when the arm and fingers are about to close on it, the hand's DS would let the arm adapt correctly to the disturbances and move toward the new location of the object. The fingers, however, would continue closing, as their DS is independent of the state of the object. If, in contrast, the fingers depend on the position of the object, the aperture and closure of the fingers would then adapt to the object's movement and close on the new location of the object in synch with the arm movement (see figure 6.5).

Hence, to apply DSs to control for arm and hand movement, we must find a way to synchronize them. In many instances in this book, we showed that we can control with a DS that converges asymptotically to a target. To perform reach-grasp tasks, such a scheme could be exploited in two ways. One could do either of the following:

1. Learn two separate and independent DSs with state vectors as end-effector pose and finger configurations.

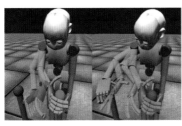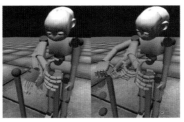

(a) Coupled   (b) Uncoupled

Figure 6.5
Reach-grasp task executions, with and without explicit coupling. The explicitly coupled execution (a) prevents premature finger closure, ensuring that given any amount of perturbation, formation of the grasp is prevented until it is safe to do so. In the implicitly coupled execution (b), the fingers close early and the grasp fails. The bottom figure shows close-ups of hand motion postperturbation with implicit (*left*) and explicit (*right*) coupling.

2. Learn one DS with an extended state vector consisting of degrees of freedom (DOF) of the end-effector pose, as well as finger configurations.

Learning two DSs would not be desirable at all, because then two subsystems (transport and preshape) would evolve independently using their respective learned dynamics. Hence, any perturbation in hand transport would leave the two subsystems temporally out of sync. This may lead to failure of the overall reach-grasp task, even when both the individual DSs will have converged on their respective goal states.

At first glance, the second option is more appealing, as one could hope to be able to learn the correlation between hand and finger dynamics, which would then ensure that the temporal constraints between the convergence of transport and hand preshape motions will be retained during reproduction. In practice, good modeling of such an *implicit* coupling in high-dimensional systems is hard to ensure when one learns the model from data. The model is as good as the data. If one is provided with relatively few demonstrations [as presented in chapter 2, in learning from demonstration (LfD), one targets fewer than ten demonstrations so that the training will be bearable to the trainer], chances are that the correlations will be poorly rendered, especially when querying the system far from the demonstrations. Hence, if the state of the robot is perturbed away from the region of the state space that was demonstrated, one may not ensure that the two systems will be properly synchronized.

Coupling two separate DSs is preferable. In the context of reach-and-grasp tasks, the DSs correspond to the hand transport (dynamics of the end-effector motion) and the hand preshape (dynamics of the finger joint motion). The transport process evolves independent of the fingers' motions, while the instantaneous dynamics followed by the fingers depends on the state of the hand.

6.3.1 Formalism of the Coupling

We start by assuming an unsynchronized system to control for the arm and the hand, whereby the arm is controlled to reach the object with a DS. We start with two DSs, $\dot{x} = f_x(x)$ and $\dot{q} = f_q(q)$, to control the finger joints. The first DS controls the end-effector in Cartesian space, $x \in \mathbb{R}^3$, and is asymptotically stable at the origin. As done in previous chapters, we place the origin on the object that we wish to grasp. The origin is, hence, on a moving frame of reference that translates with the displacement of the object. To generate a dependency between the fingers' movements and the hand transport, such that the fingers wait for the hand to reach the object, we exploit the fact that the distance to the object is known and is given by the norm of the state of the first DS (i.e., $\|x\|$). We can then create a dependency as follows:

$$\dot{q} = f_q(q - \beta\hat{q}),$$

$$\hat{q} = g(\|x\|),$$

$$q_{t+1} = q_t + \alpha\dot{q},$$

$$\beta, \alpha \in \Re.$$

(6.2)

The system in equation (6.2) results in the desired behavior—namely, that the fingers reopen when the object is moved away from the target and close on the object once the hand is on it. How fast the fingers reopen or close on the object is controlled through parameter α. By how much the fingers reopen is controlled by parameter β. The function g determines the type of coupling. The simplest coupling would be a linear dependency, such that the fingers close in proportion to how close the hand is to the object. This may be sufficient when controlling for a 1-DOF gripper, but not for a full humanoid hand, where some fingers may need to close more rapidly than others. We show next how one can learn the dynamics of each DS and the parameters of the coupling from human data. Once the dynamics is learned, the coupling can be tuned by changing the model parameters to favor either humanlike motion or fast adaptive motion to recover from quick perturbations. The stability of the system can be ensured (see exercise 6.3).

Exercise 6.3   *Consider the system described in equation (6.2).*

1. Assuming that both f_x and f_q are asymptotically stable at the origin, show that the coupled system in equation (6.2) retains the asymptotic stability of f_q.

2. Which condition must you put on $g(x)$ to ensure the stability of the coupled system?

6.3.2   Learning the Dynamics

To learn the system given in equation (6.2), one needs to gather data for training the driving hand and finger movements of the two DSs. Moreover, we also need data to estimate the open parameters that control the amplitude and speed at which the fingers reopen. If we follow a LfD scheme for robotics (see chapter 2), one can proceed to record examples of reach-and-grasp movements to train the DSs. However, as we are interested in deducing the coupling between the two dynamics, it is interesting to request data that demonstrate this coupling. One way to do so is to ask subjects to perform demonstrations under disturbances. Such an approach was followed in [135] and is illustrated in figure 6.6.

To model the natural adaptation of reach and grasp, we change the target to which subjects should reach after motion onset. We used two stationary targets, a green ball and a red ball. An on-screen target selector prompted the subject to reach and grasp one of the two balls, depending on the color shown on the screen. To start the experiment, one of the balls is switched on and the subject starts to reach toward the corresponding object. As the subject is moving the hand and preshaping the fingers to reach for the target ball, a perturbation is

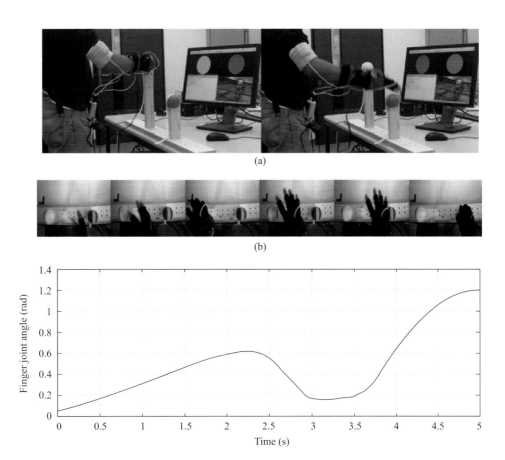

Figure 6.6
(a) Experimental setup to record human behavior under perturbations. An on-screen target selector is used to create a sudden change in the target location for reaching. (b) The motion of the fingers as seen close up from a high-speed camera at 100 fps. Note the decrease in the joint angle values (reopening of fingers) starting at the onset of perturbation.

created by abruptly switching off the target ball and lighting up the second ball. The switch across targets occurs only once during each trial, about 1 to 1.5 seconds after the onset of the motion. By then, the subject's hand has usually traveled more than half the distance to the target. The trial stops once the subject has successfully grasped the second target. To ensure that we are observing natural responses to such perturbations, subjects were instructed to proceed at their own pace, and no timing for the overall motion was enforced. Illustration of the data is shown in figure 6.7.

Visual inspection of the human data confirms a steady coupling between hand transport and fingers closing in the unperturbed situation, whereby the fingers close faster as the hand approaches the target faster, and vice versa. This coupling persists across trials and for all subjects. Most interesting is the observation that, during perturbed trials, just after the target is switched, the fingers first reopen and then close again in synch with the hand as the hand moves toward the new target (see figure 6.6b). We, hence, infer that the coupling function $g(x)$ depends on the distance of the hand to the target; therefore, we write $g(\|x\|)$.

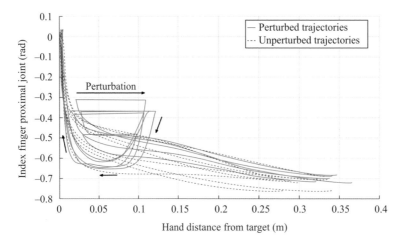

Figure 6.7
Hand-finger coordination during perturbed and unperturbed demonstrations [135].

Measurements of the dynamics of hand aperture and hand transport can be used to derive precise measurements of the correlation between hand transport and the fingers' preshape, which we then use to determine specific parameters of our model of coupled dynamical systems across these two motor programs. To proceed, we need first to build a statistical representation of the distribution of the data and then learn the DS from this distribution as described in chapter 3. For convenience, we place the attractors at the origin of the frames of reference of both the hand motion and the finger motion. This way, the hand motion is expressed in a coordinate frame attached to the object to be grasped, while the zero of the finger joint angles is placed at the joint configuration adopted by the fingers when the object is being grasped.[2]

As such, the following joint distributions are learned as separate densities:

1. $P(x, \dot{x})$: used to encode the hand's dynamics, $\dot{x} = E\{P(\dot{x}|x)\}$.

2. $P(q, \dot{q})$: used to encode the fingers' dynamics, $\dot{q} = E\{P(\dot{q}|q)\}$.

3. $P(\|x\|, q)$: encoding the coupling between the norm of the current hand position and the fingers' position; this can be used to determined the desired fingers position $\hat{q}$ by conditioning the distribution [i.e., $\hat{q}g(\|x\|) = E\{P(q|\|x\|)\}$].

The coupling function g is monotonic and zero at the origin.

The learned models can then be used in the CDS given by equation (6.2). When running the model under similar perturbations as those done during the trials, the model gives a good account of the dynamics of arm-arm movement (see figure 6.8a). Observe further that the trajectories followed by the fingers after perturbations remain within the covariance envelope of the model. This envelope represents the variability of finger motions observed during the unperturbed trials. This confirms the hypothesis that the fingers resume their unperturbed motion model shortly after responding to the perturbation. This is particularly visible when looking at figure 6.8b, which is zoomed in on the part of the trajectories during and just after the perturbation. Three demonstrations are shown. It can be seen that,

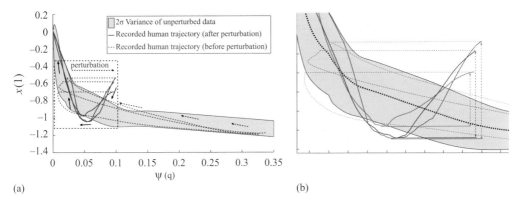

(a) (b)

Figure 6.8
(a) The data recorded from perturbed human demonstrations. The adaptation behavior under perturbation follows the same correlations between hand position and fingers as in the unperturbed behavior. (b) The region where perturbation is handled is indicated in red and zoomed, where three demonstrations (red, blue, and magenta) from the same subject are shown.

regardless of the state of the fingers q at the time of perturbation, the finger trajectories tend to follow the mean of the regressive model (which is representative of the mean of the trajectories followed by the human fingers during the unperturbed trials) before the perturbation occurs. Just after the perturbations, the fingers then reopen (trajectory goes down) and then close again (trajectory goes up).

As mentioned previously, the model has two open parameters, α and β, to control for the speed and amplitude of the coupled dynamics of finger movements to the hand movement. An illustration of the variation of the trajectories with α and β is shown in figure 6.9. Here, α modulates the speed with which the reaction to perturbation occurs. On the other hand, a high value of β increases the amplitude of reopening. figure 6.10 shows the streamlines of this system for two different α values in order to visualize the global behavior of trajectories evolving under the CDS.

To begin this discussion, we say that it was more interesting to learn the coupling separately rather than implicitly when learning a full-state Gaussian mixture model (GMM). Figure 6.11 shows a comparison of the CDS trajectories with those obtained using the single-GMM approach, where the coupling is only implicit. It shows the behavior when a perturbation is introduced only on the abscissa. Clearly, in the implicitly coupled case, the perturbation is not appropriately transferred to the unperturbed dimension q and the motion in that space remains unchanged. This behavior can be significantly different depending on the state of the two subsystems just after the perturbation. This is further demonstrated in figure 6.12, where we show the trajectories generated by the single-GMM model and the explicit CDS when initialized from different points in state space. Notice the sharp difference in the trajectories as the explicit CDS trajectories try to maintain correlation between the state-space variables and always converge from within the demonstration envelope. On the other hand, the trajectories of the single-GMM approach have no definite convergence constraint.[3] This difference is significantly important in the context of reach-grasp tasks. If the trajectories converge from the top of the envelope, it means that variable q (fingers) is converging faster than x (hand position). That is, this is premature finger closure as

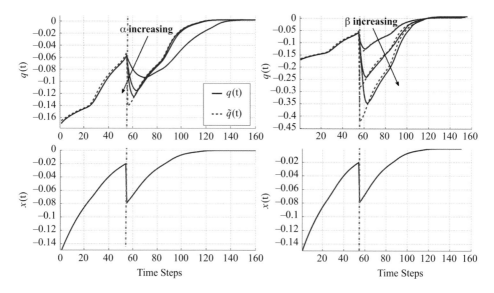

Figure 6.9
Variation of obtained trajectories with α and β. The vertical red line shows the instant of perturbation when the target is suddenly pushed away in the positive x direction. Negative velocities are generated in x in order to track $\hat{q}$. The speed of retracting is proportional to α (*left*) and amplitude is proportional to β (*right*).

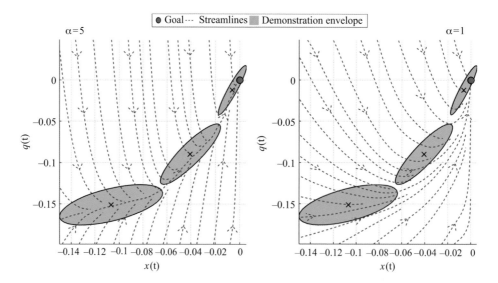

Figure 6.10
Change in α affecting the nature of streamlines. Larger α values will tend to bring the system more quickly toward the (x, q) locations seen during demonstrations.

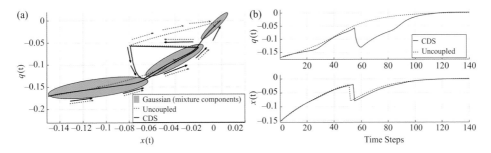

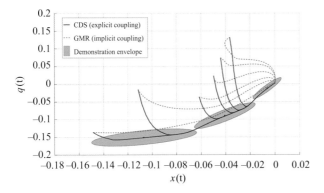

Figure 6.11
Task reproduction with explicit and implicit coupling shown in (a) state space and (b) time variation. Dotted lines show the execution of the implicitly coupled task. Note the difference in the directions from which the convergence occurs in the two cases. In the explicitly coupled execution, convergence is faster in x than in q.

Figure 6.12
Comparison between the motions obtained by a single GMM for embedding the coupling implicitly and the explicit CDS approaches. Note that the order of convergence can be remarkably different when starting at different positions in state space.

compared to what was observed during the demonstrations. If they converge from below, it means that the fingers are closing later than what was observed during the demonstrations. While the former is undesirable in any reach-grasp task, the latter is undesirable only in the case of moving/falling objects.

6.3.3 Robotic Implementation

The model was validated with the real iCub robot. The state of our system is composed of the Cartesian position and orientation of the end-effector (human/robot wrist) and of the following six finger joint angles:

- 1 for the curl of the thumb
- 2 for the index finger's proximal and distal joints
- 2 for the middle finger's proximal and distal joints
- 1 for the combined curl of the ring and little fingers.

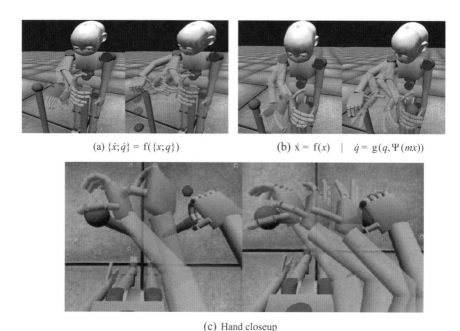

(a) $\{\dot{x}; \dot{q}\} = f(\{x; q\})$ (b) $\dot{x} = f(x) \quad | \quad \dot{q} = g(q, \Psi(mx))$

(c) Hand closeup

Figure 6.13
Reach-grasp task executions with and without explicit coupling. The explicitly coupled execution (b) prevents premature finger closure, ensuring that given any amount of perturbation, formation of the grasp is prevented until it is safe to do so. In the implicitly coupled execution (a), the fingers close early and the grasp fails. (c) shows a close-up of hand motion postperturbation with implicit (*left*) and explicit (*right*) coupling.

The Moore-Penrose inverse kinematic function to convert the end-effector pose to joint angles of the arm. In simulation, as well as on the real iCub robot, we control the 7 degrees of freedom of the arm and six finger joints at an update rate of 20 milliseconds.

Figure 6.13a illustrates a reproduction of the human experiment in which the iCub robot first reaches for the green ball. Midway through the motion, the target is switched and the robot must reach for the red ball. There, again, we compare the use of implicit coupling through an extended GMM or explicit coupling through the CDS model. The single GMM for the hand and finger dynamics fails at embedding properly the correlations between the reach-and-grasp subsystems and does not adapt the fingers' motion well to grasp for the new ball target. Lack of an explicit coupling leads to poor coordination between the finger and hand motions. As a result, the fingers close too early, causing the ball to fall. Figure 6.13b shows the same task when performed using the CDS model. The fingers first reopen following the perturbation, delaying the grasp formation, and then close according to the correlations learned during the demonstrations, leading to a successful grasp. Figure 6.13c shows the hand from the top view, clearly showing the reopening of fingers in the explicitly coupled task.

6.3.3.1 Adaptability to fast perturbations
An important aspect of encoding motion using autonomous dynamical systems is that it offers great resilience to perturbations. While we have inferred the dynamics from human data, we do not have to stick to the velocities shown by humans and exploit the

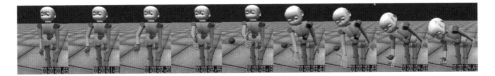

Figure 6.14
Fast adaptation under perturbation from palm-down to palm-up power grasp.

parameterization of the model through α and β to control for speed and amplitude as desired for the robotic application. Executing the task with values for α that differ from that set of human data may be interesting for robot control for two reasons:

1. Because robots can move much faster than humans, using larger values for α could exploit the robot's faster reaction times, while retaining the coupling between finger and hand motions found in human data.

2. Also, using values of α that depart from these inferred from human demonstration may allow one to generate better responses to perturbations that send the system to areas of the state space not seen during demonstrations.

We illustrate this capacity to adapt to rapid perturbations in simulation when the robot must not only adapt the trajectory of its hand, but also switch across grasp types (see figure 6.14). This requires fast adaptation from palm-up to palm-down grasp, while the target keeps moving. As shown in figure 6.14, switching to the second CDS model ensures that replanning of the finger motion is done in coordination with the hand motion (which is now redirected to the falling object). Precisely, the orientation of the hand and the finger curl are changed synchronously, causing the hand to close its grasp on the falling object at the right time.

With the real robot, we tested the ability of the system to adapt the fingers' configuration (in addition to adapting the fingers' dynamics of motion) so as to switch between pinch and power grasps (see figure 6.15). Two separate CDS models were learned for pinch-grasp of a thin object (a screwdriver, in this case) and power-grasp of a spherical object (a ball), respectively, from five demonstrations of each task during unperturbed trials.

6.4 Coupling Eye-Hand-Arm Movements[4]

We can extend the CDS for the reach-and-grasp movements, described in the previous section to enable the arm-hand system to be driven by eye movements. To this end, we add one more DS to generate the dynamics of eye movements. We then couple the arm to the eyes in a similar way to the coupling of the arm and the hand. This leads to a three-layer coupling. The dynamics can also be learned from observing the dynamics of movement in humans, using an eye-tracker for monitoring eye movements.

Coupling the eye-arm-hand system is particularly useful for adapting to disturbances along the way, such as obstacles. Adaptation of the arm posture for obstacle avoidance can be done by replacing the target with the obstacle and generating an intermediate target a few centimeters from the obstacle (see figure 6.16). While this can allow the trajectory to

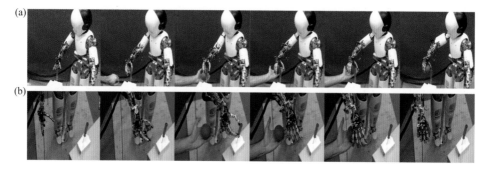

Figure 6.15
While the robot reaches for the thin object, preshaping its fingers to the learned pinch grasp, we suddenly present the spherical object in the robot's field of view. The robot then redirects its hand to reach for the spherical object in place of the first thing. (a) and (b) show the same task from the front and top views to better visualize the motion of the fingers.

naturally move away from obstacles, this does not ensure that the robot's arm will *never* collide with the obstacle. The reader is referred to chapter 9 for a discussion of ensuring the nonpenetration of the obstacle with dynamical systems.

This approach was validated in simulation and on the real iCub robot using the binocular cameras on board the robot (see figures 6.17 and 6.19). The CDS drives the gaze, arm, and hand toward the object using the pose information (in retinal and Cartesian coordinates) obtained from the vision system. As the gaze moves toward the object in every cycle of the control loop, we update the system with a more precise reestimate of the object position. Before the hand comes close to the object, the gaze fixates the object, and we get the precise information about the object position, which is crucial for successful grasping and obstacle avoidance. Our time-independent CDS automatically adapts to the reestimate of the object positions obtained from such a nonuniform resolution processing scheme. Experimental details on visual processing of the obstacle and target are given in [94].

In each run, the object to be grasped is placed at a randomly computed position within a 15-cm cube in the workspace. Figure 6.17 shows an obstacle scenario in which we test coordinated manipulation with sudden perturbations of the target object and the obstacle, respectively. The robot's end-effector avoids the obstacle when reaching for the purpose of grasping; Once the obstacle is reached, the target for the visuomotor system is changed, and the eye-arm-hand motion is directed to the object to be grasped. We use a position controller with inverse kinematics to compute the path of the arm and to adapt the arm-resting posture to be as close as possible to the desired trajectory given by the DS for the arm.

We conducted a user study where human subjects were asked to perform the same task. Analysis of the data showed that subjects ignore the obstacle when it does not obstruct the intended motion (see figure 6.18). The same principle was implemented in the robot controller. Because the eye state is the distance between the position of gaze and the position of a visual target in retinal coordinates, and the arm state is represented with respect to the position of the object in the Cartesian space, both variables are instantly updated when the perturbation occurs. The DS of the eyes adapts independent of the perturbation.

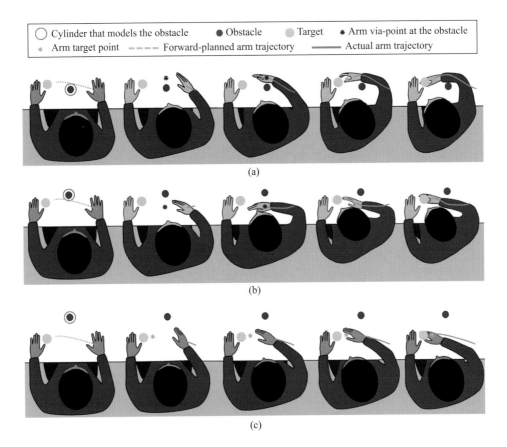

Figure 6.16
A scheme that illustrates forward planning and obstacle avoidance. After forward-integrating the CDS model, an obstacle (*dark blue disk*) is identified as an obstructing object if the estimated arm motion (*dashed orange line*) intersects with a cylinder (*dark blue circle*) that models the obstacle (certain collision); or when the cylinder lies within the area where it is very likely that it will collide with the forearm (very likely collision). If the obstacle is identified to obstruct the intended motion, then the motion of the visuomotor system is segmented from the start to the obstacle and from the obstacle to the target. When reaching to avoid the obstacle, the arm DS moves under the influence of the attractor placed at the via-point with respect to the obstacle (*dark blue star*). The direction of displacement of the via-point (anterior or ventral) is chosen to correspond to the side of the obstacle where a collision is estimated to occur: anterior side (**a**) or ventral side (**b**). If the forward-planning scheme does not detect collision with the obstacle (**c**), the visuomotor system is driven to the target object (i.e., the obstacle is ignored). The *light red star* represents the goal arm position with respect to the target object (*light red disk*). These images show execution of eye-arm-hand coordination from the start of the task (*left*) until the successful grasp completion (*right*).

—— Perturbed hand trajectory   · · · · Unperturbed hand trajectory

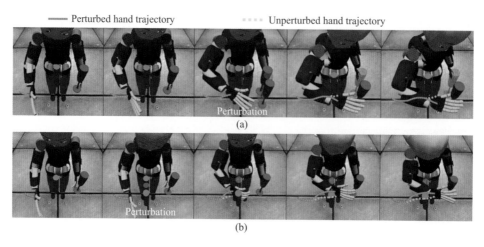

Figure 6.17
Experiments of visually guided reaching and grasping in the iCub's simulator, with the presence of the obstacle and with perturbations. The obstacle is an intermediary target for the visuomotor system, and hence obstacle avoidance is divided into two subtasks: from the start position to the obstacle (via-point) and from the obstacle to the grasping object. The photos show the execution of eye-arm-hand coordination from the start of the task (*left*) until the successful grasp completion (*right*). Figures in the upper row (a) present a scenario when the target object (red champagne glass) is perturbed during motion (perturbation occurs in the third frame from left). Visuomotor coordination when the obstacle is perturbed during manipulation is shown in the bottom row (perturbation in the second frame). The orange line shows the trajectory of the hand if there is no perturbation. The purple line is the actual trajectory of the hand from the start of unperturbed motion, including the path of the hand after perturbation, until successful grasping. In both scenarios (target perturbed and obstacle perturbed), the visuomotor system instantly adapts to the perturbation and drives the motion of the eyes and the arm and the hand to a new position of the object.

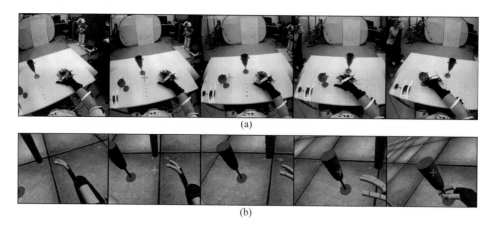

Figure 6.18
A comparison of human visuomotor coordination and visuomotor behavior of the real robot. The visuomotor coordination profile the robot produces (b) is highly similar to the pattern of coordination that was observed in the human trials (a). The photos show snapshots of the execution of eye-arm-hand coordination from the start of the task (*left*) until the successful grasp completion (*right*).

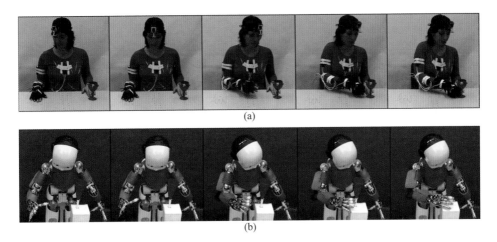

Figure 6.19
The visuomotor system ignores an obstacle when it is not relevant for manipulation (i.e., the obstacle that does not affect intended motion is not visually salient for the gaze). Analysis of the WearCam recordings from the human trials (a) reveals that subjects do not fixate the obstacle (*blue champagne glass*) in the workspace when it does not obstruct the intended reaching and grasping movement. This CDS eye-arm-hand model shows the same behavior (b), ignoring the obstacle (*green cylinder*), when the forward-planning scheme estimates that the object does not obstruct prehensile movement. The snapshots show task from the start (*left*) until completion of the successful grasp (*right*).

The behavior of the DS of the arm is modulated via the eye-arm coupling function, and the hand DS is modulated via the arm-hand coupling. Such modulation ensures that the learned profile of eye-arm-hand coordination will be preserved and that the hand will reopen as the object is perturbed away from it (see figure 6.17). Besides the anthropomorphic profile of visuomotor coordination (see figure 6.19), the gaze-arm lag allows for enough time to foveate at the object, to reestimate the object's pose, and to compute a suitable grasp configuration for the hand before it comes too close to the object.

7 Reaching for and Adapting to Moving Objects

In chapter 6,[1] we saw how dynamical systems (DSs) can be coupled to enable joint control of multiple limbs in a robot. In this chapter, we show how we can take this coupling one step further and couple the DS controlling our robot to the dynamics of an external object over which we have no control. To emphasize the need for immediate and fast replanning, we consider scenarios in which the dynamics of the object can change abruptly and the object moves very quickly.

Specifically, we tackle the problem of intercepting a moving object. We assume that all we have at our disposal is a model of the object's dynamics, but that this model may change over time. This change in estimated dynamics could be due to poor sensing, for instance. While the object is far away, we get an inaccurate prediction of its dynamics. As the object comes closer, the model is refined and the robot needs to adapt to new predictions of where the object will land. The change in dynamics may also be due to external factors. For instance, if the object is handed to the robot by a human, the interception point may change suddenly as the human repositions herself. Adapting to the speed of motion of an object is typical of industrial activities, which involve picking up objects moving on a conveyor belt.

To move in synch with the object's dynamics, we couple the robot's motion to that of a virtual object. The virtual object's dynamics takes the role of an external variable to which we can couple the robot's DS. We show that a strategy that enforces a coupling with the object's dynamics, and whereby the robot adapts to the object's velocity, improves the stability at contact by decreasing the impact force. This adaptation is a form of compliance that is different from impedance control [14]. While we can use this strategy to catch objects, it should be emphasized that we do not control explicitly for forces at contact. Compliance and force control with DS are covered in part IV.

This chapter is based on the formulation of the linear parameter varying (LPV) system for learning and modeling the control law, which was introduced in chapter 3. Formulating DSs as LPV systems allows modeling a wide class of nonlinear systems and the use of many tools from linear systems theory for analysis and control. Importantly, it allows us to model second-order control laws, and hence to control robots in acceleration, which is a necessary step to control for force at contact.

The chapter is organized as follows: section 7.1 formalizes the problem and the envisioned applications. Section 7.2 starts the technical part of this chapter by reformulating control of trajectories with a second-order DS. In section 7.3, we present an approach in which the robot's motion can be coupled to that of a moving object. We show an application for catching fast-moving objects in flight. We formulate stability constraints to ensure that the robot meets the object and, once it does so, that its velocity is aligned with the object's velocity. This allows the robot and the objects to move in synch and hence mitigates impact forces and prevent the object from bouncing off the robot's hand. Section 7.5 extends this approach in order to enable a moving object to be intercepted synchronously by two robotic arms.[2]

7.1   How to Reach for a Moving Object

When reaching for a moving object, two main issues arise. First, in contrast to a static object, a moving object changes location continuously. The chances of catching it are slim unless one can predict accurately the object's trajectory. Perfect prediction is unrealistic. However, it is possible to approximate the trajectory through first-order interpolation for short periods of time, typically using a Kalman filter. Hence, prediction will keep changing, and the motion generator will need to adapt the path continuously in response. DSs are ideally suited for this so we use one to control the robot's trajectory. To adapt to the continuous change in prediction of the target's trajectory, we locate the attractor of the DS at the predicted intercept location for the object. As this location is updated, a new trajectory is automatically recomputed.

Second, coming in contact with a moving object generates strong contact forces. If the contact forces are too strong, they may send the object flying away before one can close the grip on it. It is crucial, therefore, to control impact forces precisely, to make sure that (1) neither the system nor the object is damaged and (2) the system remains in contact with the object after the first impact. To achieve this, we opt for a strategy in which the robot does not stop once it meet the object, but rather continues moving with the object for a short period of time. This way, the impact force can be mitigated. This *soft catching* strategy is illustrated in figure 7.1.

In this chapter, we study two scenarios, illustrated in figure 7.2:

1. Unimanual interception of a flying object. In this scenario, the object is moving at a very high speed. In order to reduce the impact force, the goal of the system is to intercept the object with a velocity as close as possible to the object's velocity (see figure 7.2a).

2. Bimanual interception of a moving object. In this scenario, the object is too large for one robotic arm to grab it. Two arms are required. The goal of the system is, hence, to drive the two robot arms in synch in order for them to reach specific points on the moving object with a specific velocity. An example of this scenario is shown in figure 7.2b, where two robot arms must pick up objects moving on a conveyor belt.

In all these scenarios, the relative velocities of the robot and the object at the intercept point is close to zero. This smooths the transition from free-space motion to contact, as the

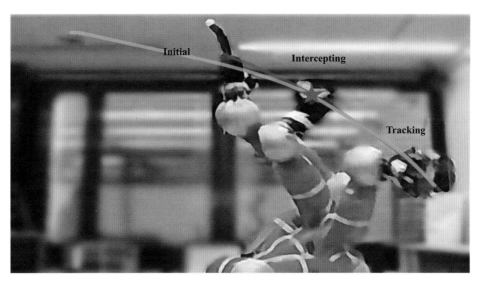

Figure 7.1
To catch a flying object and mitigate contact forces, we consider a strategy whereby the robot continues moving with the object for a little while upon intercepting it.

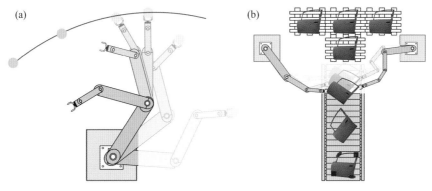

Figure 7.2
Schematic of two scenarios in which a robot may be tasked to intercept a moving object. (a) the robot catches a ball flying rapidly in the air. The challenge is to catch the object once it enters the robot's workspace, while minimizing contact force to neither damage the robot, nor send the object flying away before gripping it. (b) the object is large and requires two robot arms. The challenge is then for the two robot arms to reach the object simultaneously. Indeed, if one arm were to reach the object before the second arm got to the object, the impact would make the object spin, which is not desirable.

impact force will be zero or close to zero. To achieve this, the motion of the system must be *coupled* with the state of the object, which can be position, velocity, or acceleration of the object. As the main scope of this section is the stability of the motion generator in the free-space region and not the closed-loop motion generator, we assume that the robotic system is equipped with a low-level controller, that it is able to exactly follow the generated motion at the position/velocity levels (i.e., the measured and the commanded positions/velocity are equal).

7.2   Unimanual Reaching for a Fixed Small Object

We start by defining our second-order control law as a second-order DS that is asymptotically stable at a fixed point x^*—that is,

$$x(t^*) = x^*, \qquad\qquad \dot{x}(t^*) = 0. \qquad\qquad (7.1)$$

Assume that $x^* \in \mathbb{R}^N$ and $t^* \in \mathbb{R}^+$ are the object's location and interception time, respectively, with N the system's dimension. As we aim at controlling both the position and the velocity of the robot at contact, the DS must be a function of both variables, and the output must define the desired acceleration of the system. As introduced in chapter 3, we use a class of continuous-time LPV systems given by the following model to generate the motion of the system:

$$\ddot{x}(t) = \mathbf{A}_1(\gamma)x(t) + \mathbf{A}_2(\gamma)\dot{x}(t) + u(t). \qquad\qquad (7.2)$$

Here, $x(t) \in \mathbb{R}^N$ is the state of the DS. This corresponds, typically, to the pose of the end-effector or to the joint positions of the robot. Further, $u(t)$ is the control input vector, and $\gamma \in \mathbb{R}^{K \times 1}$ is a vector composed of the *activation parameters*. The activation parameters can be a function of time (t), the state of the system $(x(t))$, or external signals $d(t)$ [i.e., $\gamma(t, x(t), d(t))$]. In the rest of this chapter, the arguments of γ are dropped for simplicity:

$$\gamma = \begin{bmatrix} \gamma_1 & \cdots & \gamma_k. \end{bmatrix}^T. \qquad\qquad (7.3)$$

$\mathbf{A}_i(.) : \mathbb{R}^K \to \mathbb{R}^{N \times N} \; \forall i \in \{1, 2\}$ generate affine dependencies of the state-space matrices on the activation parameters and the state vectors:

$$\mathbf{A}_1(\gamma) = \sum_{k=1}^{K} \gamma_k A_1^k \quad A_1^k \in \mathbb{R}^{N \times N}$$
$$\qquad\qquad\qquad\qquad\qquad \gamma_k \in \mathbb{R}_{(0,1]}. \qquad\qquad (7.4)$$
$$\mathbf{A}_2(\gamma) = \sum_{k=1}^{K} \gamma_k A_2^k \quad A_2^k \in \mathbb{R}^{N \times N}$$

To reach the object at the desired location (x^*), we set the control input $u(t)$ in equation (7.2) as

$$u(t) = -\mathbf{A}_1(\gamma)x^*. \qquad\qquad (7.5)$$

By substituting equation (7.5) into equation (7.2), we get

$$\ddot{x}(t) = \mathbf{A}_1(\gamma)(x(t) - x^*) + \mathbf{A}_2(\gamma)\dot{x}(t). \qquad\qquad (7.6)$$

One can show (see exercise 7.3, later in this chapter) that such a control system is asymptotically stable at x^*, and hence, we get

$$\lim_{t \to \infty} \|x(t) - x^*(t)\| = 0 \qquad\qquad (7.7)$$

$$\lim_{t \to \infty} \|\dot{x}(t)\| = 0 \qquad\qquad (7.8)$$

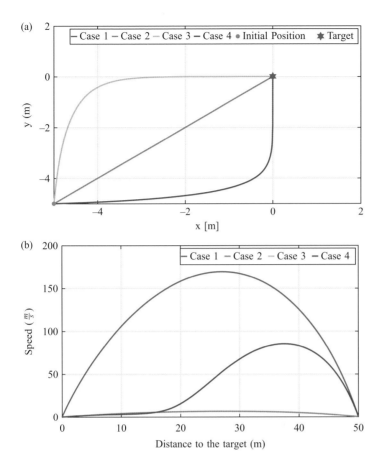

Figure 7.3
Reaching a fixed target ($x^* = 0$) with four DSs.

if equation (7.6) meets the following constraints:

$$
\begin{cases}
\begin{bmatrix} 0 & I \\ A_2^k & A_1^k \end{bmatrix}^T P + P \begin{bmatrix} 0 & I \\ A_2^k & A_1^k \end{bmatrix} \prec 0 \\
0 \prec P, \qquad\qquad P^T = P \qquad\qquad \forall k \in \{1, \dots, K\}. \\
0 < \gamma_k \leq 1, \qquad \sum_{k=1}^{K} \gamma_k = 1
\end{cases}
\tag{7.9}
$$

Example To provide an idea of the effect that particular choices of the affine functions have on the dynamics of the controller, we show four examples in figure 7.3. All these examples satisfy constraints set in equation (7.9) and are given as follows:

Case 1: $\mathbf{A}_2 = \begin{bmatrix} -10 & 0 \\ 0 & -10 \end{bmatrix}$, $\mathbf{A}_1 = \begin{bmatrix} -25 & 0 \\ 0 & -25 \end{bmatrix}$

Case 2: $\mathbf{A}_2 = \begin{bmatrix} -2 & 0 \\ 0 & -2 \end{bmatrix}$, $\mathbf{A}_1 = \begin{bmatrix} -1 & 0 \\ 0 & -1 \end{bmatrix}$

Case 3: $A_2 = \begin{bmatrix} -2 & 0 \\ 0 & -10 \end{bmatrix}$, $A_1 = \begin{bmatrix} -1 & 0 \\ 0 & -25 \end{bmatrix}$

Case 4: $A_2 = \begin{bmatrix} -10 & 0 \\ 0 & -2 \end{bmatrix}$, $A_1 = \begin{bmatrix} -25 & 0 \\ 0 & -1 \end{bmatrix}$.

Observe that the system stabilizes at the target point regardless of the values of A_1 and A_2. The amplitude of the entries in $\mathbf{A}_1$ and $\mathbf{A}_2$ defines the rate of convergence and the shape of the motion.

Programming Exercise 7.1   *The aim of this exercise is to help readers to gain a better under-standing of the DS [equation (7.6)] and the effect of the open parameters on the generated motion. This exercise is complementary to the exercises provided in chapter 3. Open MATLAB and set the directory to the following folder:*

```
1   ch7-DS_reaching/Fixed_Small_object
```

In this exercise, one can study the effect of $\mathbf{A}_1$, $\mathbf{A}_2$ on the generated motion. When calling the function (on command line) Reaching_A_Static_Target([2;2],[2;1])), you can generate motion from initial positiion [2;2] to target position [2;1]. Study the motion generated by equation (7.6) in different scenarios and answer the following questions.

1. How do $\mathbf{A}_1$, $\mathbf{A}_2$, $\forall i \in \{1, 2\}$ modulate the trajectory?

2. Modify matrices A1 and A2 such that the conditions of equation (7.9) are not met. What is the behavior of the DS?

3. Modify the initial location and target position. Do the initial location of the DS and the target position have an effect on the shape of the trajectory? Do they affect the velocity at target? Set matrices A1 and A2 so as to consider both cases: when conditions of equation (7.9) are met and not met.

7.2.1 Robotic Implementation

As presented in section 3, second-order DSs have been commonly used for approximating self-intersecting trajectories or motions for which the starting and final points coincide with each other. Another main advantage of using second-order systems [equation (7.6)] over first-order systems is that the generated motion's velocity is continuous (results in smoother trajectories).

Figure 7.4 illustrates an application of the second-order DS described previously using two KUKA IIWA robots performing point-to-point motions. The robots are tasked to press two buttons. The robots are controlled by the same DS [equation (7.6)] that was trained using example trajectories generated by optimal control. Specifically, we generated sets of the fastest kinematically feasible motions for the two robotic arms. This was generated by solving an offline optimal control problem (as shown in figure 7.4). Even though there is no direct correlation between the demonstrations and the desired task, the robots are able to accomplish the task, as the motions generated by equation (7.6) are asymptotically stable toward the desired target in space.

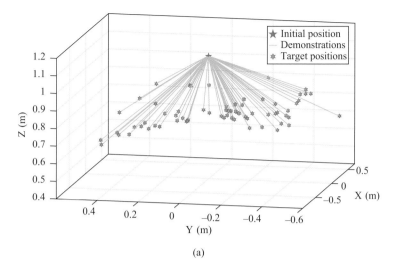

(a)

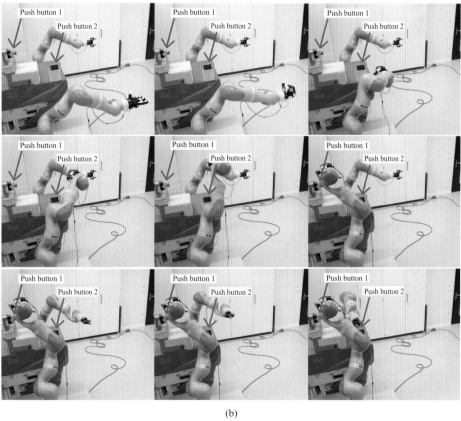

(b)

Figure 7.4
(a), The data set contains 67 demonstrations. The initial point is fixed to the robot's candle position. [100]. (b) Reaching push buttons 1 and 2 by two robotic arms. The motion generators of both robots are identical; only the desired target positions are different. [104]

Exercise 7.1   *Consider the DS presented in equation (7.6), with* $d = 2$ *and* $k = 1$. *Design* $\mathbf{A}_1$ *and* $\mathbf{A}_2$ *such that the motion generated by equation (7.6) satisfies*

$$\frac{\|x(t) - x^*\|}{\|x(0) - x^*\|} \leq e^{-10t} \quad \forall t \in \{1, \infty\}. \tag{7.10}$$

Exercise 7.2   *Consider the DS presented in equation (7.6). Given equation (7.4), derive a relation between the minimum eigenvalue of* $\mathbf{A}_i^k, \forall i \in \{1, 2\}, \forall k \in \{1, \ldots, K\}$ *and the convergence error such that the motion generated by equation (7.6) satisfies*

$$\frac{\|x(t) - x^*\|}{\|x(0) - x^*\|} \leq e^{-10t} \quad \forall t \in \{1, \infty\}. \tag{7.11}$$

Exercise 7.3   *Consider the DS presented in equation (7.6). Prove that it asymptotically converges to* $\begin{bmatrix} x^* & 0 \end{bmatrix}^{\mathrm{T}}$ *if the conditions in equation (7.9) are met.*

7.3   Unimanual Reaching for a Moving Small Object

We turn next to the problem of ensuring that our control law can successfully intercept a moving object. Our main task is to make sure that the contact with the object is stable. This means that the relative velocity between the robotic hand and the object should be zero for some period of time, one that is long enough for the fingers of the robot's hand to close on the object. Enforcing that the robot moves at the same speed as the object might not always be possible, as the motion of the object might be much faster that the maximum feasible velocity of a robotic system. Hence, to relax this constraint, we assume that the contact will be stable if the system intercepts the object at the desired location with a velocity aligned with the velocity of the object; that is,

$$x(t^*) = x^o(t^*), \qquad\qquad \dot{x}(t^*) = \rho \dot{x}^o(t^*), \tag{7.12}$$

where the state of the object is denoted by $x^o \in \mathbb{R}^N$, and $\rho \in \mathbb{R}_{[0,1]}$ is a continuous parameter, to which we refer as *softness*. Setting $\rho = 0$ leads the robot to intercept the object at the desired location, with $\dot{x}(t^*) = 0$. This corresponds to a *stiff* interception with maximal contact force. Conversely, $\rho = 1$ results in the robot intercepting the object at the desired location, with $\dot{x}(t^*) = \dot{x}^o(t^*)$. This is a *soft* interception. However, as stated previously, the latter might result in infeasible fast motion. The softness parameter must be chosen in such a way that it generates dynamically feasible trajectories while mitigating the impact forces. Now we show how we can determine the parameters of our controller to satisfy these conflicting goals.

For simplicity and brevity, we assume that the desired intercept point is located at the origin [i.e., $x^o(t^*) = 0$]. The control input $u(t)$ in equation (7.2) is defined as follows:

$$u(t) = \rho(t)\ddot{x}^o - \mathbf{A}_1(\gamma)\rho(t)x^o - \mathbf{A}_2(\gamma)\left(\rho(t)\dot{x}^o + \dot{\rho}(t)x^o\right) + 2\dot{\rho}(t)\dot{x}^o + \ddot{\rho}(t)x^o. \tag{7.13}$$

$\mathbf{A}_i(.) \in \{1, 2\}$ follows the definition given in equation (7.4) and decompose the dynamics in a set of linear DSs, modulated by the activation parameters (γ). By substituting equation (7.13) into equation (7.2), we have

$$\ddot{x}(t) = \rho(t)\ddot{x}^o(t) + 2\dot{\rho}(t)\dot{x}^o(t) + \ddot{\rho}(t)x^o(t)$$
$$+ \mathbf{A}_1(\gamma)\left(x(t) - \rho(t)x^o(t)\right) + \mathbf{A}_2(\gamma)\left(\dot{x}(t) - \left(\rho(t)\dot{x}^o(t) + \dot{\rho}(t)x^o(t)\right)\right). \tag{7.14}$$

If the conditions in equation (7.9) are met, the DS given by equation (7.14) asymptotically converges on $\left[\rho(t)x^o \quad \rho(t)\dot{x}^o + \dot{\rho}(t)x^o\right]^T$ (see exercise 7.7); that is,

$$\lim_{t\to\infty} \|x(t) - \rho(t)x^o(t)\| = 0 \tag{7.15}$$

$$\lim_{t\to\infty} \|\dot{x}(t) - (\rho(t)\dot{x}^o(t) + \dot{\rho}(t)x^o(t))\| = 0. \tag{7.16}$$

If we assume that ρ is zero, equation (7.14) generates a controller that is a composite of *reaching* and *tracking* the target object. Here, $\rho(t)$ acts as a switch from one motion type to the other. Setting $\rho = 0$ results in a pure *reaching* behavior with a DS that stabilizes at the target; that is, $\lim_{t\to\infty}\left[x(t) \quad \dot{x}(t)\right] = \left[x^o(t^*) \quad 0_{1\times d}\right]^T$. The system reaches the desired intercept point, but it stops there. It will reach the object in time if the DS [equation (7.14)] lets it move fast enough, so as to converge on an acceptable neighborhood of the desired intercept point $\rho\left[x^o(t^*) \quad 0\right]^T$ before the object gets there; that is, $\|x(t^*) - \rho x^o(t^*)\| \le \varepsilon$ and $\|\dot{x}(t^*) - \rho\dot{x}^o(t^*) - \dot{\rho}x^o(t^*)\| \le \varepsilon$, where ε is a small positive number. The robot may then stop and wait for the object. However, the robot's stopping will generate a strong impact force at contact. To ensure that the robot moves with the object's velocity (or as close as possible to it), we need to enable the robot to transition smoothly from moving toward the object to tracking the object. We could do so by setting $\rho = 1$. This would result in a tracking motion with an error decreasing asymptotically according to

$$\ddot{x}(t) - \ddot{x}^o(t) = \mathbf{A}_1(\gamma)(x(t) - x^o(t)) + \mathbf{A}_2(\gamma)(\dot{x}(t) - \dot{x}^o(t)). \tag{7.17}$$

While the motion would converge on the object's trajectory and intercept it with $\dot{x} = \dot{x}^o$ (i.e., the *velocity* constraint is satisfied), we can no longer guarantee that it will reach the object at the right location. By varying the value of the γ parameter, one can ensure that the system reaches the object not only with the right velocity, but also at the right point. This is summarized in proposition 7.1.

Proposition 7.1 *The DS given by equation (7.14) reaches the desired intercept point $(x^o(t^*) = 0_{1\times d})$ asymptotically, with a velocity aligned with that of the object, $\dot{x}(t^*) \approx \rho\dot{x}^o(t^*)$.*

Proof The intercept point is located on the object's trajectory and, as mentioned before, it is the origin (i.e., $\rho x^o(t^*) = \begin{bmatrix} 0 & \cdots & 0 \end{bmatrix}^T$). Hence, ρx^o crosses x^o at the desired intercept point. Because the system asymptotically converges on ρx^o, it intercepts the object's trajectory at the desired intercept point at $t = t^*$. Moreover, the system's velocity will be a velocity vector proportional to that of the object [i.e., $\dot{x}(t^*) = \rho\dot{x}^o(t^*)$]. $\qquad\square$

Switching from $\rho = 0$ to $\rho = 1$ is not advisable, as this may generate strong discontinuities in acceleration. We favor a smooth transition between the two. Figure 7.5 illustrates the

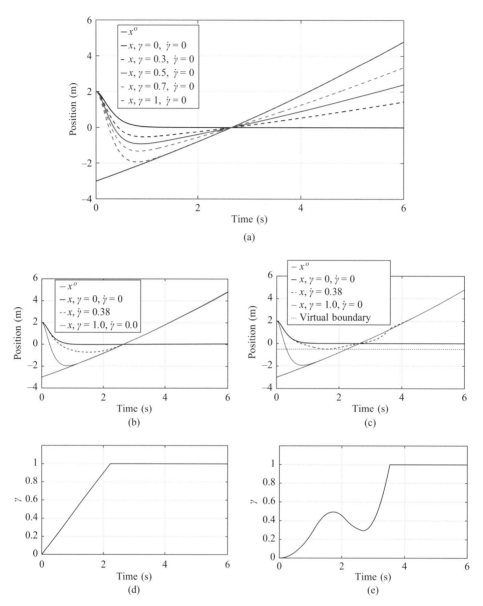

Figure 7.5
The behavior of one-dimensional (1D) DS subject to the value of ρ and $\dot{\rho}$. $\dot{\rho} = 0$ in (a). (b) and (c) show the behavior of the 1D system for constant or time-varying $\dot{\rho}$, respectively. (d) and (e) show ρ in (b) and (c), respectively. In (c), the virtual workspace constraint is satisfied by optimizing the value of ρ over time.

solutions of equation (7.14) for three values of ρ and $\dot{\rho}$: $\dot{\rho} = 0$; $\dot{\rho} = c$, where c is a constant value; and $\dot{\rho} = c(t)$, where $c(t)$ is a function of time. Observe that, for all values for ρ, $\dot{\rho}$, and $\ddot{\rho}$, the generated trajectories intercept the object's trajectory at $x(t^*) = 0$. Moreover, by increasing the value of ρ, the system's velocity is getting closer to the object's velocity: compare the black line with the purple line in figure 7.5a. Figure 7.5c shows an example in which the value of ρ is modulated online. Such adaptation may be required to slow down as one comes close to singularities or constraints in the workspace.

Programming Exercise 7.2 *The aim of this exercise is to help readers to gain a better understanding of the DS [equation (7.14)] and the effect of the open parameters on the generated motion. This exercise has three parts.*

In the first part, $\dot{\rho} = 0$. Open MATLAB and set the directory to the following folder:

```
1    ch7-DS_reaching/Moving_small_object.
```

In the first function, one can study the effect of $\mathbf{A}_1$, $\mathbf{A}_2$, and ρ on the generated motion. Run first on the command line the function Constant_rho, all the motions intercept the object at the origin.

1. Set the motion of the object to be faster than the convergence rate of the DS.

2. Modify the object's trajectory so that it does not pass through the origin (i.e., it does not pass through the desired intercept point). What happens?

3. What would happen if ρ is greater than 1, and less than zero?

4. Is there any condition where the motion generated by equation (7.14) intercepts the object at a point that is not the desired intercept point?

5. Modify the code so that the conditions of equation (7.9) are not met.

6. Change the integration step. Does this affect the behavior? For which values? Can you explain why?

7. What do high values of ρ geometrically mean?

Run, on the command line, function Constant_Drho. For this code, $\dot{\rho} = c$, where c is a constant number. With this function, one can study the effect of $\mathbf{A}_1$, $\mathbf{A}_2$, and ρ on the generated motion while the value of ρ is not constant and changes over time. In a simple case, $\rho(0) = 0.1$, $\dot{\rho} = 0.1$, the generated motion intercepts the object at the origin. By modifying the inputs to this function, answer the following questions.

1. Is it important for ρ to be monotonically increasing or decreasing?

2. What would happen if equation (7.9) are not met?

3. What does changing ρ over time geometrically mean?

Run, on the command line, the function Geometrically_constrained. In this function, $\dot{\rho} = c(t)$, where $c(t)$ is a function over time. The aim of the program is to shape $\dot{\rho}$ such that the motion of the DS avoids the virtual boundary while it moves towards the target. In this function, one can study the effect of $\mathbf{A}_1$, $\mathbf{A}_2$, and ρ on the generated motion while the value of ρ is not constant and changes with respect to geometrical constraints. By modifying the inputs to this function, answer the following questions.

1. Does the location of the virtual boundary matter? What if one placed it at $x = 0$?

2. What happens if conditions of equation (7.9) were not met?

3. What effect does changing ρ has on the angle between the DS and the virtual object's motion?

4. Is there any $\mathbf{A}_1$, $\mathbf{A}_2$ such that the constraint from the virtual boundary would not supposed be satisfied?

7.4 Robotic Implementation

The system was supposed to softly catch objects in flight using a 7-DOF robot arm, KUKA LBR IIWA, mounted with a 16-DOF Allegro hand. The output of the DS [equation (7.14)]

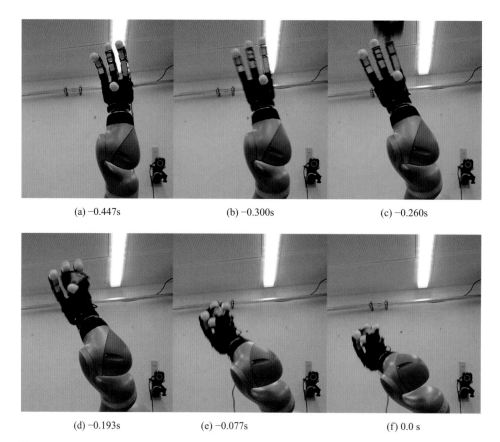

(a) −0.447s   (b) −0.300s   (c) −0.260s

(d) −0.193s   (e) −0.077s   (f) 0.0 s

Figure 7.6
Snapshots of the finger motions. The object is intercepted in (d) and caught in (f). It is important to note that the closure time for the fingers varies with the incoming object speed.

is converted into the 7-DOF joint state using velocity-based control without joint velocity integration [106]. In order to avoid high torques, the resultant joint angels are filtered by a critically damped filter. The robot is controlled in the joint position level at a rate of 500 Hz.

In order to coordinate the motions of all joints (including the arm and finger joints), the coupled DS (CDS) model, introduced in section 6, is used to generate the finger motion. This approach consists of coupling two different DSs (i.e., the end-effector motion and the finger motion). The motion of the end-effector is generated independent of the finger states, while the finger motion is a function of the state of the end-effector and the object. The metric of the coupling is the distance between the end-effector and the object ($\|x - x^o\|$). As a result, the fingers close when the object gets inside the hand, and they reopen when the object moves away.

While the object is flying, its position and orientation is tracked at 240 Hz using the Opti-Track motion-capture system from Natural point. Because the control loop is faster than the capturing system, the predicted position of the object is used as the object position in equation (7.14). Snapshots of the arm and the finger motions are shown in figures 7.6 and 7.8. In figure 7.7, the position of the end-effector generated by equation (7.14) is illustrated. A detailed experimental validation is presented in [100] and [102], as well as videos available online on the book's web page.

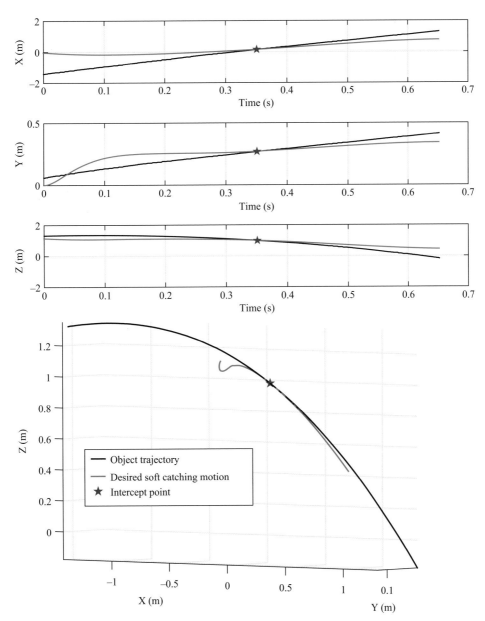

Figure 7.7
The position of the end-effector generated by the DS [equation (7.14)]. The illustrated object trajectory is the predicted trajectory of the uncaught object. This trajectory is illustrated from the first point until the stop point. As expected, the output of equation (7.14) softly intercepts the object trajectory at the desired intercept position. To stop the robot, the velocity of the robot is linearly reduced during the post-interception period (0.3*s*).

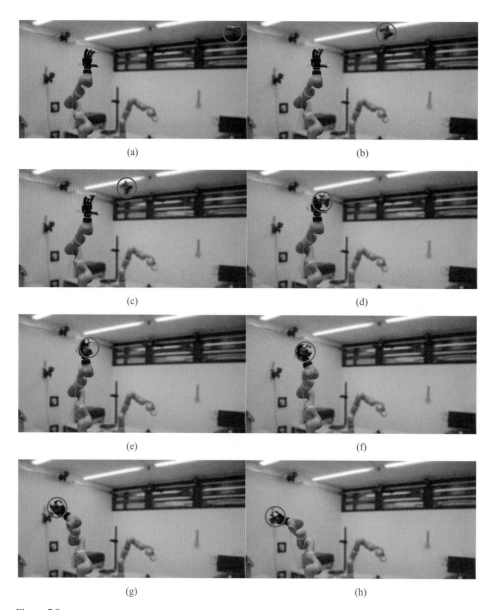

(a)   (b)

(c)   (d)

(e)   (f)

(g)   (h)

Figure 7.8
The brick is thrown. In (a), the object trajectory prediction algorithm is being initialized. (b) is approximately the first point; (e) is the interception. One can stop the robot in (g) as the fingers are closed, but it might damage the robot.

Although this approach allowed the robot to catch objects in about 80 percent of the case, there were failures. The main cause of failure was the inability of always generating an accurate joint-level motion corresponding to the desired end-effector trajectory. Because the motion is too fast, the end-effector does not accurately track the desired motion. This tracking error results in situations where the object is hit by the thumb or undesired parts of the hand and bounces away. In addition, any imprecision in tracking of the object can affect performance importantly. As stated in the introduction of this chapter, catching an object in flight requires regularly updating the measurements of the locations of the object. However, in practice, even with a very accurate, marker-based tracker, this is not possible at all times. To determine the position of the object, all the markers must be visible, but the markers often are obstructed by the robot arm or by the object itself as the object rotates while flying.

Exercise 7.4 *Given a robotic system with following kinematic constraints, consider the following scenarios:*

1. *The location* x *is kinematically reachable by the robotic system if* $0 \leq p_w(x)$, *where* $p_w(.)$ $\in \mathbb{R}^N \to \mathbb{R}$ *is a model of the reachable workspace of the robot.*

2. *The velocity* $\dot{x}$ *is feasible for the robotic system to achieve if* $\|\dot{x}\| \leq \dot{x}_{max}(q)$, *where* $\dot{x}_{max}(q)$ *is the maximum feasible velocity of the robot at the given joint configuration* (q).

3. *Similarly, the acceleration* $\ddot{x}$ *is feasible for the robotic system if* $\|\ddot{x}\| \leq \ddot{x}_{max}(q)$, *where* $\ddot{x}_{max}(q)$ *is the maximum feasible acceleration of the robot at the given joint configuration* (q).

Propose an algorithm to maximize the softness at each time step in the DS [equation (7.14)] with respect to the aforementioned kinematic constraints.

Exercise 7.5 *Consider the DS given by equation (7.14). Can one use this DS to hit a flying object? If yes, what would be the value of* ρ?

Exercise 7.6 *One can use four methods to demonstrate the desired behavior to a robotic arm in chapter 2: kinesthetic teaching, free-hand manipulation, teleoperation, and optimal control. Which one would be beneficial to use to estimate the parameters of the DS [equation (7.14)] for catching objects in flight.*

Exercise 7.7 *Consider the DS presented in equation (7.14), prove that it asymptotically converges* $\left[\rho(t)x^o \quad \rho(t)\dot{x}^o + \dot{\rho}(t)x^o\right]^T$ *if the conditions in equation (7.9) are met.*

7.5 Bimanual Reaching for a Moving Large Object

This section turns to the problem of intercepting a moving object with two robotic arms moving in coordination, Such dual, and by extension multiple, robotic systems is advantageous in that it extends the workspace of a single arm and makes it possible to manipulate heavy or large objects that would other be infeasible. A plethora of applications in smart factories or smart buildings would benefit from such strategies. Examples include grabbing boxes transported by cart or a conveyor belt, or simply handed by humans to robots (see figure 7.2b).

To accomplish bimanual interception of a moving object, one needs to break down the problem into two subproblems. First, make sure that the robots are synchronized with one another, such that if one robot should get delayed, the other robot should wait for it before catching the object. Second, the two robots should synchronize their movements with that of the moving object.

The first level of control requires that the robot move in coordination with each other. This is necessary not only to ensure that the systems simultaneously intercept the object, but also to avoid collisions between them while they adapt to the motion of the moving object. The second level imposes position and velocity constraints to ensure proper interception as described in the unimanual case presented in the previous section: all the robots must intercept the object at the desired location with a velocity aligned with that of the object. Thus we have two constraints, which we denote as *position* and *velocity* constraints, given by

$$x_i(t^*) = x_i^o(t^*), \qquad\qquad \dot{x}_i(t^*) = \rho \dot{x}_i^o(t^*) \qquad \forall i \in \{1, \ldots, N_R\}, \qquad (7.18)$$

where $x_i \in \mathbb{R}^N$, x_i^o is the state of the ith robot and the desired point on the object that it should reach, and N_R is the number of robotic systems. The object's state can be determined by tracking the N_R reaching points on the object:

$$x^o(t) = \sum_{i=1}^{K} x_i^o(t). \qquad (7.19)$$

To achieve the three-way coordination described here, we adopt the idea of a virtual object. This virtual object will coordinate the motion of the robots with each other and coordinate the robots with the real object. It is a duplicate of the real object, and it also entails duplicates of the reaching points that are virtually connected to the end-effector of the robots via spring-damper terms. The motion of the virtual object is coordinated and aligned to the motion of the real object such that it eventually reaches the latter. We set an LPV-based DS for the virtual object as follows:

$$\ddot{x}^v(t) = A_1^v x^v(t) + A_2^v \dot{x}^v(t) + u^v(t), \qquad (7.20)$$

where $A_i^v \in \mathbb{R}^{N \times N}$ $\forall i \in \{1, 2\}$, and $x^v \in \mathbb{R}^N$ is the state of the virtual object. The virtual object (and consequently the arms) must simultaneously intercept the object at the feasible reachable points. To achieve this, we define the following control input $u^v(t)$:

$$u^v(t) = \frac{1}{N_R + 1} \left(\rho(t)\ddot{x}^o - A_1^v \rho(t)x^o - A_2^v(\rho(t)\dot{x}^o + \dot{\rho}(t)x^o) + 2\dot{\rho}(t)\dot{x}^o + \ddot{\rho}(t)x^o + \sum_{j=1}^{N_R} U_j \right)$$

$$- \frac{N_R}{N_R + 1}\left(A_1^v x^v(t) + A_2^v \dot{x}^v(t) \right). \qquad (7.21)$$

Similar to what we saw in section 7.3, $x^o \in \mathbb{R}^N$, and $\rho \in \mathbb{R}_{[0,1]}$ is the same continuous softness parameter. We further assume that the origin of each of the robot controllers is set at the desired intercept point $x^o(t^*) = 0$. U_j is the interaction effect of the tracking controller of the jth end-effector on the virtual object as follows:

$$U_j = \ddot{x}_j(t) + A_{1j}(\gamma_j)(x_j^v(t) - x_j(t)) + A_{2j}(\gamma_j)(\dot{x}_j^v(t) - \dot{x}_j(t)), \qquad (7.22)$$

where $\mathbf{A}_{ij}(.)$, $\forall i \in \{1, 2\}$, and $\forall j \in \{1, \dots, N\}$, act similarly to equation (7.4) and generate affine dependencies of the state-space matrices on the activation parameters (γ_j), defined as

$$\mathbf{A}_{1j}(\gamma) = \sum_{k=1}^{K_j} \gamma_{kj} A_{1kj}$$

$$\hspace{3cm} \forall j \in \{1, \dots, N_R\}. \hspace{2cm} (7.23)$$

$$\mathbf{A}_{2j}(\gamma) = \sum_{k=1}^{K_j} \gamma_{kj} A_{2kj}$$

Here, $x_j^v \in \mathbb{R}^N$ is the state of jth point on the virtual object and is such that $x^v(t) = \sum_{i=1}^{K} x_i^v(t)$. The desired motion of the jth end-effector ($x_j(t)$) is calculated based on the tracking error between the state of the jth point on the virtual object ($x_j^v(t)$) and the end-effector ($x_j(t)$):

$$\ddot{x}_j(t) = \ddot{x}_j^v(t) + \mathbf{A}_{1j}(\gamma_j)(x_j(t) - x_j^v(t)) + \mathbf{A}_{2j}(\gamma_j)(\dot{x}_j(t) - \dot{x}_j^v(t)). \hspace{1cm} (7.24)$$

Substituting equation (7.21) into equation (7.20), the dynamic of the virtual object becomes

$$\ddot{x}^v(t) = \frac{1}{N_R + 1} \Big(\rho(t)\ddot{x}^o + A_1^v(\gamma)(x^v - \rho(t)x^o) + A_2^v(\gamma)\big(\dot{x}^v - \rho(t)\dot{x}^o - \dot{\rho}(t)x^o\big)$$

$$+ 2\dot{\rho}(t)\dot{x}^o + \ddot{\rho}(t)x^o + \sum_{j=1}^{N_R} U_j \Big). \hspace{1.5cm} (7.25)$$

It is shown in exercise 7.11 that if none of the three DSs are perturbed, the motion of the virtual object, and consequently of the robotic systems generated by equations (7.25) and (7.24), respectively, asymptotically converge to the real object; that is,

$$\lim_{t \to \infty} \|x_j(t) - x_j^v(t)\| = 0 \hspace{1.5cm} \lim_{t \to \infty} \|\dot{x}_j(t) - \dot{x}_j^v(t)\| = 0 \hspace{1cm} (7.26)$$

$$\lim_{t \to \infty} \|x^v(t) - \rho(t)x^o(t)\| = 0 \hspace{1cm} \lim_{t \to \infty} \|\dot{x}^v(t) - (\rho(t)\dot{x}^o(t) + \dot{\rho}(t)x^o(t))\| = 0 \hspace{0.5cm} (7.27)$$

if there are P^v, P_j, Q^v, Q_j such that

$$\begin{cases} 0 \prec P^v, \hspace{0.3cm} 0 \prec P_j \hspace{3cm} 0 \prec Q^v, \hspace{0.3cm} 0 \prec Q_j \\[2mm] \begin{bmatrix} 0 & I \\ A_2^v & A_1^v \end{bmatrix}^T P^v + P^v \begin{bmatrix} 0 & I \\ A_2^v & A_1^v \end{bmatrix} \prec -Q^v \\[4mm] \begin{bmatrix} 0 & I \\ A_{2j}^k & A_{1j}^k \end{bmatrix}^T P_j + P_j \begin{bmatrix} 0 & I \\ A_{2j}^k & A_{1j}^k \end{bmatrix} \prec -Q_j \hspace{0.3cm} \forall j \in \{1, \dots, K_r\}. \\[4mm] 0 \leq \gamma_{kj} \leq 1, \end{cases} \hspace{1cm} (7.28)$$

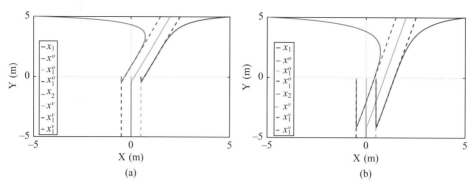

Figure 7.9

Reaching and intercepting a moving object by two agents. (a) $\rho = 0.1$; (b) $\rho = 0.9$. $\mathbf{A}_{2j} = \begin{bmatrix} -20 & 0 \\ 0 & -20 \end{bmatrix}$, $\mathbf{A}_{1j} = \begin{bmatrix} -100 & 0 \\ 0 & -100 \end{bmatrix}$, $\forall j \in \{1, 2\}$, and $A_2^v = \begin{bmatrix} -4 & 0 \\ 0 & -4 \end{bmatrix}$, $A_1^v = \begin{bmatrix} -4 & 0 \\ 0 & -4 \end{bmatrix}$. Both robots reach the virtual object and converge the motion of the real object in coordination and synch.

If we set the softness parameter $\rho(t) = \dot{\rho}(t) = 0$, equation (7.25) generates asymptotically stable robot motions toward the predicted intercept point; that is, the coordination between the robots is preserved but the coordination between the robots and the object is lost. On the other hand, if $\rho(t) = 1$ and $\dot{\rho}(t) = 0$, equation (7.25) generates asymptotically stable motions toward the real object even though the object's motion is not accurately predicted; that is, perfect *coordination* with the object.[3] However, in this case, there is no guarantee that the virtual object intercepts the real object at the predicted intercept point and within the robots' workspace. In this case, the coordination between robots and object is lost. Thus, one needs to vary ρ between zero and 1 such that it is 1 only when the object is at the vicinity of the predicted intercept point. Figure 7.9 illustrates an example of two systems reaching a moving object. The motion of the systems are coordinated such that both of them simultaneously intercept the object at the desired location.

Programming Exercise 7.3   *The aim of this exercise is to help readers to get a better understanding of the virtual object–based DS [equations (7.24), (7.25)] and the effect of the open parameters on the generated motion. Open MATLAB and set the directory to the following folder:*

```
ch7-DS_reaching/Moving_large_object
```

In this exercise, one can study the effect of $\mathbf{A}_{1i}$, $\mathbf{A}_{2i}$, $\forall i \in \{1, 2\}$, A_1^v, A_2^v and ρ on the generated motion. Run, on the command line, the function Bimanual_reaching. Both robots and the virtual object intercept the object at the origin. Modify the input to this function or the parameters inside the function and answer the following questions.

1. What would happen if the motion of the object was much faster than the convergence rate of the DS of

    (a) Just one of the robots

    (b) Both robots, but not the virtual object

    (c) Both robots and the virtual object

2. *What would happen if the object did not pass through the origin (i.e., it did not pass through the desired intercept point)?*

3. *What would happen if ρ were higher than 1, or lower than zero?*

4. *Is there any condition where the motion generated by equations (7.24) and (7.25) intercepts the object at a point that is not the desired intercept point?*

5. *Does integration step affect the behavior? Why?*

6. *What effect do high values of ρ have on the dynamics of the two DSs?*

7. *What happens when $A_1^v = A_2^v = 0$?*

8. *What happens when the virtual object was initialized very far from the robotic system?*

9. *What happens when $\mathbf{A}_{11} = 100\mathbf{A}_{12}$ and $\mathbf{A}_{21} = 20\mathbf{A}_{22}$?*

7.6 Robotic Implementation

The approach was implemented on a dual-arm platform consisting of two 7-DOF robotic arms: a KUKA LWR 4+ and a KUKA IIWA mounted with a 4-DOF Barrett hand and a 16-DOF Allegro hand. The distance between the base of the two robots is $\begin{bmatrix} 0.25 & 1.5 & -0.1 \end{bmatrix}^T$ meters. The robot implementation involves converting the output of the DS [equation (7.24)] into a 7-DOF joint state (for each arm) using a velocity-based control method without joint velocity integration [106]. In order to avoid high torques, the resulting joint angles are filtered by a critically damped filter. The robot is controlled at a rate of 500 Hz. The fingers are controlled with joint position controllers. All the hardware involved (e.g., arms and hands) are connected to and controlled by one 3.4-GHz i7 PC. The position of the *feasible reaching points* of the objects are captured by an OptiTrack motion-capture system from the natural point at 240 Hz. Because the control loop is faster than the motion-capture system, the predicted position of the object is used as the object position in equation (7.25), when the current position of the object is available.

This empirical validation section is divided into three parts that highlight the motion generator's ability to do the following: (1) to coordinate the multiarm systems; (2) to adapt the two arms' motions in coordination so as to reach and grab a large moving object, when introducing unpredictability in the object's motion (such as by having the object be carried by a blindfolded human); and (3) for very fast adaptation of bimanual coordination to intercept a flying object, without using a predefined model of the object's dynamics. A detailed experimental validation is presented in [101, 104, 100], as well as in videos available on the book's web page.

7.6.1 Coordination Capabilities

The first scenario is designed to illustrate the coordination capabilities of the arms with each other and with the object. Arm-to-arm coordination capabilities are shown by setting ρ to 0, favoring arm-to-arm coordination. As the human operator perturbs one of the robot arms, the virtual object is perturbed as well, resulting in a stable, synchronous motion of the other unperturbed arm (figure 7.10). Because the motion generator is a centralized controller based on the virtual object's motion, there is no master/slave arm; thus, when any of

Figure 7.10
Snapshots of the video illustrating the coordination of the arms in free space. The real object is outside the workspace of the robots; hence, the coordination parameter ρ_i is close to 0 and arm-to-arm coordination is favored. The human operator perturbs one of the arms, which leads the other arm to move in synch following the motion of the virtual object attached to the two end-effectors.

the robots are perturbed, the others will synchronize their motions accordingly. The coordination of the arms with the object is shown by moving the object inside the workspace of the robots. The object used is a large box ($60 \times 60 \times 40$ cm) held by a human operator. The edges of the box are specified as the feasible reaching points. When the box is inside the joint workspace of the robots, the operator changes the orientation and position of the box to show the coordination capabilities between the robots and the object (figure 7.11).

7.6.2 Grabbing a Large Moving Object

In this second scenario, the same object is used. Yet, now the operator holds the box while walking toward the robots. Once the end-effectors are less than 2 cm away from the feasible reaching points, finger closures of the hands are triggered and the box is successfully grabbed from the human. As can be seen in figure 7.13, the operator can be even blindfolded to achieve unpredictable trajectories and avoid the natural reactions of the humans to help the robots. When the human operator carrying the box is approaching the robots, the virtual object converges to the box and follows it until the desired interception points are reached.

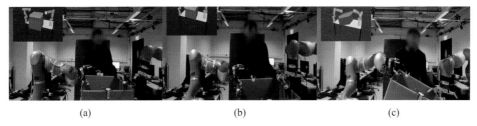

(a)   (b)   (c)

Figure 7.11
Snapshots of the coordination capabilities between the arms and a moving/rotating object. The real object is inside the workspace of the robots; hence, the coordination parameter ρ is close to 1 and the arm-to-object coordination is favored. The top right figures show the real-time visualization of the robots, and the virtual (green) and real object (blue).

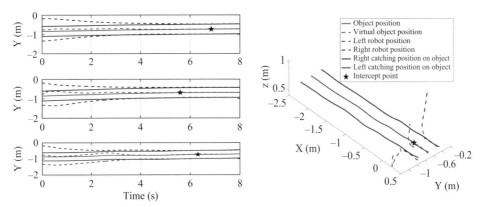

Figure 7.12
Examples of the position of the end-effectors generated by the DS [equation (7.25)]. Only the trajectories along the y-axis is presented. The illustrated object trajectory is the predicted trajectory of the uncaught object. The prediction of the box's trajectory requires some data to be initialized and uses almost all of the first 0.2 m of the object in x. As expected, the outputs of equation (7.25) first converge on the desired intercept position as ρ is a small value, and then it softly intercept the object's trajectory and follow the object's motion. The robots are stopped if the object is not moving or the fingers are closed.

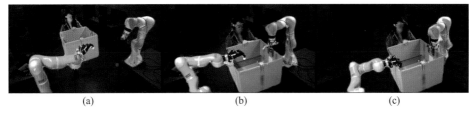

(a)   (b)   (c)

Figure 7.13
Snapshots of the robots' motion when reaching for a moving object carried by a blindfolded operator. (a), (b) and (c) onset of object trajectory's prediction (c) arms have intercepted the object and the fingers have closed on the object.

Figure 7.14
Snapshots of the arms reaching for a fast-moving object. To keep from damaging the robot's hands, they do not close on the object when they intercept the object.

The fingers close and the box is grabbed from the human. An example of the desired robot trajectory and the box trajectory is shown in figure 7.12. As expected, the end-effectors converge on the box and continue to track its motion. As initially $\rho = 0$, the virtual object asymptotically converges on the desired reaching position. While the box is approaching the robots, ρ starts increasing and finally reaches 1 when the object is in the workspace of the robots. Hence, equation (7.25) generates asymptotically stable motions toward the real object instead of the intercept point. Consequently, the prediction of the intercept point does not play a vital role in grabbing the box.

7.6.3 Reaching for Fast-Flying Objects

The third scenario is designed to show the coordination between the robots and a fast-moving object, where a rod (150×1 cm) is thrown to the robots from 2.5 m away, resulting in approximately 0.56s flying time. The distance between the base of the two robots is reduced to $\begin{bmatrix} 0.25 & 1.26 & -0.1 \end{bmatrix}^T$ meters. Due to an inaccurate prediction of the object trajectory, the feasible intercept points need to be updated and redefined during the reach. The new feasible intercept point is chosen in the vicinity of the previous one to minimize the convergence time.

As the motion of the object is fast and the predicted reaching points are not accurate, the initial value of ρ is set to 0.5 to decrease the convergence duration of the robots to the real object. Snapshots of the real robot experiments are shown in figure 7.14. Visual inspection of the data and video confirmed that the robots follow the motion of the object in a coordinated manner and intercept it at the vicinity of the predicted feasible intercept point.

Exercise 7.8 *Given two robotic arms driven by the virtual object–based DS [equations (7.25) and (7.24)], propose two DSs to generate the softness variable ρ such that $\rho \approx 1$ in the following scenarios:*

- *If and only if the object is inside the workspace of the robots*
- *If and only if the distance between the object and the desired intercept point is less than 0.5 cm.*

Note that the object can move toward or away from the robots.

Exercise 7.9 *Consider two robotic arms driven by the virtual object–based DS [equation (7.25)]. However, only one of the arms is able to follow equation (7.24) and the other mechanically fails during the task execution (i.e., $\ddot{x}_2 = \dot{x}_2 = 0$). Study the motion of the virtual object and the other robotic arm. Can they intercept the object at the desired intercept point?*

Exercise 7.10 *Given two robotic arms driven by the virtual object–based DS [equations (7.25) and (7.24)] and $K_j = 1, j \in \{1, 2\}$. Show that if one satisfies the conditions of theorem 4.10, the robotic arms converge on the desired reaching points on the virtual object?. Assume that the virtual object is initialized in the middle of two arms. What is the motion of the virtual object?*

Exercise 7.11 *Considering the DS presented in equations (7.25) and (7.24), prove that equations (7.26) and (7.27) hold if the conditions presented in equation (7.28) are met.*

8 Adapting and Modulating an Existing Control Law

In the previous chapters, we have shown various techniques to learn a control law from a set of training data points. This learning was done once for all, offline, based on examples from a full set of training trajectories. There are, however, many occasions when it would be useful to be able to train the system again, such as to enable a robot to take a different approach path toward a target. Often, the changes apply only to a small region of the state space. Hence, it would be useful to be able to retrain the controller by modifying the original flow only locally.

This chapter shows how one can learn to modulate an initial (nominal) dynamical system (DS) to generate new dynamics. We consider modulations that act *locally* to preserve the generic properties of the nominal DS (e.g., asymptotic or global stability). We further show how such modulations can be made explicitly dependent on an external input and illustrate the usefulness of such a concept with a few examples where the speed is modulated to enter in contact with a surface.

We start the chapter with a description of the properties required for the modulation in section 8.1. We then introduce several methods to learn and construct the modulated functions for internal and external signals in section 8.2 and section 8.3, respectively. Finally, we consider a scenario where a robotic system should stably contacts a surface in section 8.4. This chapter is accompanied by hands-on programming exercises to exemplify the overall algorithms. Readers are highly encouraged to download the source code, vary the parameters and analyze the parameters' effects on the generated system.[1]

8.1 Preliminaries

Let us assume that we have at our disposal a nominal DS $\dot{x} = f(x)$, with $x, \dot{x} \in \mathbb{R}^N$, which is asymptotically stable at a fixed-point attractor $x^* \in \mathbb{R}^N$. Such nominal dynamics can either be learned, following methods offered in chapter 3, or hard-coded by the user. We can modulate this nominal DS to generate new dynamics, $g(x)$, by multiplying $f(x)$ by a continuous matrix function $M(x) \in \mathbb{R}^{N \times N}$. The modulated dynamics is then given by

$$\dot{x} = g(x) = M(x)f(x). \tag{8.1}$$

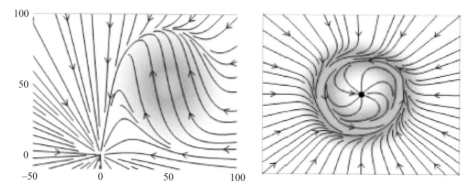

Figure 8.1
Example of modulated dynamical systems (DS). The nominal DS is a linear system. (Left) A local rotation around the point x^o generates a nonlinear dynamics locally. (Right) The local modulation results in a limit cycle around the origin.

Making the modulation a function $M(x)$ of the state $x \in \mathbb{R}^N$ allows us to activate it differently in different regions of the state space. The activation is hence *local*. The simplest activation one can think of is a rotation that can be written as

$$M(x) = I + (R - I)e^{-\|x - x^o\|}, \tag{8.2}$$

with $R \in \mathbb{R}^{N \times N}$ being a rotation matrix and $x^o \in \mathbb{R}^N$, and the rotation acts locally around the point x^o. The effect of the modulation vanishes exponentially from that point. Figure 8.1 illustrates two examples of such a local modulation generating a curvy motion to a nominal linear DS: $\dot{x} = Ax$.

8.1.1 Stability Properties

One can show (see Exercises 8.2 and 8.4) that if $M(x)$ is full rank $\forall x \in \mathbb{R}^N$ and locally active in a compact set that does not include the attractor, the modulated dynamics has the same equilibrium points as the nominal dynamics. If the modulated DS preserves boundedness (see exercise 8.3), it preserves the attractors of the nominal DS. This condition, however, is not sufficient to prevent the appearance of limit cycles.

These conditions put the emphasis on generating local modulations that would preserve the general stability properties of the nominal DSs. Conversely, one can also use modulations to generate instabilities. Consider, for instance, a modulation of the form $M(x) = (1 - e^{-\|x - x^o\|})$. This modulation is locally active but does not have full rank everywhere. It cancels the flow at point x^o, making it a spurious attractor.

One can further generate an unstable DS through the following modulation: $M = \begin{pmatrix} 1 - \gamma(x, x^o) & 0 \\ 0 & 1 - \gamma(x, x^o) \end{pmatrix}$, with $\gamma(x, x^o) = (1 - e^{-\sigma\|x - x^o\|})$ and $0 < \sigma$. This generates a local repulser, whose influence is modulated by the factor σ. Numerically, the influence vanishes away from the repulser and the flow reverts to the nominal DS. This is illustrated in Figure 8.2.[2]

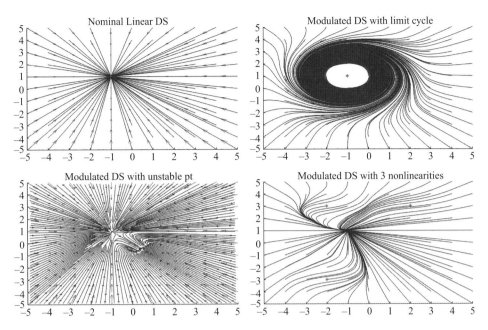

Figure 8.2
Example of modulated DSs. The nominal DS is a linear system (top left). A local rotation around the point ∗ generates a limit cycle locally (top right). A repulsion can be generated locally at the ∗ point (bottom left). Finally, a series of local modulations through three local modulator points is shown (bottom right).

One can further generate explicitly limit cycles and local nonlinearities, as illustrated in figure 8.2.

One way to generate local modulation and to ensure that the modulation matrix remains full rank is to generate a local rotation of the nominal dynamics. Let $\phi(x) = \gamma(x, x^o)\phi$ denote a state-dependent rotation. This results in a smoothly decaying rotation that fully rotates the dynamics by the angle ϕ only at $x = x^o$. The modulation function is then defined as the associated rotation matrix:

$$M(x) = \begin{bmatrix} \cos(\phi(x)) & -\sin(\phi(x)) \\ \sin(\phi(x)) & \cos(\phi(x)) \end{bmatrix}. \tag{8.3}$$

This principle will be reused extensively for landing on a surface (see section 8.4) and for obstacle avoidance (see chapter 9), by generating a rotation to let the flow align with the tangential plane to the object's surface. The rotation is valid locally only when approaching the surface.

8.1.2 Parametrizing a Modulation

The properties enunciated thus far remain even when we scale the modulation. For instance, assume that the modulation is a local rotation as in equation (8.2), multiplying it by a positive scalar [i.e., $g(x) = \lambda M(x)f(x)$]. In this case, $\forall \lambda \in \mathbb{R}^+$ results in faster rotation.

Another parameter that we can choose is the center of the modulation x^o in equation (8.2). As long as we make sure that the effect of the modulation does not include the attractor, we can preserve the stability. Doing this requires choosing an appropriate scaling of the exponential decreasing function, such that its effect vanishes numerically close to the attractor.

Now we have at our disposal a set of parameters that can be learned to fit a modulated matrix at the desired location with a desired strength, without affecting the stability of the system. In brief, if we choose to represent the modulation matrix as a local rotation, we can simply choose the speed, direction, center, and local region of the rotation. In the next section, we are going to discuss an approach to learning and estimating the parameters of the modulation function.

Programming Exercise 8.1   *The aim of this exercise is to help the readers in getting a better understanding of the modulated DSs [equation (8.2)] and the effect of the open parameters on the generated motion. Open MATLAB and set the directory to the following folder:*

```
ı          ch8-DS_modulated
```

Open ch8_ex1_2.m file. This code generates the first top left example shown in figure 8.2. The goal of this exercise is to allow readers to understand the effect of the parameters of the modulation function on the local modulation.

1. Construct a modulation that preserves the stability at the attractor for all the values of the rotation matrix.

2. Which type of modulations lead to a limit cycle around the origin, as illustrated in figure 8.2? Program such a modulation.

3. Construct a modulation function that makes an unstable nominal system stable.

Exercise 8.1   *Consider the nominal DS $\dot{x} = Ax$ with $A = \begin{pmatrix} -1 & 0 \\ 0 & -1 \end{pmatrix}$. Construct a matrix $M(x)$ that is locally active but generates a limit cycle.*

Exercise 8.2   *Show that if $M(x)$ is full rank for all x, the modulated dynamics has the same equilibrium points as the nominal dynamics.*

Exercise 8.3   *Show that if the nominal dynamics is bounded and $M(x)$ is locally active in a compact-set $\chi \subset \mathbb{R}^N$. Then the modulated dynamics is bounded.*

Exercise 8.4   *Consider a system $\dot{x} = f(x)$ that has a single equilibrium point. Without loss of generality, let this equilibrium point be placed at the origin. Assume further that the equilibrium point is stable and the modulated dynamics is bounded and has the same equilibrium point as the nominal dynamics. Show that, if χ does not include the origin, the modulated system is stable at the origin.*

Programming Exercise 8.2 *The aim of this exercise is to help readers to get a better understanding of the modulated DSs illustrated in figure 8.2. Open again MATLAB and set the directory to the following folder:*

```
ch8-DS_modulated
```

and then open

```
ch8_ex1_2
```

This file generates the first top left example shown in figure 8.2. Expand this code to do the following:

1. *Create two local rotations at points* $x^1 = [2, 1]^T$ *and* $x^2 = [3 - 2]^T$.
2. *Generate two limit cycles on two separate parts of the state space.*

8.2 Learning an Internal Modulation

The previous section introduced the conditions to generate a modulation that either preserves the stability at an attractor of the nominal DS or allows new attractors to be generated. We also showed how these modulations could be parameterized. Here, we show how these parameters can be learned.[3]

8.2.1 Local Rotation and Norm-Scaling

As seen previously, rotations (or any other orthogonal transformations) always have full rank. It is possible to define and parameterize rotations in any dimension, but we will focus mainly on two-dimensional (2D) and three-dimensional (3D) systems for illustration of the approach here. For increased flexibility, a scaling of the speed of the DS can be achieved by multiplying the rotation matrix by a scalar. Let $R(x)$ denotes a state-dependent rotation matrix, and let $\kappa(x)$ denote a state-dependent scalar function strictly larger than -1. We then construct a modulation function that can locally rotate and speed up or slow down the dynamics as follows:

$$M(x) = (1 + \kappa(x))R(x). \tag{8.4}$$

Both κ and R should vary in a continuous manner across the state space. In a continuous system, the inclusion of speed-scaling does not influence the stability properties, although it may do so in discrete implementations, so care should be used not to allow $\kappa(x)$ to take large values. Also, note that κ has been given an offset of 1 so that with $\kappa(x) = 0$, the nominal speed is retained. This is useful when modeling κ with local regression techniques such as Gaussian process regression (GPR), as will be done in section 8.2.2.1. Rotations in arbitrary dimension can be defined by means of a 2D rotation set and a rotation angle ϕ. In 2D, the fact that the rotation set is an entire $\mathbb{R}^2$ means that a rotation is fully defined by the rotation angle only. Hence, the parameterization in that case is simply $\theta_{2d} = [\phi, \kappa]$. In 3D, the rotation plane can be compactly parameterized by its normal vector. Hence, the parameterization in 3D is $\theta_{3d} = [\phi \mu_R, \kappa]$, where μ_R is the rotation vector (the normal of the rotation set). Parameterization in higher dimensions is possible, but it requires additional parameters for describing the rotation set.

8.2.2 Gathering Data for Learning

Assume that a training set of M observations of x and $\dot{x}$ is available: $\{x^m, \dot{x}^m\}_{m=1}^{M}$. To exploit this data for learning, it is first converted to a data set consisting of input locations and corresponding modulation vectors: $\{x^m, \theta^m\}_{m=1}^{M}$. To compute the modulation data, the first step is to compute the nominal velocities, denoted by $^o\dot{x}^m, \forall m \in \{1 \ldots M\}$. This is done by evaluating the nominal dynamics at $x^m, \forall m \in \{1 \ldots M\}$ in the trajectory data set. Each pair $\{^o\dot{x}^m, \dot{x}^m\}$ then corresponds to a modulation parameter vector θ^m. How this parameter vector is computed depends on the structure and parameterization chosen for the modulation function. For example, algorithm 8.1 describes the procedure for computing the modulation parameters for the particular choice of modulation function [equation (8.4)]. Parameter vectors for each collected data point are computed this way, and paired with the corresponding state observations, they constitute a new data set: $\{x^m, \theta^m\}_{m=1}^{M}$. Regression can now be applied to learn $\theta(x)$ as a state-dependent function.

Algorithm 8.1 Procedure for converting 2D or 3D trajectory data to modulation data.

Require: Trajectory data $\{x^m, \dot{x}^m\}_{m=1}^{M}$
1: **for** $m = 1 \to M$ **do**
2:     Compute nominal velocity: $^o\dot{x}^m = f(x^m)$
3:     Compute rotation vector (3D only): $\mu^m = \frac{\dot{x}^m \times {^o\dot{x}^m}}{\|\dot{x}^m\| \| {^o\dot{x}^m}\|}$
4:     Compute rotation angle: $\phi^m = \arccos \frac{\dot{x}^{m^T} {^o\dot{x}^m}}{\|\dot{x}^m\| \| {^o\dot{x}^m}\|}$
5:     Compute scaling: $\kappa_m = \frac{\|\dot{x}^m\|}{\| {^o\dot{x}^m}\|} - 1$
6:     3D: $\theta^m = [\phi^m \mu^m, \kappa^m]$, 2D: $\theta^m = [\phi^m, \kappa^m]$
7: **end for**
8: **return** Modulation data $\{x^m, \theta^m\}_{m=1}^{M}$

8.2.2.1 GP-Modulated Dynamical Systems

GPR is a regression technique which, in its standard form, can model functions with input of arbitrary dimension and scalar outputs. The basic GPR equations are reviewed briefly in section B.5 of appendix B. The behavior of GPR depends on the choice of covariance function $k(\cdot, \cdot)$. In this section, we use the squared exponential covariance function, defined by

$$k(x, x') = \sigma_f \exp\left(-\frac{x^T x}{2l}\right),$$

where $l, \sigma_f > 0$ are scalar hyperparameters that can be set to predetermined values or optimized to maximize the likelihood of the training data.

The modulation is based on encoding the parameter vector of the modulation function with Gaussian processes (GPs). The data set from section 8.2.2 is used as a training set for the GP, where the positions x^m are considered as inputs and the corresponding modulation parameters θ^m are considered as outputs. Note that because θ is multidimensional, one GP per parameter is needed. This can be done at little computational cost if the same hyperparameters are used in each GP. A vector of scalar weights can be precomputed as follows:

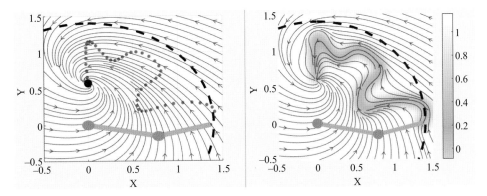

Figure 8.3
Nominal DS learned with SEDS based on training data trajectory illustrated in red/plain line (*left*). The DS does not follow well the dynamics. When fitting the dynamics with LMDS and using the fit to modulate the DS, we obtain a much better fit (*right*).

$$\alpha\left(x_*\right) = \left[K_{XX} + \sigma_n^2 I\right]^{-1} K_{Xx_*}.$$ (8.5)

Prediction of each entry of θ then only requires computing a dot-product: $\hat{\theta}^j\left(x_*\right) = \alpha\left(x_*\right)^T \Theta^j$, where Θ^j is a vector of all the training samples of the j th parameter of θ.

8.2.2.2   Enforcing Local Modulation

The particular choice of GP with zero mean and with the squared exponential covariance function results in all elements of θ going to 0 in regions far from any training data. Hence, for the modulation to be local, it should be parameterized such that $M \to I$ as $\theta \to 0$. This is the case for the rotated and speed-scaled modulations, which encode the rotation angle in the norm of a subvector of θ. Also, when the speed factor κ goes to 0, the speed of the reshaped dynamics approaches the nominal speed. Consequently, the modulation function does go to identity, but there is no strict boundary outside of which M is exactly equal to I. To make the modulation locally active in the strict sense, the entries of $a\left(x^*\right)$ in equation (8.5) should be smoothly truncated at some small value. To this end, one can use a sinusoidal signal computing the truncated weights $\alpha'\left(x^*\right)$ as follows:

$$\alpha'\left(x^*\right) = \begin{cases} 0 & \alpha\left(x^*\right) < \underline{\alpha} \\ \frac{1}{2}\left(1 + \sin\left(\frac{2\pi\left(\alpha\left(x^*\right) - \underline{\alpha}\right)}{2\rho} - \frac{\pi}{2}\right)\right)\alpha\left(x^*\right) & \underline{\alpha} \le \alpha\left(x^*\right) \le \underline{\alpha} + \rho. \\ \alpha\left(x^*\right) & \underline{\alpha} + \rho < \alpha\left(x^*\right) \end{cases}$$ (8.6)

An illustrative example of modulation applied on toy 3D data is given in figure 8.4.

It should be noted that this particular choice of truncation function is not critical; it could surely be replaced by other methods without perceivably affecting the resulting dynamics. The computation of the reshaping parameters $\hat{\theta}^j(x^*)$ at a query location x^* can hence be summarized as follows:

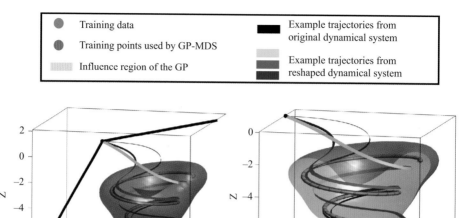

Figure 8.4
Left: Example of reshaped dynamics in a 3D system. The colored streamtapes represent example trajectories of the reshaped dynamics. The streamtapes in black represent trajectories that do not pass through the reshaped region of the state space, and hence retain the straight-line characteristics of the linear system that is used as the nominal dynamics here. The green streamtubes are artificially generated data representing an expanding spiral. Points in magenta represent the subset of this data that was selected as significantly altered (corresponding to a level set of the predictive variance in the GP). The colored streamtapes are example trajectories that pass through the reshaped region. Right: Same as left, but zoomed in, and the influence surface has been sliced to improve visibility of the training points and the trajectories.

- Compute $\alpha'(x^*)$ according to equation (8.5).
- Compute the truncated weights according to equation (8.6).
- Compute the predicted parameters $\hat{\theta}^j(x^*) = \alpha'(x^*)^T \Theta^j$.

Another example of application on the data set of handwriting motion, which has already been used in chapter 3, is shown in figure 8.5.

> **Programming Exercise 8.3**   *The aim of this exercise is to help readers to get a better under-standing of the examples illustrated in figure 8.3. Open MATLAB and set the directory to the following folder:*
>
> ```
> ch8-DS_modulated/Local_Modulation
> ```
>
> *Then run*
>
> ```
> modulating_dynamical_systems.m
> ```
>
> *This file generates a GUI that generates a local rotation based on user's input. Modify this code to generate other modulations, such a rotation in the opposite direction and with smaller angles that varies over time.*

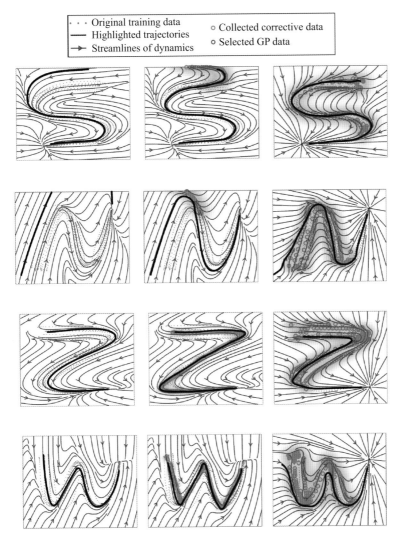

Figure 8.5
Left column: Demonstrated trajectories and the resulting nominal DS model learned for the letters S, N, Z, and W, using SEDS (see chapter 3). **Middle column**: A learned modulation is used to improve various aspects of the SEDS models. In the case of S and N, the favorable starting region of the state space is achieved. In the case of Z and W, the modulation with a fine length scale is used to sharpen the corners of the letters. **Right column**: The nominal training data are provided to learned the modulation and added to a standard linear DS as nominal dynamics.

Programming Exercise 8.4   *The aim of this exercise is to help readers to get a better under-standing of the modulated DS [equation (8.4)] and the effect of the open parameters on the generated motion. Open MATLAB and set the directory to the following folder:*

```
1    ch8_DS_modulated/Learning_modulations
```

Matlab file locally_modulating_dynamical_systems.m allows you to draw some new trajectories that will generate a local modulation. Test the effect of changing the parameters of the kernel on the precision of the reconstruction of the training trajectories. Answer the following questions:

1. *Would it be possible to have an S-shape trajectory while*

    (a) *the target is in the middle of the S-shape trajectory?*

    (b) *the target is far away from the S-shape trajectory?*

2. *Does the direction of the demonstrated trajectories matter?*

3. *What happens if the nominal DS is very stiff or unstable?*

8.2.3   Robotic Implementation

One application of this work to modify the dynamics of movement for a robot tasked to insert plates into slots in a dish rack was presented in [85] (also see figure 8.6). The system starts with a nominal DS that does not have the proper orientation. Through kinesthetic teaching via tactile feedback, a set of additional trajectories could be generated, which was then used to train a modulation to achieve proper orientation. For details, refer to [85].

To perform this task, the robot needs to grasp the plates, transport them from an arbitrary starting location to the slot, and insert them with the correct orientation. In this example, proper orientation is achieved by keeping the end-effector orientation fixed. The grasping is completed by manual control of the Barrett hand by a human operator.

As nominal dynamics, a Cartesian DS model corresponding to a standard place-type motion trained from trajectories recorded from humans is used. Example trajectories from this system are illustrated in figure 8.6a. As can be seen, the general motion pattern is appropriate for the task, but trajectories starting close to the dish rack tend to go to straight toward the target, colliding with the rack on the way. This model can be improved by locally reshaping the system in the problematic region.

A Cartesian impedance controller is used to follow the trajectory generated by the DS. Because the teaching procedure takes place iteratively as the robot is performing the task, it is necessary to inform the system when a corrective demonstration is being given. This is achieved through the use of an artificial skin module mounted on the robot. The idea is to achieve accurate tracking in combination with compliant motion when necessary for corrective demonstrations. This is achieved by multiplying the feedback-component of the controller with a positive scalar that is inversely proportional to pressure detected on the artificial skin. Let $\tau_{PD} \in \mathbb{R}^7$ denote the vector of joint torques coming from the Cartesian impedance controller, and $\tau_G \in \mathbb{R}^7$ denote the gravity compensation torques. Then the control torque τ commanded to the robot joints is:

$$\tau = \psi \tau_{PD} + \tau_G, \tag{8.7}$$

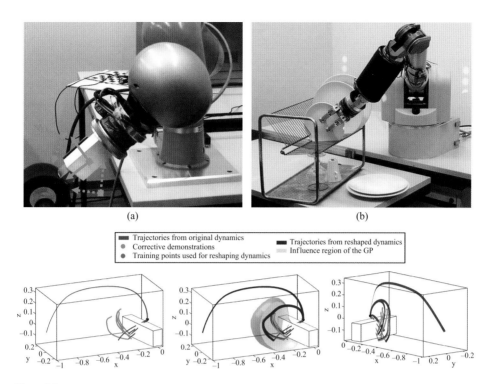

Figure 8.6

Top: A Barrett WAM 7-DOF arm performs a plate-stacking task. Trajectories resulting from the nominal dynamics from a set of starting points (bottom left) would lead to failure due to lack of correct orientation. Trajectories starting close to the rack tend to collide with it. Provided corrective training data delivered through physical guiding of the robot is shown in green. Bottom middle: Resulting reshaped system. The gray-shaded region illustrates the region of influence of the GP and is computed as a level set of the predictive variance. Bottom right: Reshaped system from a different point of view. Note the sparse selection of the training data.

where $\psi \in [0, 1]$ is a truncated linear function that is equal to 1 when there is no pressure on the skin, and 0 when the detected pressure exceeds a predetermined threshold value. As an effect of equation (8.7), the resistance to perturbations is decreased when the teacher pushes the arm to deviate its trajectory during a corrective demonstration.

The teaching procedure was initialized by starting the robot from a problematic point (one that would result in collision with the rack if the teacher did not intervene). The teacher then would physically guide the robot during its motion so as to prevent collision with the rack. The data were recorded and used by the modulation to reshape the dynamics according to the incoming data. This procedure was repeated from a few problematic points to expand the reshaped region of the state space. Corrective demonstrations were provided during four trajectories starting in the problematic region, resulting in the training data visible in Figures 8.6b and 8.6c. A total of $1,395$ data points were collected. With the length scale $l = 0.07$, signal variance $\sigma_f = 1$ and signal noise $\sigma_n = 0.4$, and selection parameter values $\overline{J}^1 = 0.1, \overline{J}^2 = 0.2$, only 22 training points needed to be saved in the training set.

The region in which the dynamics are reshaped is illustrated as a gray surface in figure 8.6b. As can be seen in figures 8.6b and 8.6c, the dynamics were successfully reshaped to avoid collision with the edge of the rack. The total computational time (GP prediction followed by reshaping) took about 0.04 ms, two orders of magnitude faster than required for the control frequency of 500 Hz. The program was written in C++ and executed on an average desktop with a quadcore Intel Xeon processor. On this particular machine, the maximum number of training points compatible with our control frequency is just over 2,000. A detailed experimental validation is presented in [85].

8.3   Learning an External Modulation[4]

In the previous section, we showed how the parameters of the modulation function can be learned to locally reshape the behavior of the system with respect to the state of the system (e.g., the state of a robot). However, in many tasks, it is necessary to be able to react to sensory inputs in a task-specific manner (i.e., force/torque or vision sensors). The goal of the externally modulated dynamical system (EMDS) is to provide a modulation formulation that allows one to learn reactions to sensory events such as contact detection with tactile sensing arrays or force-torque sensors.

Let $s \in \mathbb{R}^M$ be an M-dimensional external signal, independent of the state of the DS. In EMDS, the dynamics are reshaped by a modulation field $M(x, s)$. The form of the dynamics follows the same reshaping structure as equation (8.1):

$$\dot{x} = M(x, s)f(x), \tag{8.8}$$

where $M(x, s) \in \mathbb{R}^{N \times N}$ is a continuous matrix that modulates the nominal dynamics $f(x)$, as a function not only on the system's state, but also on an external signal.

As the resulting DS is not autonomous, we cannot expect the same stability properties as in the case of the autonomous modulation formulation. However, by constructing the modulation matrix appropriately, one can achieve guaranteed boundedness and convergence of the dynamics by ensuring that M is full rank and locally active. It can easily be derived by adapting the proofs in section 8.1 to include the external signal such that the following points are true:

- The reshaped dynamics has the same equilibrium point as the nominal DS.
- The reshaped dynamics is bounded.
- If the nominal DS is locally stable, the reshaped system is also locally stable.
- The reshaped dynamics has the same equilibrium point as the nominal DS.
- If the nominal DS is locally asymptotically stable, the reshaped system is also locally asymptotically stable.

To this end, the modulation field $M(x, s)$ has only a few design constraints. In the following section, we introduce one possible way to design and parameterize this function.

8.3.1   Modulating, Rotating and Speed-Scaling Dynamics

As shown in section 8.1, the modulation function can be defined as a composition of a speed-scaling and a rotation matrix. Rotations always have full rank, so the modulated dynamics has the same equilibrium point(s) as the nominal dynamics. Besides, any vector can be expressed as a rotation and scaling of another non-null vector, which justifies this choice of representation for the modulation matrix.

It is always possible to represent a modulation function compactly as a parameter vector $\theta \in \mathbb{R}^L$, where $L \geq N$ depends on the chosen parameterization and the dimension of the state

$x \in \mathbb{R}^N$. Complex reshaping of the nominal DS can then be achieved by using nonlinear regression to learn a function mapping from the state to this parameter vector.

The rotation angle ϕ can always be recovered from θ, as the norm of a subvector of θ (angle axis representation) or as an independent element of θ. Hence, given a learned function from the state to the reshaping parameter vector, we can find the rotation angle as a function of the state, $\phi(x)$. In EMDS, we let the external signal s modulate the rotation angle and the speed scaling before reconstructing M and applying the modulation to the nominal DS:

$$\theta(x, s) = h_s(s)[\phi(x)\mu_R, \kappa(x)], \tag{8.9}$$

with μ_R as the rotation vector defining a rotation axis. The modulation function $M(x, s)$ is defined as

$$M(x, s) = (1 + \kappa(x, s))R(x, s), \tag{8.10}$$

with $R(x, s)$ as the rotation matrix associated with the rotation vector $\phi(x)\mu_R$, and hence, full rank by construction. So does $M(x, s)$, and therefore all the stability properties are guaranteed for any values of x and s. The mappings $\phi(x) : \mathbb{R}^N \to [-\pi, \pi]$ and $\kappa(x) : \mathbb{R}^N \to \mathbb{R}^+$ are continuous functions from a robot state to a rotation angle and a speed-scaling, respectively. To preserve the stability and convergence properties, the state-dependent maps $\phi(x)$ and $\kappa(x)$ should be locally active. The parameters are also influenced by the continuous external activation function $h_s : \mathbb{R}^M \to [0, 1]$, which depends on the external signal s.

It is worth noting that the local property ensures boundedness and local asymptotic stability for any chosen modulation matrix. It may thus be useful to keep locality even when it is not required by the desired dynamics, but only for the provided stability purposes.

As an illustrative example, consider the following linear nominal dynamics:

$$\dot{x} = -Ax = -\begin{bmatrix} 10 & 0 \\ 0 & 10 \end{bmatrix} x. \tag{8.11}$$

Let the following continuous function $h_x : \mathbb{R}^2 \to \mathbb{R}$ describe the influence of the modulation and impose the locally active property:

$$h_x(x) = \begin{cases} 0 & \text{if } \|x\| < 0.08 \\ 50 \cdot \|x\| - 4 & \text{if } 0.08 \leq \|x\| < 0.1 \\ 1 & \text{if } 0.1 < \|x\| < 0.7 \\ -20 \cdot \|x\| + 15 & \text{if } 0.7 \leq \|x\| < 0.85 \\ 0 & \text{otherwise.} \end{cases} \tag{8.12}$$

The value of $h_x(x)$ is visible as a grayscale for the examples shown in figure 8.7.

The external signal s influences the local modulation according to the following activation function $h_s : \mathbb{R} \to [0, 1]$:

$$h_s(s) = \begin{cases} 1 & \text{if } s < 0.0 \\ 1 - 10s^3 + 15s^4 - 6s^5 & \text{if } 0.0 \leq s \leq 1.0 \\ 0 & \text{otherwise.} \end{cases} \tag{8.13}$$

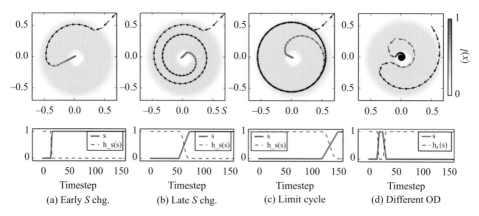

(a) Early S chg.   (b) Late S chg.   (c) Limit cycle   (d) Different OD

Figure 8.7
Top: Examples of the DS modulated by an external signal. **Bottom:** Corresponding profiles of the external signal s and the function $h_S(s)$, which inhibits the rotational modulation. The DS is the same in (a)-(b), but the external signal's profile is different in each plot. (c) Example of a modulation of the DS, using $\phi_c = 90°$. The system does not converge while s is 0 that is, the external activation function $h_S(s)$ is 1. If that is the case, the system stays in a limit cycle. (d) An example with different nominal dynamics and a modulation that also applies speed-scaling. OD stands for the original (nominal) dynamics.

The modulation function in two dimensions is defined as the following rotation matrix without speed-scaling:

$$M(x,s) = \begin{bmatrix} cos(\Phi(x,s)) & sin(\Phi(x,s)) \\ -sin(\Phi(x,s)) & cos(\Phi(x,s)) \end{bmatrix}. \tag{8.14}$$

Introducing the local activation function $\phi(x) = h_x(x)\phi_c$, with $\phi_c \in [-\pi, \pi]$ being a constant angle, the rotation angle $\Phi(x,s)$ is given by

$$\Phi(x,s) = h_s(s)h_x(x)\phi_c. \tag{8.15}$$

This results in a spiraling behavior where and when the DS is modulated. When the external signal is active, the rotation is inhibited, and thus the system converges much faster than the nominal DS on a straight line.

The resulting dynamics are given in figures 8.7(a) and (b) using $\phi_c = 81°$ and different arbitrary profiles of external signal s. The evolution in time of the external signal s, and consequently of the activation function $h_S(s)$, is drawn on the plots below the evolution of the DS in 2D, and the value of s is also represented in the top figures by the color of the arrows. When the signal s becomes high, the activation function $h_S(s)$ goes to 0 and the system switches from spiraling to the nominal linear dynamics, converging rapidly. The resulting behavior can be used, for instance, to switch between searching and reaching motions on a robot. In figure 8.7(a), s is activated early, and so are the dynamics, changing from spiraling to reaching directly. In figure 8.7(b), s is activated late and at a slower rate, and hence the system follows the modulated dynamics, spiraling for a long time until gradually changing. We also provide an example where the system is not globally asymptotically stable in figure 8.7(c), setting the maximum modulation angle ϕ_c to $90°$. When the external signal is 0, the DS goes into a limit cycle. Boundedness is enforced, however, thanks to the locally active property.

In figure 8.7(d), we illustrate another sample behavior of our modulated system using different nominal dynamics [with $A = \begin{bmatrix} 0.05 & 0.2 \\ -0.2 & 0.05 \end{bmatrix}$ in equation (8.11)], a maximum modulation angle $\phi_c = 160°$, and a signal s varying between 0 and 1. The modulation also applies speed-scaling of factor 3, visible on the top image from the length of the arrows changing with s. Depending on s, the direction of the rotation is changed.

8.3.2 Learning the External Activation Function

Using the design of the modulation function presented here, it is possible to retrieve a normal modulation function by removing the dependency on the external signal; that is, by replacing $h_s(s)$ with 1 in equation (8.15) (i.e., never inhibiting the local modulation). Conversely, an EMDS can be created by associating an existing modulation function with the function $h_s(s)$.

Therefore, an EMDS can be based on a modulation function learned in the same way presented in section 8.2, using GPR or any arbitrary local learning algorithm. The external signal activation function $h_s(s)$ can then be provided or learned separately to form the EMDS. To sum up, one way to learn a complete EMDS from scratch with training data can be the following procedure:

1. Learn a DS, the nominal dynamics, from demonstration data (e.g., with SEDS as introduced in section 3.

2. Learn a modulation function from other demonstration data to represent different dynamics, expressed as a modulation of the nominal DS.

3. Learn the function $h_s(s)$.

The function $h_s(s)$ mapping the external signal to the activation of the modulation can either be hard-coded or learned, with the constraint that its values lie between 0 and 1. To learn the function $h_s(s)$, we go through a short learning phase, during which a teacher manually selects the desired behavior while the task is being executed. During this phase, the teacher chooses how much to inhibit the local modulation of the modulated system. The teacher chooses continuous values between 0 (deactivated modulation function) and 1 (activated modulation function).

The recorded data are used to train a GPR model using the squared exponential covariance function. The kernel's hyperparameters are determined manually by using prior

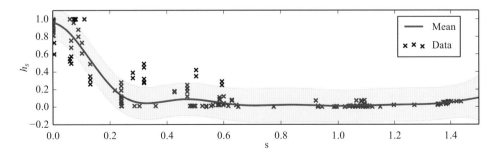

Figure 8.8
Learning $h_s(s)$ from a set fo demonstration data. The blue envelope around the mean represents the variance of the GP function.

knowledge from the training data. An example of a learned function $h_s(s)$ and the training data can be seen in figure 8.8.

The value of h_s starts to increase from 0 to 1 when s is below 0.2. During the run time, we use the stored GP model to predict the value of h_s, given input s.

Programming Exercise 8.5   *The aim of this exercise is to help readers to get a better understanding of the modulated DS [equations (8.10) and (8.15)] and the effect of the open parameters on the generated motion. Open MATLAB and set the directory to the following folder:*

```
1            ch8-DS_modulated/Externally_learning_modulations
```

The file externally_modulating_dynamical_systems.m allows you to define a modulation function around a specific point. The external signal is defined as the Euclidean distance between the target and the state ($h_s = \|x - x^*\|$*), such that*

$$h_s = \begin{cases} 1 & \|x - x^*\| = 0 \\ -\dfrac{Cof}{1+e^{-k\|x-x^*\|}} + a & 0 < \|x - x^*\| \leq Upper \\ 0 & Upper < \|x - x^*\|. \end{cases} \tag{8.16}$$

where the positive scalars a *and* k *are defined in such a way that the aforementioned function is continuous and monotonically decreasing.*

Answer the following questions:

1. What values should be set for the scalars Upper *and* Cof *such that the modulated function is not activated for any* $x \in \mathbb{R}^2$*?*

2. What would happen if h_s *was not a monotonically decreasing function?*

3. What happens if the nominal DS is very stiff or unstable?

4. Apart from $\|x - x^*\|$*, can you think of other external variables?*

8.3.3   Robotic Implementation

Consider a task of going from point A to point B for the robot end-effector with the desired dynamics, while there are obstacles with unknown positions, on the path. The external variables are the time since last contact and the angle of the last contact. We aim at learning how to avoid obstacles depending on information from the collision (in this case, the force direction during contact).

For this purpose, we encode the nominal dynamics and the modulated dynamics as two opposite velocity fields in a central region, where the experiment is taking place. Both dynamics converge on the target when going far enough away from that region. The first is directed perpendicular to the direction between initial and target frames, in a horizontal plane. The second is directed in the opposite direction (see figure 8.9).

A whole range of dynamics is reachable by changing the activation of the modulation. For instance, by setting the activation to 0.5, the resulting trajectory is a straight line. The angle of the deviation can be adjusted by modifying the activation value between 0 and 1.

To learn the mapping $h_s(s)$, eight demonstrations are shown with different collision angles. the model is learned with GPR and the output is such that when no collisions occur,

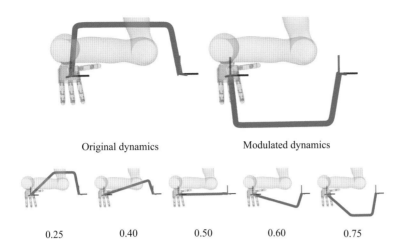

Original dynamics   Modulated dynamics

0.25   0.40   0.50   0.60   0.75

Figure 8.9
The RGB frames on the left and right correspond on the starting point and the target, respectively. Seen from above, in green, are trajectories of the DS. *Top:* The nominal and modulated dynamics. *Bottom:* The resulting dynamics with different levels of activation h_S. With $h_S = 0.50$, the dynamics is in a straight line.

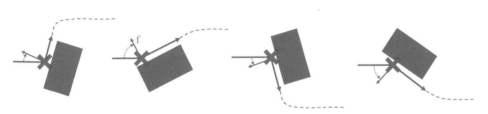

Figure 8.10
Schematic of the demonstrated trajectories. The collision point and the force sensed during the collision are marked in red. The end-effector follows the shape of the objects after contact, and then continues again in a straight line after a few seconds.

the robot moves in a straight line [i.e., $h_S(s) = 0.5$]. After a collision, the end-effector adjusts its trajectory depending on the collision angle (see figure 8.10). If the angle is small, the robot does a large detour, hence picking an extreme value of the activation function (0 or 1, depending on the direction of avoidance).

The task is executed by activating the modulation depending on the input s (figure 8.11a), or fixing the value of the activation to 0.5, hence ignoring external signals for comparison purposes (see figure 8.11b). When the external signal is ignored, the end-effector moves in a straight line. Even though, the robot is compliant, when ignoring the external signal, the robot does not avoid the obstacles. The friction in the joints prevents it from being really deviated from its trajectory, and the robot displaces the obstacles while colliding with them. When using information from the external signal, the robot adapts its trajectory after each collision and navigates between the obstacles. Depending on the collision angle, the robot adapts the avoidance trajectory. Therefore, it sometimes slides along the object (here with the second obstacle), or moves away from it (first and third obstacles). A detailed experimental validation is presented in [138].

(a) $h_s(s)$ learned (b) $h_s(s) = 0.5$ (fixed)

Figure 8.11
Experiment b: Evolution of the obstacle avoidance task with the controller activated (a) or not (b). The end-effector reaches from left to right.

8.4 Modulation to Transit from Free Space to Contact

In the previous two sections, we have been methods by which we could modify a nominal *first-order* DS through a multiplicative function $M(x)$. This allowed us to change the direction of the flow in certain regions of the state space and to create a dependency through an external input. In this section, we go one step further and show how we can extend this modulation to change the direction of the flow locally so as to safely make contact with a surface. To this end, we extend the modulated DS formulation to *second-order* DS, used already in section 3.5.1 in chapter 3. We introduce the idea of having a function $\Gamma(x)$, measuring the distance to the surface to slow down the flow as we approach the surface. This concept will be reused extensively in chapter 9, when performing obstacle avoidance.

8.4.1 Formalism

For many manipulation tasks in robotics, making stable contact is essential. For example, consider a scenario in which a robot is tasked to reach a table to wipe it (see figure 8.12). For wiping to be effective, the robot must remain in contact once it reaches the surface and moves swiftly over it. The challenge is to do this without stopping the robot as it makes its first contact. The speed must hence be modulated prior to contact to prevent the force at contact to let the robot bounce off the surface during the transition. To achieve this, we create a *transition* region before the surface highlighted as the gray/greenish region in figure 8.12.

Definition 8.1 *A contact is called stable if the impact happens only once and the system remains in contact with the surface after the impact.*

Suppose that the contact surface is nonpenetrable and passive and that we have at our disposal a C^∞ function $(\Gamma(x): \mathbb{R}^N \to \mathbb{R})$, which conveys a notion of distance to the surface. Let us assume, further, that the surface and the level curves of Γ enclose a convex region, and that this function monotonically increases with respect to the shortest distance between x and the surface. Hence, $\Gamma(x) = 0$ if and only if x is on the contact surface. Moreover, $e^i \in \mathbb{R}^N$, $\forall i \in \{1, \dots, N\}$ forms an orthonormal basis in $\mathbb{R}^N$, with $e^1(x) = \frac{\nabla \Gamma(x)}{\|\nabla \Gamma(x)\|}$ pointing to the normal of the surface and $\|\nabla \Gamma(x)\| \neq 0$, $\forall x \in \mathbb{R}^N$. The task space can then be categorized into two regions: a free-space when $0 < \Gamma(x)$ and a contact region when $\Gamma(x) = 0$

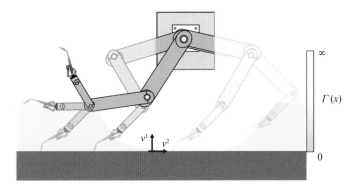

Figure 8.12
A robot is tasked to swipe swiftly along a surface. To ensure that the contact is stable, we introduce a modulation to slow down the robot when approaching the surface through a distance function $\Gamma(x)$. Far from the surface, the function is zero and has no influence.

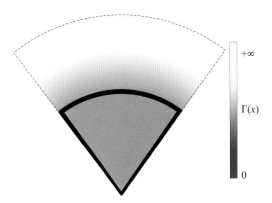

Figure 8.13
The distance function can be used to partition the space into a free-space region and a contact region, where the function $\Gamma(x) = 0$, illustrated with the black curve.

(see figure 8.13). The value of Γ is proportional to the intensity of the modulation. Close to the surface, the modulation is maximal. Far away, at infinity, the modulation is negligible and the robot moves with the nominal DS. If *Gamma* decreases exponentially, the nominal motion is modulated only within a short region, known as the *transition* region, close to the surface. The robot's motion is then modulated only shortly before and when making contact with the surface.

Let us further assume that the impact is perfectly elastic; that is, the coefficient of restitution (COR) is one. In this case, the normal velocities[5] of the system before and after impact are equal in amplitude but pointing in opposite directions. Hence, to achieve a stable contact, the normal velocity of the system at the contact must be zero; that is,

$$e^1(x)^T \dot{x}(t^*) = 0, \tag{8.17}$$

where t^* is the time when the robot makes contact with the surface.

We use a second-order nominal DS $f(x, \dot{x}, t)$ to move the robot in free space and asymptotically stable to a fixed target (x^*) located above the surface (i.e., $0 < e^{1T}x^*$). The nominal acceleration is nonzero everywhere except at the target [i.e., $f^T(x,\dot{x})f(x,\dot{x}) \neq 0 \ \forall (x,\dot{x}) = \mathbb{R}^{N \times N} - \{x^*, 0\}$]. Such a system can be learned as described in chapter 3.

We introduce a modulation through the function $M(x,\dot{x})$ and control the desired acceleration of the robot through

$$\ddot{x} = M(x,\dot{x})f(x,\dot{x},t). \tag{8.18}$$

Notice that both nominal DS and modulation depend on both state and velocity.

$M(x,\dot{x}) \in \mathbb{R}^{N \times N}$ reshapes the nominal DS such that it aligns its flow with the contact surface. To perform this alignment, we must control, separately, the motion in the normal and tangential directions to the surface. We decompose the motion of the DS into a frame of reference attached to the surface, by defining the modulating function as follows:

$$M(x,\dot{x}) = Q \Lambda Q^T \quad Q = \begin{bmatrix} e^1 & \cdots & e^d \end{bmatrix} \quad \Lambda = \begin{bmatrix} \lambda_{11}(.) & \cdots & \lambda_{1N}(x,\dot{x}) \\ \vdots & & \vdots \\ \lambda_{N1}(.) & \cdots & \lambda_{NN}(x,\dot{x}) \end{bmatrix}, \tag{8.19}$$

where $e^i \ \forall i \in \{1,\ldots,N\}$ forms an orthonormal basis in $\mathbb{R}^N$, and $\lambda_{ij}(x,\dot{x}) \ \forall i,j \in \{1,\ldots,N\}$ gives the entries of Λ, where i is the row number and j is the column number. The motion direction, tangential and normal to the surface, can be controlled through the scalar values $\lambda_{ij} \ \forall i,j \in \{1,\ldots,N\}$. For example, by setting $\lambda_{1j}(x,\dot{x}) = 0 \ \forall j \in \{1,\ldots,N\}$, the acceleration of the robot normal to the surface will be zero (i.e., $e^{1T}\ddot{x} = 0$). Moreover, by setting $\lambda_{ii}(x,\dot{x}) = 1$, $\lambda_{ij}(x,\dot{x}) = 0 \ \forall i,j \in \{1,\ldots,N\}$, $i \neq j$, the nominal DS drives the robot in the e^ith direction. We exploit this property and limit the influence of the modulation function to a region in a vicinity of the surface denoted as the *transition* region.

Given that we have at our disposal function $\Gamma(x)$ to measure the distance to the surface, we set the transition region to be all points such that $0 < \Gamma(x) \leq \rho$, $\rho \in \mathbb{R}_{>0}$. Outside this region, to avoid undesirable modulation, the modulation exponentially decreases as a function of the distance to the surface. To modulate locally the dynamics of the DS given by equations (8.18) and (8.19), we set

$$\lambda_{ij}(x,\dot{x}) = \begin{cases} \lambda_{ij}(x,\dot{x}) & \text{if } \Gamma(x) \leq \rho \\ \left(\lambda_{ij}(x,\dot{x}) - 1\right) e^{\frac{\rho - \Gamma(x)}{\sigma}} + 1 & \text{if } i = j \ \rho < \Gamma(x) \\ \lambda_{ij}(x,\dot{x}) e^{\frac{\rho - \Gamma(x)}{\sigma}} & \text{if } i \neq j \ \rho < \Gamma(x) \end{cases} \tag{8.20}$$

$\forall i,j \in \{1,\ldots,d\}$, where $0 < \sigma$ defines the speed at which the modulation vanishes in the free motion region, and, ρ defines the region of the influence of the modulation function. If $\rho < \Gamma(x)$, the robot is far from the contact surface and $\Lambda = I_{d \times d}$ (i.e., the robot is driven solely by the nominal DS).

To project the modulated DS onto the basis directions, λ_{ij} is defined as follows:

$$\lambda_{ij}(x,\dot{x}) = \left({}_{e^i}A_1(\lambda)e^i(x)^T x(t) + {}_{e^i}A_2(\lambda)e^i(x)^T \dot{x}(t) + {}_{e^i} u \right) \frac{f(x,\dot{x},t)^T e^j}{f(x,\dot{x},t)^T f(x,\dot{x},t)}, \tag{8.21}$$

where

$$\frac{|e^{1T}\dot{x}_0|}{e^{1T}x_0} \leq \omega. \tag{8.22}$$

By substituting equation (8.21) into equation (8.18) and assuming that $\Gamma(x) \leq \rho$, the modulated DS projected onto the basis directions becomes

$$e^1(x)^T \ddot{x}(t) = {}_{e^1}\mathbf{A}_1(\lambda) e^1(x)^T x(t) + {}_{e^1}\mathbf{A}_2(\lambda) e^1(x)^T \dot{x}(t) + {}_{e^1} u$$

$$\vdots \tag{8.23}$$

$$e^d(x)^T \ddot{x}(t) = {}_{e^d}\mathbf{A}_1(\lambda) e^d(x)^T x(t) + {}_{e^d}\mathbf{A}_2(\lambda) e^d(x)^T \dot{x}(t) + {}_{e^d} u,$$

where ${}_{e^i}u$, $\forall i \in \{1, \ldots, N\}$ are the control inputs tailored for each direction.[6] Here, ${}_{e^j}\mathbf{A}_i(.) : \mathbb{R}^K \to \mathbb{R} \; \forall i \in \{1, 2\}$ and $j \in \{1, \ldots, N\}$ are the affine dependencies of the state-space matrices on the activation parameter and the state vectors:[7]

$$\begin{aligned}
{}_{e^j}\mathbf{A}_1(\lambda) &= \sum_{k=1}^{{}_{e^j}K} {}_{e^j}\lambda_k \, {}_{e^j}A_{k1} \\
&\qquad\qquad\qquad \forall j \in \{1, \ldots, N\}, \qquad {}_{e^j}A_{k1}, \, {}_{e^j}A_{k2}, \, {}_{e^j}\lambda_k \in \mathbb{R}. \\
{}_{e^j}\mathbf{A}_2(\lambda) &= \sum_{k=1}^{{}_{e^j}K} {}_{e^j}\lambda_k \, {}_{e^j}A_{k2}
\end{aligned} \tag{8.24}$$

To make smooth contact with the surface and avoid bouncing—in other words, to satisfy equation (8.17)—we create a forward command u_{e^1}, aligned with the normal to the surface as follows:

$$_{e^1}u = -\dot{x}^T \nabla e^1(x)^T \dot{x} - {}_{e^1}\mathbf{A}_1(\lambda) e^1(x)^T x(t) + {}_{e^1}\mathbf{A}_1(\lambda) \Gamma(x). \tag{8.25}$$

By substituting equation (8.25) into equation (8.23), the motion generated at the normal direction will be

$$e^1(x)^T \ddot{x}(t) = -\dot{x}^T \nabla e^1(x)^T \dot{x} + {}_{e^1}\mathbf{A}_2(\lambda) e^1(x)^T \dot{x}(t) + {}_{e^1}\mathbf{A}_1(\lambda) \Gamma(x). \tag{8.26}$$

It can be shown (see exercise 8.6) that the motion generated by equation (8.26) converges to $\begin{bmatrix} 0 & 0 \end{bmatrix}^T$; that is,

$$\lim_{t \to \infty} \|\Gamma(x(t))\| = 0 \tag{8.27}$$

$$\lim_{t \to \infty} \|e^1(x)^T \dot{x}(t)\| = 0 \tag{8.28}$$

if the following constraints are satisfied:

$$\begin{cases}
\begin{bmatrix} 0 & I \\ _{e^1}A_{k2} & _{e^1}A_{k1} \end{bmatrix}^T P + P \begin{bmatrix} 0 & I \\ _{e^1}A_{k2} & _{e^1}A_{k1} \end{bmatrix} \prec 0 \\[2ex]
0 \prec P, \qquad\quad P^T = P \qquad\qquad\qquad \forall k \in \{1, \ldots, K\}, \\[2ex]
0 < {}_{e^1}\lambda_k \leq 1, \qquad \sum_{k=1}^{K} {}_{e^1}\lambda_k = 1
\end{cases} \tag{8.29}$$

Interestingly the constraint imposed in equation (8.29) is similar to the one introduced in chapter 3 when learning LPV-based DS. These open parameters could hence be learned/estimated for the contact task at hand.

8.4.2   Simulated Examples

To illustrate the concepts introduced before, we show a few examples, in figure 8.14, of 2D simulations where the DS makes contact with either a circle or an ellipsoid. As it can be seen in figure 8.14b, the system can reach the surface with zero velocity. In these simulations, we start with a linear nominal DS with no specific attractor. Hence, as we do not control for the contact location, the trajectories contact the surface at different points.

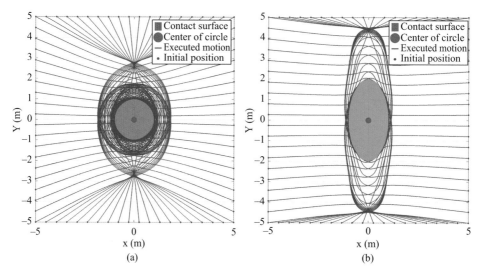

Figure 8.14
Reaching a convex shaped surface with equation (8.26). In (a), the surface is an ellipse, and in (b), it is a circle. As can be seen, the motion generated by equation (8.26) aligns with the curvature of the surface such that the normal velocity is zero at contact.

Programming Exercise 8.6   *The aim of this exercise is to help readers to get a better understanding of the DS [equation (8.26)] and how it shapes the motion of the robot with respect to the curvature of the surface and the effect of the open parameters on the generated motion. This exercise has two parts.*

Open MATLAB and set the directory to the following folder:

```
ı   ch8-DS_modulated/Surface_contact
```

The code contains two functions with which, one can study the effect of $\mathbf{A}_1$, $\mathbf{A}_2$ on the generated motion. Run the command line Surface_Circle([2;2],1) , all trajectories adapt to the surface of an ellipse. By studying the motion generated by equation (8.26) and modifying the parameters of the equation in the matlab code, answer the following questions:

1. Do the values of A_1, A_2, $\forall i \in \{1, 2\}$ have an effect on the motion of the robot? What would happen if $A_1 = 100A_1$, $A_2 = 20_{e^2}A_2$, $A_1 = 0.01A_1$, $A_2 = 0.2A_2$ or $A_1 = A_1$, $A_2 = A_2$?

2. Consider these two extreme cases:

- $A_1 = 0$
- $A_2 = 0$

Can you guess what the behavior of the robotic system would be? Verify your intuition with the code.

3. Does the radius or the location of the circle have an effect on the generated motion? Is there any radius that results in an unstable contact?

4. Is there any case in which the generated motions do not adapt to the ellipsoid or the circle? If yes, which are those, and why? If no, why?

8.4.3 Robotic Implementation

A real-world evaluation of this approach was done using a 7-DOF robotic arm (KUKA IIWA) to swipe various rigid surfaces. The robot is controlled in joints at a rate of 200 Hz. The output of the DS [equation (8.26)] is converted into joint state using the

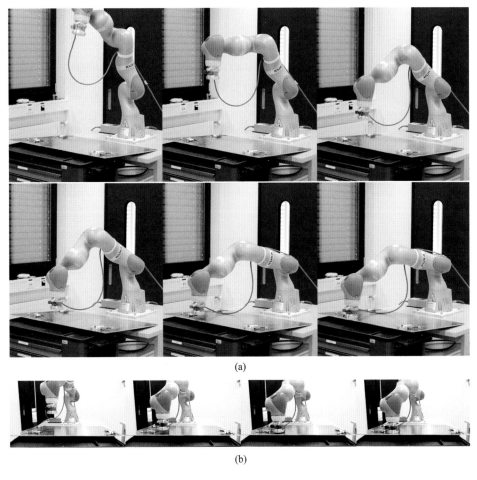

(a)

(b)

Figure 8.15
In (a) and (b), snapshots of the motion of the robot while reaching a rigid surface. In (b), the close-up snapshots depict the end-effector motion. Both the surface and the tool are metallic and rigid.

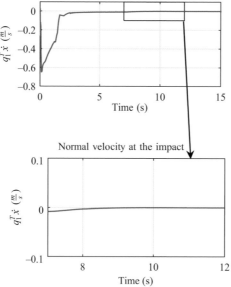

Figure 8.16
The location of the surface is precisely measured. In the bottom figures, the normal velocity of the robot at the impact region is illustrated.

damped least-squares inverse kinematic solver. The contact surface is approximated by a plan (i.e., $\nabla e^1(x) = 0$).

$$\mathbf{A}_2 = -40 I_{3 \times 2}, \; \mathbf{A}_1 = -\begin{bmatrix} 400 & 0 & 0 \\ 0 & 400 & 0 \\ 4000 & 4000 & 400 \end{bmatrix}.$$

Two experimental setups were used to showcase the performance of the system. In the first, both surface and tool are metallic and rigid. In the second, the surface is a metallic fender and the tool is made of plastic. The snapshots of the motion execution in both experimental setups are shown in figures 8.15 and 8.16. A visual inspection of video and the measurements confirms that the robot stably makes contact with the surface. A detailed experimental validation is presented in [100].

Exercise 8.5 *Let's consider a scenario where the contact surface is a planar surface defined by $\Gamma = e^{1T} x$. By using equation (8.23) and $\mathsf{d} = 2$, a robotic arm can reach the surface with zero velocity. In order to control the contact position, one can define $_{e^2} u = -_{e^2} \mathbf{A}_1 (\lambda) e^{2T} x^*$, where $x^* \in \mathbb{R}^2$ is the desired contact point. By this, one can control the motion of the robot such that it stably makes contact with the surface at the desired location. However, controlling the contact position would be possible only for flat/planar surfaces. What would happen if the surface is not planar and it is, for example, an ellipse?*

Exercise 8.6 *Considering the DS presented in equation (8.23), prove that it asymptotically converges $\begin{bmatrix} 0 & 0 \end{bmatrix}^T$ if the conditions in equation (8.29) are met.*

9 Obstacle Avoidance

In the previous chapters, we always have assumed that the control law was valid throughout the entire state space. This is not the case, however, when there are obstacles along the path or when the workspace is limited, as is common when controlling the movement of a robotic arm. While the workspace of a mobile robot may be limited due to the presence of external obstacles, the workspace of the robotic arm is constrained by joints' limits, which create dead-locks. For instance, a humanoid robot is limited in its movement because each of its body parts creates an obstacle for the other parts (e.g., the arms may collide with one another and with the legs when the robot bends down to pick up an object).

In this chapter, we show that, in order to address this problem, one can locally modulate the dynamical system (DS) to contour obstacles or to remain within a given workspace. Importantly, while doing so, we can preserve some of the inherent properties of the DS, such as convergence on a given target.

To recall, in chapter 8, we saw how one can modulate an initial DS $f(x)$ through a state-dependent function $M(x)$. The new system becomes $g(x) = M(x)f(x)$. When $M(x)$ is a full-rank matrix and is nonzero at the target, $g(x)$ inherits the asymptotic stability of the initial DS. In this chapter, we exploit this property and construct modulations that prevent a DS to penetrate boundaries delineating regions that cannot be traveled through. We start with a modulation that allows for avoiding convex obstacles, and then extend it to concave obstacles and to multiple obstacles in movement. We show that the formulation can also be used to enforce a flow to move inside a volume, which would be useful to ensure that the path stays within the robot's workspace (to cite just one instance).

As the method assumes that the shape of the obstacle is known, we discuss how, in practice, one can estimate the obstacles' shape at run time from a point-cloud rendered by proximity sensors (e.g., Lidar, RGB-D cameras), and how the obstacle's shape can be learned through scanning of the object (as discussed in section 9.1.9).

Obstacle avoidance is particularly useful for controling robots in joint space to avoid self-collision. The boundary of free space in joint space is not fixed, however; it evolves with joint motion. It is also very nonlinear. This is a problem ideally suited for machine learning. In section 9.2, we show how such a moving boundary can be learned by knowing the kinematics of the robotic systems. The learned boundary can then be used at run time in conjunction with DSs to prevent two robot arms from intersecting with each other, while they rapidly modify their trajectories to track moving targets.[1]

9.1   Obstacle Avoidance: Formalism

Following the notation used in this book, let $x \in \mathbb{R}^N$ be the state of the robotic system, and let us assume that the controller follows a nominal linear DS. We further assume that the system is asymptotically and globally stable at a single attractor x^*:

$$f(x) = -(x - x^*). \tag{9.1}$$

9.1.1   Obstacle Description

We assume that the obstacle is known. Moreover, we assume that we have an explicit description of the obstacle's shape, and that it is given by a closed-form functional $\Gamma(x)$: $\mathbb{R}^N \setminus \mathcal{X}^i \mapsto \mathbb{R}_{\geq 1}$, with $\Gamma(x) = 1$ the isoline surrounding the obstacle.

Each isoline of Γ conveys a measure of distance to the obstacle. $\Gamma(x)$ is continuous and continuously differentiable (C^1 smoothness). Hence, the gradient at each point in space gives us the normal and tangent to the obstacle.

We distinguish three regions:

Exterior points: $\mathcal{X}^e = \{x \in \mathbb{R}^N : \Gamma(x) > 1\}$

Boundary points: $\mathcal{X}^b = \{x \in \mathbb{R}^N : \Gamma(x) = 1\}$ (9.2)

Interior points: $\mathcal{X}^i = \{x \in \mathbb{R}^N \setminus (\mathcal{X}^e \cup \mathcal{X}^b)\}$.

By construction, $\Gamma(\cdot)$ increases monotonically while moving away from the obstacle. An illustration is given in figure 9.1.

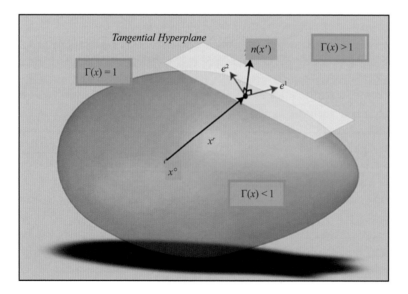

Figure 9.1
A C^1 function $\Gamma(x)$ determines the outer shape of the obstacle and a measure of the distance to the obstacle. Its isoline, which contours the obstacle, has value 1. The isolines have positive and increasing values away from the obstacle, hence providing a measure of the distance to the obstacle. All isolines inside the obstacles have values strictly inferior to 1. The gradient of $\Gamma(x)$ can be used to determine at each point x^r the normal $n(x^r)$ and tangential plane to the obstacle (formed by vectors e_1, e_2 here).

9.1.2 Modulation for Obstacle Avoidance

We follow the approach described in chapter 8 and apply a modulation to our nominal DS $f(x)$ in order to take into account the obstacle. We construct the modulating matrix as follows:

$$\dot{x} = \mathbf{M}(x)f(x) \qquad \text{with} \quad \mathbf{M}(x) = \mathbf{E}(x)\mathbf{D}(x)\mathbf{E}(x)^{-1}. \tag{9.3}$$

The modulation matrix $\mathbf{M}(\cdot)$ is constructed through eigenvalue decomposition, using as the basis of the eigenvectors the normal $n(x)$ and tangents to the obstacle, as follows:

$$\mathbf{E}(x) = [\mathbf{n}(x)\ \mathbf{e}_1(x)\ ..\ \mathbf{e}_{d-1}(x)]. \tag{9.4}$$

The tangents $\mathbf{e}_{(\cdot)}(x)$ form a $d-1$–dimensional orthonormal basis to the gradient of the distance function $d\Gamma(x)/dx$.

We can use this basis to express the modulation in the frame of reference of the obstacle, which greatly simplifies the computation. The matrix of eigenvalues can then be shaped such that the flow is canceled along the normal direction once it reaches the boundary of the obstacle:

$$\mathbf{D}(x) = \mathbf{diag}\,(\lambda_n(x), \lambda_e(x), \dots, \lambda_e(x)), \tag{9.5}$$

where the eigenvalues $\lambda_{(\cdot)}(x)$ determine the amount of stretching in each direction.

We set that the eigenvalue associated with the first eigenvector decreases to become zero on the obstacle's hull. This cancels the flow in the direction of the obstacle and ensures that the robot does not penetrate the obstacle's surface. The eigenvalues along the tangent direction increase the speed in the other directions:

$$0 \leq \lambda_n(x) \leq 1 \qquad \lambda_e(x) \geq 1 \tag{9.6}$$

$$\lambda_r(x|\Gamma(x) = 1) = 0 \quad \underset{\Gamma(x)}{\arg\max}\ \lambda_e(x) = 1 \quad \lim_{\Gamma(x) \to \infty} \lambda_{(\cdot)}(x) = 1.$$

By construction, the isolines $\Gamma(x)$ [see equation (9.2)] give a measure of the distance to the obstacle's surface. We can, hence, set the following:

$$\lambda_n(x) = 1 - \frac{1}{\Gamma(x)} \qquad \lambda_e(x) = 1 + \frac{1}{\Gamma(x)}. \tag{9.7}$$

9.1.3 Stability Properties for Convex Obstacles

Theorem 9.1 *Consider an obstacle in $\mathbb{R}^N$ with boundary $\Gamma(x) = 1$ with respect to the center of the obstacle $x^o \in \mathscr{X}^i$, as given in equation (9.2). Any trajectory $x(t)$ that starts outside the obstacle (i.e., $\Gamma(x(0)) \geq 1$) and evolves according to equation (9.3) will never penetrate the obstacle [i.e., $\Gamma(x(t)) \geq 1, t = 0 \dots \infty$].*

Proof see appendix D.   □

Interpretation: When we combine such a modulation with a linear DS set with a single attractor, this leads the system to move around the obstacle toward the attractor. An example of such a modulation around a circular obstacle is given in figure 9.2.

Observe that in equation (9.7), the first eigenvalue decreases rapidly to become zero at the obstacle's boundary, while the eigenvalues along the tangential directions grow inversely

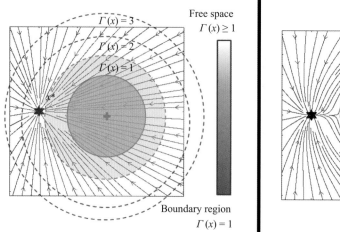
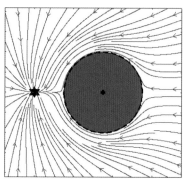

Figure 9.2
Left: A linear DS asymptotically stable at a single attractor is modulated through a circular obstacle. The value of the Γ function are overlaid. *Right:* The resulting modulated DS does not penetrate the obstacle while still asymptotically stable at the target.

proportionally to the decrease of the first eigenvalue. This leads the DS to accelerate along these tangential directions. This effect may be undesirable for some applications. For instance, when the robot avoids a human or a fragile object, one would prefer the robot to slow down and decrease the overall norm of its velocity, rather than accelerate. Conversely, in some situations, accelerating may be desirable. For instance, if the robot must avoid a dangerous obstacle, speeding up to move away as soon as possible would be a suitable action.

To generate different responses depending on the application is possible by modulating the rate of change of the eigenvalues. For instance, one could wish to preserve the overall magnitude of the velocity and set that the sum of the eigenvalues remains constant. Similarly, one can also ensure that the magnitude of the velocity is preserved in certain directions (see exercise 9.1).

Programming Exercise 9.1   *The aim of this exercise is to help readers to get better acquainted with the obstacle avoidance algorithm and the effect of the open parameters on obstacle modulation. Open MATLAB and set the directory to the following folder:*

```
1        ch9-obstacle_avoidance
```

Open file called "ch9_ex1". This file allows you to change the parameters of the modulation function and to generate a local modulation. Do the following:

1. Set the eigenvalues such that the norm of the velocity is preserved when going around the obstacle. Repeat to let the robot slow down or accelerate when moving around obstacles.

2. Change the shape of the obstacle to create an ellipse and set the reference point to ensure that the flow is not stuck on the obstacle's boundary.

3. Set a linear DS to make the obstacle move and observe the effect on the modulation over time. For this step, you must change the original code and generate the flow over time of the DS.

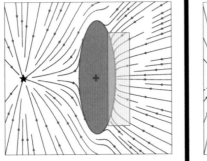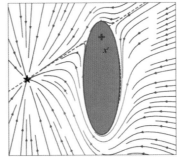

Figure 9.3
Left: A linear DS, asymptotically stable at a single attractor, is modulated through an oval obstacle. one spurious attractor exists in the middle of the obstacle as the flow of the nominal DS vanishes and there is no tangential component. A large number of trajectories lead to this spurious attractor, as highlighted in the orange region. *Right:* Introducing a reference point within the obstacle that is off center breaks the symmetry and allows the flow to rotate around the obstacle. There is still one spurious attractor, but only one single trajectory leads to it (illustrated by the dashed line).

Limitations: The modulation, given by equation (9.4), ensures that the flow will never penetrate the obstacle. However, it does not prevent the appearance of spurious attractors. It is easy to see this. Consider a flow that is moving directly perpendicular to the obstacle; the flow will vanish entirely because there is no tangential component. While this may seem to be a minor issue, in practice numerous trajectories can lead to this spurious attractor. An example of such an issue is shown in figure 9.3. Another issue is that this modulation requires the obstacle to be convex. Next, we show how a simple change to this modulation can avoid the problem illustrated in figure 9.3 and avoid some concave obstacles.

9.1.4 Modulation for Concave Obstacles

Recall that to ensure that the modulation preserves the stability of the nominal DS at the attractor, it must be full rank and should not apply to the attractor. Full rank can be achieved if matrix $E(x)$ is invertible. The vectors composing the columns of E do not need to be orthogonal. It is sufficient that they be linearly independent. We, hence, relax the orthogonality constraint and define $r(x)$ as a *reference direction* in place of the normal vector $n(x)$. This change of variable allows us to drag the flow in specific directions when coming close to the obstacle. This avoids the problem illustrated in figure 9.3, where the flow was rotating in the direction of the spurious attractor and stopping there. It also allows us to get out of concavities when avoiding concave obstacles.

The new matrix E becomes

$$E(x) = [\mathbf{r}(x) \; \mathbf{e}_1(x) \; .. \; \mathbf{e}_{d-1}(x)], \quad \mathbf{r}(x) = \frac{x - x^r}{\|x - x^r\|}, \tag{9.8}$$

and the tangents $\mathbf{e}_{(\cdot)}(x)$ form a $d - 1$–dimensional orthonormal basis to the gradient of the distance function $d\Gamma(x)/dx$.

The eigenvalues must remain the same to generate the same effect—namely, vanishing flow at the boundary and transfer of motion along the tangent directions. We, hence, set the following:

$$\lambda_r(x) = 1 - \frac{1}{\Gamma(x)} \qquad \lambda_i(x) = 1 + \frac{1}{\Gamma(x)}, i = 1, ..., d - 1, \tag{9.9}$$

where $\lambda_r(x)$ is the eigenvalue associated with the first column of E.

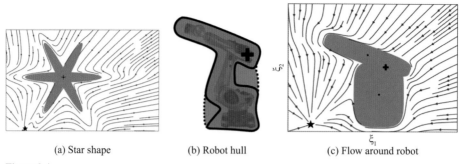

(a) Star shape   (b) Robot hull   (c) Flow around robot

Figure 9.4
A *star-shaped* obstacle is such that there exists at least one central point, such that all rays originating from this point cross the object's boundary only once. (a) An ideal star-shaped obstacle. (b) An obstacle that is not star-shaped, such as a robot arm. To make it star-shaped, one can expand the convex hull (yellow areas). (c) A modulated DS around the robot's obstacle. The robot is represented with two separate convex elliptic obstacles with a common reference point (illustrated with a cross in the drawing.)

The reference direction $\mathbf{r}(x)$ is a unitary vector pointing from a reference direction x^r inside the obstacle to the current position x. The position of reference point x^r is constrained by the fact that the basis $\mathbf{E}(x)$ needs to be invertible, and hence that reference direction $\mathbf{r}(x)$ must be linearly independent of the tangent vectors $\mathbf{e}_{(\cdot)}(x)$. As the reference direction changes, this must be true for any position x. This condition is ensured for a reference point within a convex obstacle $x^r \in \mathscr{X}^i$. A good choice for the reference point is often the geometric center x^c, but any point within the obstacle is valid.

9.1.4.1 Shape of concave obstacles

The shape of the obstacle is also constrained by the condition of $\mathbf{E}(x)$ having full rank from equation (9.3). Convex and concave obstacles for which there exist a reference point x^r inside the obstacle, from which all rays cross the boundary only once (figure 9.4a) fulfill this condition. These obstacles are referred to as *star-shaped*. For such an obstacle, one can construct $\Gamma(x)$ as follows. If x^c is the center of the obstacle, the distance function is given by

$$\Gamma(x) = \sum_{i=1}^{d} \left(\|x_i - x_i^c\|/R(x) \right)^{2p} \quad \text{with} \quad p \in \mathbb{N}_+, \tag{9.10}$$

with $R(x)$ the distance from a reference point x^r within the obstacle to the surface of the obstacle ($\Gamma(x) = 1$) in direction $r(x)$ (figure 9.4a).

Considering star-shaped obstacles is not as limiting as one may think, as many objects which we handle in daily life are *star-shaped*, such as bottles, sofa, closet with door open. When faced with nonstar-shaped concave obstacles, one solution is to complement the hull, as illustrated in figure 9.4b. One solution is to pick a reference point x^r and then build convex shapes that include the reference point and surround each robot part (see figure 9.4b).

Note, finally, that a combination of convex obstacles can generate concave obstacles (as discussed in section 9.1.7.1). The method can then still be applied, so long as this combination shares a common nonempty set of points.

9.1.5 Impenetrability and Convergence

The modulated system with a single obstacle is certain to never penetrate the obstacle. The system is also sure to retain convergence to the attractor for all trajectories except one. Indeed, as discussed previously, with a linear nominal DS, there is one trajectory $\mathscr{X}^s$ that leads to a saddle point (the black dashed line in figure 9.3). The saddle point is located on the obstacle's boundary, which arises when the trajectory $\mathscr{X}^s$ points in the direction of the reference point. At the intersection with the obstacle, the modulated dynamics vanishes as both tangent and normal velocity direction have zero magnitude. This is a saddle point, as any small deviation in either direction will lead the flow to converge back on the attractor x^*. The trajectory leading to the saddle point is such that

$$\mathscr{X}^s \subset \mathbb{R}^N = \{x \in \mathbb{R}^N : \mathbf{f}(x) \parallel \mathbf{r}(x), \|x - x^*\| > \|x - x^r\|\}. \tag{9.11}$$

Theorem 9.2 *Consider a time invariant, linear DS $\mathbf{f}(x)$, with a single attractor at x^*, as in equation (9.1). The DS is modulated according to equations (9.3) and (9.8) around an obstacle with boundary $\Gamma(x) = 1$. Any motion $x(t)$, $t = 0 \ldots \infty$ that starts outside the obstacle and does not lie on the saddle point trajectory [i.e., $\{x(0) \in \mathbb{R}^N \setminus \mathscr{X}^s : \Gamma(x(0)) > 1\}$] never penetrates the obstacle and converges on the attractor [i.e., $x(t) \to x^*$, $t \to \infty$].*

Proof see section D.4 in a appendix D. □

> **Exercise 9.1** *The modulation generated by the eigenvalues results in a change of magnitude along the various basis directions, as described earlier. The increase in velocity along the tangent direction is bounded.*
>
> - *What is its upper bound?*
> - *Where does it occur?*

9.1.6 Enclosing the DS in a Workspace

A simple inversion of the distance function allows one to enclose the DS inside an obstacle (illustrated in figure 9.5). A linear DS with a fixed point attractor is forced to remain within the bounded volume. Convergence to the attractor is preserved.

Using an enclosing space can be useful in many applications. It can force a robot to stay within a region of the space. For instance, a robot welcoming customers in a shopping mall could be tasked to remain always within the main hall by delineating the boundaries virtually. It can also delimit the robot's free joint space. This is particularly useful when controlling articulated robot arms that may intersect with one another, as presented in section 9.2. However, the approach presented here is limited to enclosing volumes that can be represented as star-shaped obstacles. When controlling in very high dimensions and joint space, this can no longer be guaranteed. In section 9.2, we present an extension that allows one to learn such complex boundaries and to move inside the volume by using a DS as a reference trajectory but solving the constraints due to the boundaries through a quadratic program at run time.

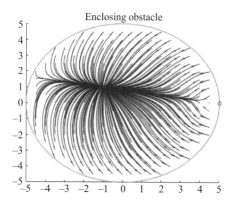

Figure 9.5
An inversion of the distance function enables the flow here to be enclosed within a circle of diameter 5.

Programming Exercise 9.2   *Open MATLAB and set the directory to the following folder:*

```
1              ch9-Obstacle_avoidance
```

Open file called "ch9_ex2.m" This file generates a circular workspace in which is embedded a nominal linear DS, as illustrated in figure 9.5.

1. Modify the shape of the workspace to approximate the workspace of a two-joint end-effector, assuming that the first joint is constrained to move between ±90 degrees and the second joint between 0 and 45 degrees.

2. Change the modulation to ensure that near the boundary, the direction of movement accelerates and rotates clockwise.

9.1.7 Multiple Obstacles

In the presence of multiple obstacles, the nominal DS is modified by taking the weighted mean of the modulated DS $\dot{x}^o$ created by each obstacle $o = 1 \ldots N^o$ separately for magnitude $\|\dot{x}^o\|$ and direction $\mathbf{n}^{\dot{x}^o}$ (figure 9.6).[2] The modulated DS for each obstacle o is given as $\dot{x}^o = \|\dot{x}\|^o \mathbf{n}^{\dot{x}^o}$. To balance the effect of each obstacle and to ensure that, on the boundary of

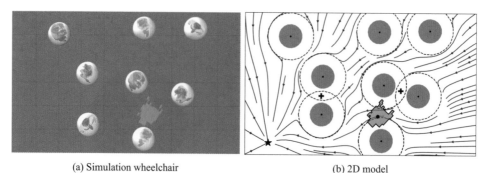

(a) Simulation wheelchair   (b) 2D model

Figure 9.6
The wheelchair (orange) tries to avoid a human crowd represented by circular obstacles (a). To account for the wheelchair's geometry, a margin around the obstacle (dashed line) is added (b).

each obstacle, the influence of the other obstacles vanish, we enforce that :

$$\sum_{k=1}^{N^o} w^o(x) = 1 \quad \text{and} \quad w^o(x \in \mathscr{X}^{b,\hat{o}}) = \begin{cases} 1 & o = \hat{o} \\ 0 & o \neq \hat{o}, \end{cases} \tag{9.12}$$

where N^o is the number of obstacles and $\hat{o}$ denotes the boundary of obstacle o.

It is recommended to set the weight w^o to be inversely proportional to the distance measure $\Gamma^o(x) - 1$ (notice that each obstacle can have its own distance measure Γ^o):

$$w^o(x) = \frac{\prod_{i \neq o}^{N^o} (\Gamma^i(x) - 1)}{\sum_{k=1}^{N^o} \prod_{i \neq k}^{N^o} (\Gamma^i(x) - 1)} \quad o = 1..N^o. \tag{9.13}$$

The magnitude is evaluated by the weighted mean as follows:

$$\|\dot{\bar{x}}\| = \sum_{o=1}^{N^o} w^o \|\dot{x}^o\|. \tag{9.14}$$

We compute the deflection from the original DS with respect to the unitary vector $\mathbf{n}^f(x)$, aligned with the original DS. We define the function $\kappa(\cdot) \in \mathbb{R}^{N-1}$ that projects the modulated DS from each obstacle onto a $(d-1)$–dimensional hypersphere with radius π.[3]

$$\kappa(\dot{x}^o, x) = \arccos\left(\mathbf{n}_1^{\dot{x}^o}\right) \frac{\left[\hat{\mathbf{n}}_2^{\dot{x}^o} \; .. \; \hat{\mathbf{n}}_d^{\dot{x}^o}\right]^T}{\sum_{i=2}^{d} \hat{\mathbf{n}}_i^{\dot{x}^o}}, \quad \hat{\mathbf{n}}^{\dot{x}^o} = \mathbf{R}_f^T \mathbf{n}^{\dot{x}^o}.$$

The orthonormal matrix $\mathbf{R}_f(x)$ is chosen such that the initial DS $\mathbf{f}(x)$ is aligned with the first axis $[x_1 \; 0]^T = \mathbf{R}_f(x)^T \mathbf{f}(x)$, with $\mathbf{R}_f(x) = \left[\mathbf{n}^f(x) \; \mathbf{e}_1^f(x) \; \ldots \; \mathbf{e}_{d-1}^f(x)\right]$. The vectors $\mathbf{e}_{(\cdot)}^f(x)$ are chosen so as to form an orthonormal basis.

The weighted mean is evaluated in this κ-space:

$$\bar{\kappa}(x) = \sum_{o=1}^{N^o} w^o(x) \kappa^o(\dot{x}, x). \tag{9.15}$$

The direction vector of the modulated DS $\dot{\bar{x}}$ is then expressed back in the original space:

$$\bar{\mathbf{n}}(x) = \mathbf{R}_f(x) \left[\cos \|\bar{\kappa}(x)\| \quad \frac{\sin \|\bar{\kappa}(x)\|}{\|\bar{\kappa}(x)\|} \bar{\kappa}(x)\right]^T. \tag{9.16}$$

With equation (9.14), the final velocity is evaluated as

$$\dot{\bar{x}} = \bar{\mathbf{n}}(x) \|\dot{\bar{x}}\|. \tag{9.17}$$

9.1.7.1 Intersecting obstacles

In the case of intersecting obstacles, the algorithm given previously is applicable, too, but convergence occurs only if there exists one common reference point $x^{r,o}$ among all the obstacles. Hence, there must exist a common region. This is always the case for two intersecting obstacles. For more intersecting obstacles, it happens in special cases (figure 9.7). The convex obstacles must form a star shape as a group (see the concave obstacle discussed in section 9.1.4.1 earlier in this chapter). Furthermore, if several obstacles share one reference point x^r, there exists only one common, saddle point trajectory.

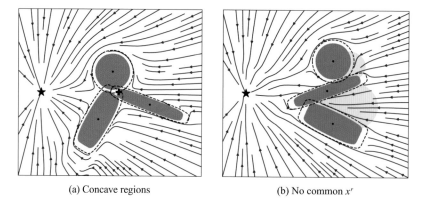

(a) Concave regions   (b) No common x^r

Figure 9.7
Several intersecting convex obstacles can be avoided if all the obstacles have a common region pairwise. (a).
No global convergence can be observed otherwise, and convergence to a local minimum (yellow) occurs (b). A
star-shaped hull can be created to exit those concave regions, as in figure 9.4.

> **Exercise 9.2**   *Prove some of the steps necessary to guarantee the impenetrability and convergence of a linear DS system with obstacle modulations, as described in equations 9.12–9.17.*
>
> *a) Impenetrability: Prove that the multiobstacle avoidance system will never penetrate any of the obstacles.*
>
> *b) Convergence: Prove that no other stationary point is created than the one generated by the saddle point trajectory.*

9.1.8   Avoiding Moving Obstacles

While many obstacles in real life are static (e.g., walls, doors, furniture), many others are not; indeed, some of them may even move fairly rapidly. For instance, if the robot is a robotic arm tasked to pick up objects from a conveyor belt next to a human coworker, it is to avoid humans at all times as both reach for objects on the same conveyor belt. It may also have to avoid other robots working on the same conveyor belt. If the robot is an autonomous car, it must avoid other vehicles on the road.

When avoiding a moving obstacle, the key is to be able to estimate its velocity so as to determine where it moves. Assuming that one can estimate the translational and rotational velocity of the moving object, one can add this to our obstacle modulation to take into account the direction of movement of the obstacle and move likewise. To achieve this, we use a trick introduced already several times throughout this book—namely, to change the origin of the system so as to simplify computation. Here, we place it on the obstacle's center x^c through a change of variable $\tilde{x} = x - x^c$. The first derivative of this new variable can then be computed by taking into account the translational and rotational velocities of the obstacle, $\dot{x}^0, \omega^0$, respectively:

$$\dot{\tilde{x}} = \dot{x}^o + \dot{\omega}^o \times \tilde{x}. \tag{9.18}$$

The nominal DS $f(x)$ is then modulated as follows:

$$\dot{x} = M(x)(f(x) - \dot{x}^o) + \dot{x}^o. \tag{9.19}$$

Exercise 9.3 *The modulation generated by equation (9.19) has two unfortunate side effects. See if you can find a way to solve the following problems:*

1. The normal component of the DS on the surface of the moving obstacle is equal to the obstacle's velocity in this direction. As a result, a robot controlled by such a DS would be pulled along with a passing obstacle, even though the robot may be static and the robot is not moving toward the robot. How could you cancel out this side effect to make sure that an obstacle passing the robot tangentially does not affect the robot's motion?

2. Because the velocity of the obstacle is additive in equation (9.19), the attractor may be moved. What do you need to change to ensure that the attractor remains untouched? (This, of course, comes at the cost of being hit by an obstacle at that point!)

3. Return to the code used in programming exercise 9.11. Modify the modulation to allow the obstacle to move with a linear motion sliding next to the robot-point-mass and moving toward the attractor. Observe the induced movement of the robot and that of the attractor. Apply the solutions you proposed described in questions 1 and 2 above.

9.1.9   Learning the Obstacle's Shape

The proposed obstacle avoidance algorithm requires a user to provide an analytical formulation of the outer surface of the obstacles. If the space is not crowded and you can easily avoid obstacles without moving close to them, then the best tactic is to approximate them with convex shapes such as ellipsoids. If space is a concern, then having a tighter envelope is preferable. One will have to make sure, however, that the obstacle or group of obstacles remains star-shaped.

The algorithm requires one to have an explicit function that surrounds the object and gives a measure of distance. Such an envelope can be learned using Support vector regression (SVR) (readers unfamiliar with SVR should refer to section B.4 of appendix). Readers can test this approach using the code offered in 9.3.

Programming Exercise 9.3 *Open MATLAB and set the directory to the following folder:*

```
1        ch9-Obstacle_avoidance/construct_obstacles
```

This folder allows you to hand-draw an object's surface and fit it with SVR to generate the Γ function. You can also load a point-cloud gathered from a real surface, and generate a DS that slides along the surface.

When provided with the three-dimensional (3D) model of the object, one may compute a smooth convex envelope (also known as a *convex bounding volume*) that fits tightly around the object. This bounding volume (BV) can be used (instead of the object's shape) to perform obstacle avoidance. When solely the point cloud description of the object is available, one may use one of the estimation techniques to approximate the BV. For example, in [11], the BV is approximated using a set of spheres and tori. To use this method, one first needs to find the relevant patch (either sphere or torus) of the BV that corresponds to the current position of the robot. Then, based on the analytical formulation of that patch, one

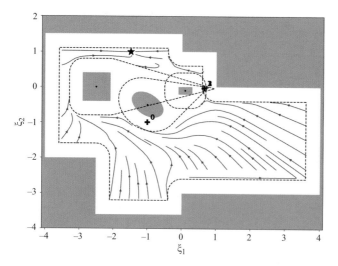

Figure 9.8
A known office environment is described by a collection of convex objects with a common boundary to create star-shaped obstacles. We overlay the dynamics of the robot in this environment.

can compute the dynamic modulation matrix as described before. Recall that our obstacle avoidance module only requires C^1 smoothness of the.

When doing obstacle avoidance in a dynamic environment, it is hardly possible to generate the BVs from the output of the vision system in real time. Thus, it is necessary to generate a library that stores the analytical formulations of different objects. At run time, this library can be used in conjunction with an object's recognition module to map the shape onto the object. When the objects are concave, one option is to store a library of analytical convex envelopes. One can then use this analytical descriptor of the envelopes both to detect the object and to construct an obstacle avoidance module for concave objects by concatenating convex shapes with a common intersection to make star-shaped concave envelopes. Figure 9.8 illustrates how a set of obstacles can be fit with a series of convex objects incrementally.

In the presence of fast, unknown moving obstacles, the object recognition phase may not provide the agility required to avoid the obstacle (especially when there is a large library of objects). In these situations, it might be more adequate to replace the object recognition phase with an automatic BV generator algorithm. Generating a simple BV (e.g., an ellipsoid) around the point cloud of an obstacle can be done very quickly. If the object moves very rapidly, it is recommended to set a safety margin, such as by adding an offset to the isoline such that the isoline of value 1 at the boundary of the obstacle corresponds to a surface around the obstacle within a safe distance.

Furthermore, when there are many obstacles in the working space of the robot, it may not be necessary (and also computationally feasible) to track all the obstacles all the time. Because the modulation decreases as the distance to the obstacle increases, one could ignore all obstacles for which the associated modulation matrices are close to identity. By taking into account the obstacles that are locally relevant, the processing time for the vision systems could decrease significantly. However, this will be at the cost of imposing a small discontinuity in the robot velocity when an obstacle is added or removed from the set of relevant obstacles. By setting a small threshold, this discontinuity practically becomes negligible.

9.2 Self-Collision, Joint-Level Obstacle Avoidance

Self-collision avoidance is one of the main challenges in multiarm manipulation. In this section, we present one solution for self-collision avoidance in a bimanual, two-arm robotic system. It is important to note that the approach is very fast, running at a rate of < 2 ms. Speed is indeed of the essence, as self-collision may arise suddenly and needs to be avoided immediately to prevent damage.

In the previous section, we introduced a reactive obstacle avoidance algorithm that relied on computing minimum distances between the robot's state and the boundary to be avoided. Moreover, this boundary was represented as sphere/swept-spheres/polygons. While there were no limitations in the dimension of the problem and the method presented in the first part of this chapter can be extended to joint space, this is possible only if the boundary to feasible joint space can be represented by simple polygons. This is the case for a single robot arm, but it is no longer the case when we want to model self-collision of multiarm/joint system. In this case, the boundary is complex and moves as the joint moves. A reactive method as presented earlier would no longer be guaranteed to be free of local minima.

9.2.1 Combining Inverse Kinematic and Self-Collision Constraints

To avoid collisions between the joints of the arms in a two-robotic-arm system, we use an Inverse Kinematic (IK) solver, which considers the kinematic constraints of each robot, but also self-collision constraints. Assuming that the robots' bases are fixed with respect to one another, we explore the joint workspace of the robots by having them move randomly in space. This allows us to get a sense of the regions in joint space that may lead to collisions and those that do not.

To determine how the boundary between feasible (safe) and infeasible (collided) configurations changes depending on the joint configuration, we must construct a continuous map from the joint configuration to the boundary of our binary classification problem. We follow a similar notation as introduced at the beginning of the chapter and assume that we can bound the infeasible joint space region through a continuous and continuously differentiable function $\Gamma(q^{ij}) : \mathbb{R}^{d_{q_i}+d_{q_j}} \to \mathbb{R}$, where $q^{ij} = [q^i, q^j]^T \in \mathbb{R}^{d_{q_i}+d_{q_j}}$ are the joint angles of the ith and jth robots, respectively. We define $\Gamma(.)$ such that

Collided configurations: $\Gamma(q^{ij}) < 1$

Boundary configurations: $\Gamma(q^{ij}) = 1$ (9.20)

Free configurations: $\Gamma(q^{ij}) > 1.$

Equation (9.20) provides constraints that must be met when solving the IK. To solve this, we propose the following quadratic program:

$$\underbrace{\arg\min_{\dot{\mathbf{q}}} \frac{\dot{\mathbf{q}}^T W \dot{\mathbf{q}}}{2}}_{\text{Minimize expenditure}} \qquad\qquad (9.21a)$$

subject to:

$$\underbrace{\mathbf{J}(\mathbf{q})\dot{\mathbf{q}} = \dot{\mathbf{x}}}$$ (9.21b)

Satisfy the desired end-effector motion

$$\underbrace{\dot{\boldsymbol{\theta}}^{-} \leq \dot{\mathbf{q}} \leq \dot{\boldsymbol{\theta}}^{+}}$$ (9.21c)

Satisfy the kinematic constraints

$$-\nabla \Gamma^{ij}(q^{ij})^{T} \dot{q}^{ij} \leq \log(\Gamma^{ij}(q^{ij}) - 1)$$

$$\underbrace{\forall (i,j) \in \{(1,2),(1,3),\ldots,(N_R - 1, N_R)\},}$$ (9.21d)

Do not penetrate the collision boundary

where, $\mathbf{q} = [q^1, \ldots, q^{N_R}]^T \in \mathbb{R}^{D_\mathbf{q}}, D_\mathbf{q} = \sum_{i=1}^{N_R} D_{q_i}$ for N_R robots;[4] W is a block diagonal matrix of the positive definite matrices; $\mathbf{J} = diag(J_1, \ldots, J_{N_R})$ is a block diagonal matrix of the Jacobian matrices; $\dot{\mathbf{x}} = \begin{bmatrix} \dot{x} & \ldots & \dot{x}_{N_R} \end{bmatrix}^T \in \mathbb{R}^{N_\mathbf{n}}$, $d_\mathbf{n} = N_R N$ is the desired velocity given by a task-space motion generator; and $\dot{\boldsymbol{\theta}}^i = \begin{bmatrix} \dot{\theta}^i_1 & \ldots & \dot{\theta}^i_{N_R} \end{bmatrix}$ $\forall i \in \{-, +\}$ and $\dot{\theta}^+_i \in \mathbb{R}^D$ and $\dot{\theta}^-_i \in \mathbb{R}^D$ are conservative lower and upper bounds of the joint limits.

Figure 9.9 illustrates the shape of our boundary, $\Gamma(.)$ in a 2D toy in order to highlight its role in the quadratic program. While the robots are in a configuration far from the boundary, $\log(\Gamma^{ij}(q^{ij}) - 1) > 0$, which relaxes the inequality constraints, and the robots accurately follow the desired end-effector's trajectory. When the robots' configuration approaches the boundary, $\log(\Gamma^{ij}(q^{ij}) - 1)$ becomes negative, hence activating the constraint [equation (9.21d)]. This forces the joint angles to move away from the boundary.

Satisfying constraints to avoid collision and to acchieve kinematic feasibility have higher priority than following the desired end-effector's motion. We, thus, give a higher penalty to equations (9.21c) and (9.21d), than to equation (9.21b).

Equation (9.21) is a convex quadratic program (QP) with equality and inequality constraints. This can be solved either via a standard nonlinear programming solver such as Nlopt [62] or a constrained convex optimization solver such as CVXGEN [98].

In the following section, we describe a data-driven approach to learn a sparse, continuous, and continuously differentiable model for $\Gamma(.)$ from a data set in the order of millions of feasible and infeasible joint configurations.

9.2.2 Learning an SCA Boundary

As per equation (9.20), the self-collision avoidance (SCA) boundary function $\Gamma(q^{ij})$ should be continuous and continuously differentiable. Interestingly, the problem can be formulated as a binary classification problem: $y \leftarrow \text{sgn}\left(\Gamma(q^{ij})\right)$ for $y \in \{+1, -1\}$, where collided joint configurations belong to the negative class (i.e., $y = -1$) and noncollided configurations belong to the positive class (i.e., $y = +1$). When $\Gamma(q^{ij}) = 1$, q^{ij} is at the boundary of the positive class; that is, the self-collision boundary (denoted by the black line in figure 9.9).

We consider two robots with 7 degrees of freedom (DOF) each. The manifold in which the multiarm joint-angle vector lies is $q^{ij} \in R^{14}$. Employing q^{ij} as the feature vector for a

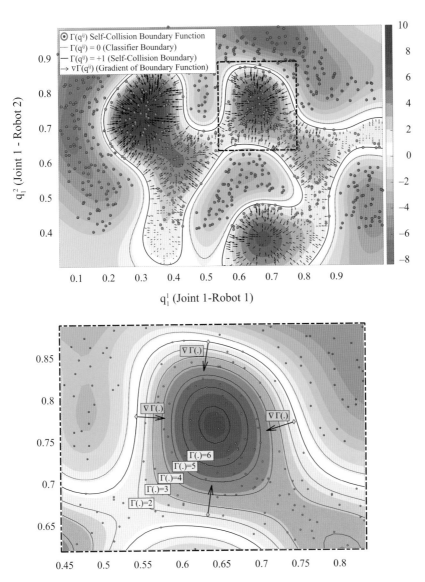

Figure 9.9

$\Gamma(q^{ij})$ function for a 2D toy problem. Assume two robots with 1-DOF, each corresponding to each axis (i.e., $q^{ij} = [q_1^1, q_2^1]$). The green data points represent collision-free robot configurations ($y = +1$), while the red data points represent collided robot configurations ($y = -1$). The background colors represent the values of $\Gamma(q^{ij})$; refer to the colorbar for exact values, where the blue area corresponds to collision-free robot configurations ($\Gamma(q^{ij}) > 1$), and the red area to collided configurations ($\Gamma(q^{ij}) < 1$). The arrows inside the collision-free region denote $\nabla\Gamma(q^{ij})$.

classification problem can be problematic for several reasons. First, many machine learning algorithms rely on computing distances/norms in Euclidean space, assuming that the features are independent and identically distributed from an underlying distribution. Hence, a Euclidean norm applied on $q^{ij} \in \mathbb{R}^{D_1+D_2}$ is merely an approximation of the actual distance in the $\mathbb{R}^{D_1+D_2}$ manifold. In fact, a proper distance metric for joint angles [i.e., $d(q_1^{ij}, q_2^{ij})$, where $q^{ij} \in \mathbb{R}^{D_1+D_2}$] is nonexistent. For this reason, instead of learning the SCA decision boundary function $\Gamma(.)$ on the joint-angle data q^{ij}, this solution learns $\Gamma(.)$ on the 3D

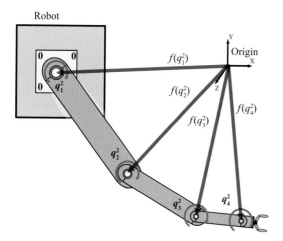

Figure 9.10

Illustration of $F(q^2)$, 3D Cartesian positions [e.g., $F(q_1^2) : S^1 \rightarrow \mathbb{R}^3$] of the individual joint angles in the multiarm system reference frame used to learn the SCA boundary function $\Gamma(F(q^2))$, where $F(q^2) = \{F(q_1^2), \ldots, F(q_2^4)\}$, assuming that we have two robots with 4 DOF each.

Cartesian representation of the joint angles $F(q^{ij})$. As illustrated in figure 9.10, $F(q^{ij})$ is a vector composed of the 3D Cartesian positions of all joints for the ith and jth robots, computed via forward kinematics. The *feature vector* for a dual-arm robotic system is thus $F(q^{ij}) \in \mathbb{R}^{3D_{q_i} + 3D_{q_j}}$. We posit that by using $F(q^{ij})$ instead of q^{ij}, we can achieve a better trade-off between model complexity and error rate. Moreover, as the output of $\Gamma(.)$ is expected to be a scalar [equation (9.20)], no extra computation is necessary, as $\Gamma(q^{ij}) \equiv \Gamma(F(q^{ij}))$. The linear inequality constraints in equation (9.21d) require $\nabla\Gamma(.)$ to be continuous; this can also be provided by either support vector machines (SVMs) or neural networks (NNs). This solution favors the use of SVMs for two main reasons: (1) Learning about an SVM is a convex optimization problem, and hence, we can always reach a *global optimum*, whereas NNs rely on heavy parameter tuning and multiple initializations in order to avoid *local minimum* solutions; (2) SVMs yield sparser models than NNs for high-dimensional, nonlinear classification problems, leading to better run times at the prediction stage.

We follow the kernel SVM formulation and propose to encode $\Gamma\left(F(q^{ij})\right)$ as the SVM decision rule. By omitting the sign function and using the radial basis function (RBF) kernel $k\left(F(q^{ij}), f(q_n^{ij})\right) = e^{\left(-\frac{1}{2\sigma^2}||F(q^{ij}) - f(q_n^{ij})||^2\right)}$, for a kernel width σ, $\Gamma\left(F(q^{ij})\right)$ (from here on the superscripts (ij) are dropped for simplicity) takes the following form:

$$\Gamma\left(F(q^{ij})\right) = \sum_{n=1}^{N_{sv}} \alpha_n y_n k\left(F(q^{ij}), f(q_n^{ij})\right) + b$$

$$= \sum_{n=1}^{N_{sv}} \alpha_n y_n e^{\left(-\frac{1}{2\sigma^2}||F(q^{ij}) - f(q_n^{ij})||^2\right)} + b \tag{9.22}$$

for N_{sv} support vectors, where $y_i \in \{-1, +1\}$ are the positive/negative labels corresponding to noncollided/collided configurations; $0 \leq \alpha_i \leq C$ are the weights for the support vectors,

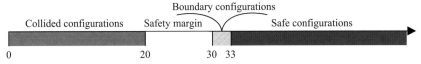

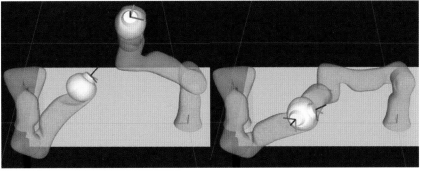

(b) "boundary configuration"   (c) "collided configuration"

Figure 9.11
Examples of the collided/boundary configurations of a *dual-arm setting* with an offset of $X_{off} = [0.0, 1.3.0.34]m$ between their bases.

which must yield $\sum_{n=1}^{N_{sv}} \alpha_n y_n = 0$; and $b \in \mathbb{R}$ is the bias for the decision rule. $C \in \mathbb{R}$ is a penalty factor used to trade off between maximizing the margin and minimizing classification errors. Given C and σ, α_i and b are estimated by solving the dual optimization problem for the *soft-margin* kernel SVM. Equation (9.22) naturally yields a continuous gradient:

$$\nabla \Gamma \left(F(q^{ij}) \right) = \sum_{n=1}^{N_{sv}} \alpha_n y_n \frac{\partial k \left(F(q^{ij}), f(q_n^{ij}) \right)}{\partial F(q^{ij})}$$

$$= \sum_{n=1}^{N_{sv}} -\frac{1}{\sigma^2} \alpha_n y_n e^{\left(-\frac{1}{2\sigma^2} \|F(q^{ij}) - f(q_n^{ij})\|^2 \right)} \left(F(q^{ij}) - f(q_n^{ij}) \right). \tag{9.23}$$

Although $\nabla \Gamma (F(q^{ij}))$ already satisfies the constraints imposed by equation (9.21d), it lives in a $3d_{q_i} + 3d_{q_j}$–dimensional space, $\nabla \Gamma \left(F(q^{ij}) \right) \in \mathbb{R}^{3d_{q_i} + 3d_{q_j}}$. We must then project this gradient onto its corresponding $\mathbb{R}^{d_{q_i} + d_{q_j}}$ joint-space; that is, $\nabla \Gamma (q^{ij}) \in \mathbb{R}^{d_{q_i} + d_{q_j}}$ with the following expansion:

$$\nabla \Gamma \left(q^{ij} \right) = \frac{\partial \Gamma \left(F(q^{ij}) \right)}{\partial F(q^{ij})} \cdot \frac{\partial F(q^{ij})}{\partial q^{ij}}, \tag{9.24}$$

with the first term equivalent to equation (9.23) and the second term being the Jacobian of each 3D joint position with regard to each joint angle $J(q^{ij}) = \frac{\partial F(q^{ij})}{\partial q^{ij}}$ for which we have a closed-form solution.

9.2.3   SCA Data set Construction

In order to learn $\Gamma\left(F(q^{ij})\right)$, we must first generate a data set capable of identifying the *self-collision boundary*. We consider a dual-arm setting, with each arm being a KUKA 7-DOF element (figure 9.11). We simplify the representation of the robot's structure by fitting spheres to each joint and its adjoining physical structure. We thus generate a discrete representation of the multiarm robotic system as a set of spheres $S^{ij} = \{s_1^i, \ldots, s_7^i, s_1^j, \ldots, s_7^j\}$. By using spheres as a geometric representation of a joint, the distance computation between joints is simplified. As the distance from any point in a sphere to the nearest obstacle is lower-bounded to $d(c) - r$, where c is the center of the sphere and r its corresponding radius. Further, the lower bound between the two spheres is the distance between their centers (c_k^i) minus the sum of their respective radii (r_k^i), for the kth spheres of the ith robot. For example, given s_5^1 and s_7^2, the lower-bounded distance between them is $d(s_5^1, s_7^2) = d(c_5^1, c_7^2) - (r_5^1 + r_7^2)$.

To identify a collision in the dual-arm system, we compute the pairwise distances of the centers of the set of spheres of the ith robot (S^i) with regard to the set of spheres of the jth robot (S^j) and find the minimum distance $\min[d(c_{k*}^1, c_{k*}^2)]$. We then define a label for each robot configuration S^{ij} as

$$
y(S^{ij}) = \begin{cases} -1 & \text{if} \quad \min[d(c_{k*}^1, c_{k*}^2)] < (r_{k*}^1 + r_{k*}^2) \\ +1 & \text{if} \quad b_- \leq \min[d(c_{k*}^1, c_{k*}^2)] \leq b_+ \\ \emptyset & \text{if} \quad \min[d(c_{k*}^1, c_{k*}^2)] > b_+, \end{cases} \tag{9.25}
$$

where r_{k*}^i corresponds to the radius of the kth sphere, and b_-, b_+ correspond to minimum-/maximum distances of the safe boundary. Specifically, a joint configuration is collided (i.e., labeled as $y = -1$), when the $\min[d(c_{k*}^1, c_{k*}^2)]$ between the centers of the closest spheres is less than the sum of the radii of the corresponding spheres [i.e., $(r_{k*}^1 + r_{k*}^2)$]. In practice, we set the spheres to a fixed radius of 10 cm, hence $(r_{k*}^1 + r_{k*}^2) = 20$ cm. Given that virtually any robot configuration where $\min[d(c_{k*}^1, c_{k*}^2)] > (r_{k*}^1 + r_{k*}^2)$ can be considered a "noncollided" configuration, we would end up with a heavily unbalanced data set of collided/noncollided data points. We thus, introduce a decomposition of the noncollided robot configurations into boundary, labeled as $y = +1$, and safe configurations, which are not labeled $y = \emptyset$. If $\min[d(c_{k*}^1, c_{k*}^2)]$ lies within a safety margin, denoted by b_- and b_+, the robots are very close to each other, but still safe (see figure 9.11). We empirically found $b_- = 30$ cm and $b_+ = 33$ cm to be safe boundaries for our dual-arm setting. Hence, a noncollided configuration is in fact a boundary configuration, as all of the safe configurations are filtered out. This has a geometric meaning, rather than finding the margin between collided and safe configurations, our boundary function will model the tighter margin between noncollided and boundary configurations. From herein, we consider boundary configurations as the noncollided configurations.

To generate the positive ($y(S^{ij}) = +1$) and negative samples ($y(S^{ij}) = -1$) for our SCA data set, we sample from all the possible motions of the robots in their respective workspaces and apply equation (9.25) to each configuration. To explore all possible joint configurations q^{ij}, we systematically displace all of the joints of both robots by 20 degrees each. Joints q_1^i, q_3^i, q_5^i, and q_7^i have a range of ±170 degrees, whereas joints q_2^i, q_4^i and q_6^i have a range of ±120 degrees. Given the 20-degree sampling resolution, this leads to 18 samples

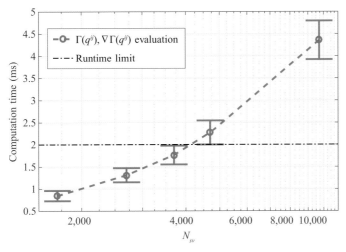

Figure 9.12
Comparison of run-time computational cost for evaluating $\Gamma(F(q^{ij}))$ and $\nabla\Gamma\left(F(q^{ij})\right)$ on the specified hardware for a dual-arm setting. The presented run times are of $\approx 2k$ control loop cycles of a self-collision avoidance test. The maximum allowable $N_{sv} \leq 3k$ in order to comply with the 2-ms limit.

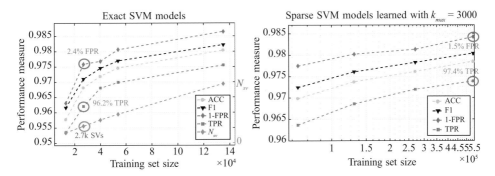

Figure 9.13
Performance Comparison of learning **exact SVM models** on randomly sub-sampled data sets vs **sparse SVM models** on larger chunks of the data set. Each model was evaluated on the test set, which contains 2.7 million unseen sample robot configurations. (left) With the **random sub-sampling method**, using the 2nd model ($N_{sv} = 2,7k$), one can achieved $FPR \approx 2.4\%$ and $TPR \approx 96.19\%$ within the desired $2ms$ runtime limit. (right) With a sparse SVM model trained on 540k points, we can achieve $FPR \approx 1.45\%$ and $TPR \approx 97.4\%$ $k_{max} = 3,000$.

for the former group and 12 for the latter. Hence, the total number of possible configurations for one arm[5] is $18^3 \cdot 12^3$, which would lead to $\approx 1e14$ possible joint configurations for a dual-arm setting. However, using our systematic sampling of collided and boundary robot configurations, we gathered a balanced data set of approximately ≈ 5.4 million data points, ≈ 2.4 million belonging to the collided configuration class $y = -1$ and the rest to the noncollided configurations $y = 1$.

9.2.4 Sparse Support Vector Machines for Large Data Sets

The training time of a kernel SVM has a complexity of $\approx \mathcal{O}(N_M^2 D)$, where N_M is the number of samples and D is the dimension of the data points. The prediction time, on the other hand, depends on the number of support vectors N_{sv} learned through training. In practice,

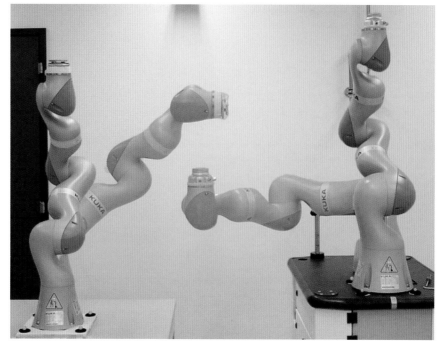

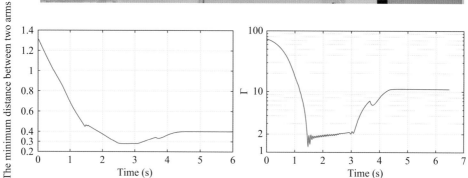

Figure 9.14
Self-collision avoidance test. (top photo) the initial and the final configurations (shown in semi-transparent image) are depicted. Bottom left and right figures show the minimum distance between the arms and the value of Γ during the motion execution. $\Gamma(\cdot)$ is never < 1, indicating that the motion of the arms are safe.

N_{sv} tends to increase linearly with the amount of training data N_M. More specifically, for a kernel SVM $N_{sv}/N_M \to \mathscr{B}_k$, where $\mathscr{B}_k$ is the smallest achievable classification error by the kernel k (i.e., in a nonseparable classification scenario), to achieve 5 percent error, at least 5 percent of the training points must become support vectors. This is a nuisance when large training sets are involved, as in this application. $N_{sv} \gg$ signifies a dense solution for representing the hyperplane of the classifier margin $\mathbf{w} = \sum_{i=1}^{N_{sv}} \alpha_i y_i \Phi(f(q_i^{ij}))$. Naturally, the denser the solution, the more computationally expensive it is at run time. This makes dense SVMs infeasible for real-time robot control. To achieve fast adaptation for both the desired end-effector positions and self-collisions, the IK solver must run at a rate of 2 ms at *most*.

Note that during this cycle, prior to solving equation (9.21), equations (9.22) and (9.24) must both be evaluated.

Given the desired control rate (2 ms), the specific hardware used to control the robots (i.e., 3.4-GHz i7 PC with 8 GB of random access memory) and the kinematic specifications of each robot, one needs to define a computational budget for our SCA boundary function. This budget translates to defining a limit of the maximum allowable N_{sv} for our SVM representation of $\Gamma(F(q^{ij}))$. Figure 9.12 shows a plot of various computation times.[6]

To comply with the 2-ms run-time requirement, the computational budget is $N_{sv} \leq 3k$. Given the size of our data set, it is not feasible to train SVM models that typically optimize for the dual arms. To address this problem, one solution could be reformulating the SVM optimization problem using the Cutting Plane Subspace Pursuit (CPSP) method introduced by [61]; it directly estimates a solution to the hyperplane with a strict bound on the number of support vectors, $\mathbf{k_{max}}$. In short, the CPSP method approximates a sparse hyperplane by expressing it in terms of a set $\mathbf{B} = \{\mathbf{b}_1, \ldots, \mathbf{b}_{\mathbf{k_{max}}}\}$ of basis vectors $\mathbf{b}_i \in \mathbb{R}^{3d_{q_i}+3d_{q_j}}$ (not necessarily training points), as $\mathbf{w} = \sum_{i=1}^{\mathbf{k_{max}}} \alpha_i y_i \Phi(\mathbf{b}_i)$. The optimization algorithm to estimate this new value of $\mathbf{w}$ then focuses on *pursuing* such a subspace through the fixed-point iteration approach for RBF kernels. The learned basis vectors $\mathbf{B}$ and α_is can be directly used in equations (9.22) and (9.23).

9.2.5   Robotic Implementation

The proposed approach has been successfully used to control the multiarm reach-to-grab motions as described in chapter 7, section 5. The approach is also evaluated in tasks where the robots must track moving targets that go inside each other's workspaces, and tasks where robots are stationary but a human applies external forces; in both of these scenarios, the robots are capable of avoiding each other. We report here one experiment, where the robots are tasked to move in straight lines to reach fixed targets inside the workspace of the other robot, as shown in figure 9.14. We can also see how the value of $\Gamma(.)$ is updated online during the motion execution with respect to the robot configurations and when it is < 2, based on equation (9.21d), the robots are moved away from each other. Movies illustrating the robotic experiments presented in this chapter are available on the book's website.

IV COMPLIANT AND FORCE CONTROL WITH DYNAMICAL SYSTEMS

10 Compliant Control

The previous chapters of this book have been devoted to using dynamical systems (DSs) to generate trajectories for position or velocity control of robots. However, we did not control for forces. Force control is necessary for many tasks, such as the precise and fine manipulation of objects, and for mitigating risks if the robot enters in contact with humans.

In section 8.4 of chapter 8, we took a first step toward controlling for force, using second-order DSs. We showed that, when the robot was making contact with a surface, we could bound the forces to ensure that the contact remained stable, but we could not control explicitly for the force applied to the surface. In this chapter, we show how DSs can be combined with impedance control, an approach traditionally used to perform robust torque control, to perform torque control. In chapter 11, we will show an extension of this to control explicitly for force applied at contact.

Impedance control can be shaped in such a way as to enable robots to absorb interaction forces and dissipate energy at contact. This is particularly useful when contacts may be undesired as the result of unexpected disturbances, such as bumping into an object inadvertently, and to make robots safer in the presence of humans. It is, however, often difficult to determine beforehand what the right impedance would be for a given task. For this reason, many methods have been offered to learn what is the right impedance to apply, when, and where. This chapter reviews some of these and shows how learning of variable impedance can be done in conjunction with learning a DS based control law.

This chapter is organized as follows. Section 10.1 discusses the need to provide compliance when in contact with the environment. This is followed by an introduction to compliant control architectures for controlling robots once they are already in contact with an object, presented in section 10.2. In section 10.3, we offer two approaches to learn compliant behaviors. At last, in section 10.4, we show how this framework can be combined with DSs.[1]

10.1   When and Why Should a Robot Be Compliant?

Consider a system driven by the following equation:

$$m\ddot{x} + k(x - x^*) = 0, \quad x(0) = x_0. \tag{10.1}$$

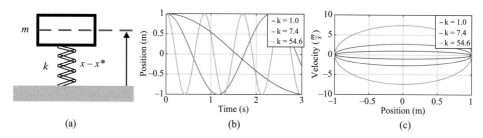

Figure 10.1
An example of a point-mass robotic system approximated by a mass-spring system is illustrated in (a). The mass's state (x) indicates the state of the robot. (b) and (c) illustrate the behavior of the system for three stiffness coefficients, $m = 1$, $x_0 = 1$, and $x^* = 0$. As can be seen, the peak velocity happens at $x = x^* = 0$, which is monotonically increasing by the stiffness coefficient.

This is the Newtonian equation of a mass m subjected to a spring with constant k. Observe that this is a linear, time-invariant, second-order DS; $x \in \Re$ is the state of the system, $x^* \in \mathbb{R}$ a fixed point of the system, and x^* is a stationary point. If the system is initialized at x^*, it does not move. However, a slight perturbation would send the system into an infinite oscillatory behavior.

Equation (10.1) describes the behavior of a mass-spring system attached to x^*. As shown in figure 10.1, the system oscillates around x^* without losing any energy. Peak velocity occurs at $x = x^*$. The frequency of the oscillation, and hence the velocity of the system at $x = x^*$, are a function of k and m. Here, k is known as the stiffness of the system. The larger k is, the stiffer the system and the higher the peak velocity and oscillation frequency (see figure 10.1c).

Observe that equation (10.1) is similar to that introduced in Chapter 3 to describe the dynamics of a robot arm tasked to hit a golf ball located at x^*. If we do not control the interception speed, the robot will hit the target at a high speed. While this may be desirable to send a ball flying into the air, if the target is a fragile vase or a wall, this may cause damage to either the object or the robot. To modulate the force at contact with the target, one must control k.

One option to prevent infinite oscillations and control the speed of the system is to add a *damping* term to equation (10.1):

$$m\ddot{x} + d\dot{x} + k(x - x^*) = 0, \qquad x(0) = x_0, \tag{10.2}$$

where $d \in \mathbb{R}^+$. As illustrated in figure 10.2c, damping significantly alters the behavior of the system. For constant mass and stiffness, increasing the damping coefficient changes the characteristics of the system from lossless (blue line) to overdamped (purple line).

The speed of the system at $x = x^*$ can be precisely controlled through a choice of damping and stiffness coefficients. One can show that the mass-damping-spring system stably converges to the equilibrium point x^* in minimum time if $k = \omega^2$ and $d = 2\omega$, where $\omega \in \mathbb{R}^+$ is the natural frequency of the system. This is called a *critically damped system*. Hence, the larger the damping, the slower the speed at target, but this comes at the cost of a higher rising/settling time.

In robotics, one often wishes to control both the velocity of the robot (to determine the traveling time and the direction of motion) and the force at contact. However, it is physically impossible to control force and displacement separately, when applied to the same control

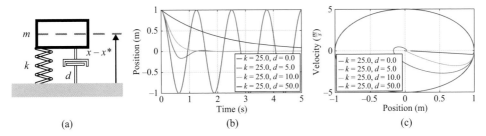

Figure 10.2
(a) illustrates a mass-spring-damper system (10.2) attached to x^*. The mass's state is indicated by x. In (b) and (c), the behavior of the system for four damping coefficients is exemplified while $m = 1$, $x^* = 0$ and $x_0 = 1$. As can be seen in (c), the damping coefficient has a significant effect on the response of the system; the system can be overdamped (purple), or lossless (blue).

dimension when applied in the same direction. To address the issue and be able to relate velocity to force control and vice versa, robotics has endorsed two approaches, known as *impedance* and *admittance* control [55]. The term *impedance* is inspired by a similar concept in electricity. Electric impedance relates to the ratio of voltage to current in a circuit. In control, impedance indicates how much a system resists a harmonic force (i.e., the ratio of the force to the resulting velocity). *Admittance* is the inverse of impedance; it indicates the ratio of the force, as the input, to the resulted velocity/displacement. For instance, the impedance and admittance of the mass-spring-damper system [equation (10.2)] with $x^* = 0$ under a harmonic force $f = f_0 e^{i\omega t}$ in the frequency domain are

$$Z_m(i\omega) = (i\omega m + d - \frac{k}{i\omega}) \tag{10.3a}$$

$$Y_m(i\omega) = \frac{i\omega}{k + i\omega d - \omega^2 m}, \tag{10.3b}$$

where i is an imaginary unit. As in steady-state behavior $s = i\omega$, equation (10.3) is equivalent to

$$Z_m(s) = (ms + d + \frac{k}{s}) \tag{10.4a}$$

$$Y_m(s) = \frac{s}{k + sd + s^2 m} \tag{10.4b}$$

in the Laplace domain, where s is the Laplace variable. If we take a closer look at equations (10.3) and (10.4), we can make the following observations:

• Equations (10.3) and (10.4) indicate that impedance and admittance are functions of not only stiffness, but also mass and damping. Moreover, impedance (admittance) of a mass-spring-damper system is positively (negatively) correlated with the mass, stiffness, and damping of the system.

• The impedance and admittance of the system are also functions of the input frequency. As an example, the impedance of spring and mass components are negatively and positively correlated with the input frequency, respectively, and the impedance of a damper is uncorrelated with the frequency. These correlations are highlighted in figure 10.3b, which

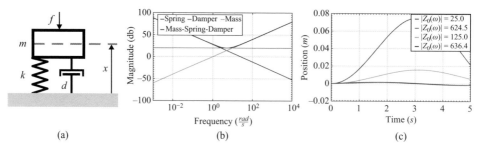

Figure 10.3
(a) illustrates a mass-spring-damper system under external force f. The mass's state is indicated by x. The frequency response of a mass ($m = 1$), a spring ($k = 25$), and a damper ($d = 10$) component and a mass-spring-damper system (with the same coefficients) are shown in (b). In (c), the behavior of a mass-spring-damper for different impedances ($m = 1$, $d \in \{25, \ 125\}$ and $k \in \{1, \ 625\}$) is exemplified. The response of the systems with similar impedance values (the red and purple lines in figure 10.3c) are almost the same.

illustrates the impedance of a mass, damper, and spring component, both separately and together, for different frequencies.

• The frequency response of the mass-spring-damper (purple line in figure 10.3b) indicates that the impedance of the system has one global-optimal working frequency (i.e., the highest admittance and compliance).

• Higher values of impedance result in lower displacements with respect to a constant exerted force. On the other hand, if the impedance is zero (infinite admittance), the system swiftly moves in the presence of external forces. This is also highlighted in figure 10.3c, which illustrates the response of the mass-spring-damper [equation (10.2)] to $f = sin(t)$ for different impedance values. As can be seen, the lower impedance is, the higher displacement will be (i.e., compare the blue and purple lines).

In robotics, impedance (accordingly admittance) characterizes how the robot reacts to external forces. These forces do not have to be real. One can generate *virtual* forces to produce a specific behavior. For instance, one can generate a spring and a damper system that is *virtually* connected to the desired target point. This can be used to drive the robot to a desired target, as introduced at the beginning of this chapter. Note that the desired target may not need to be a fixed point. As we will see in this chapter, *the desired target can be the nominal motion of the system, driven by a DS*. This target becomes a reference trajectory, and the impedance[2] indicates how swiftly a system converges to the desired reference trajectory.

The concept of impedance is closely related to the notion of compliance. A system is said to be compliant if it obeys an external input. In impedance control, the system may comply more or less to disturbances generated by external forces. The system is highly compliant if its impedance is low. Conversely, it is stiff if its impedance is high.

Exercise 10.1   *Consider a second-order, one-dimensional (1D) DS, which can be used to hit a target at* $x = x^*$, *with desired velocity* $\dot{x} = \dot{x}^*$ *at* $t = t^*$. *Consider first the undamped system for the DS using equation (10.1) and then consider DS following equation (10.2). Is the damping term required?*

What if the DS needs to pass through two points on the space with two different velocities (i.e., $x = x_1^*$ with $\dot{x} = \dot{x}_1^*$ at $t = t_1^*$ and $x = x_2^*$ with $\dot{x} = \dot{x}_2^*$ at $t = t_2^*$). Can this scenario be accomplished? Support your answer with analytical and numerical results.

Exercise 10.2 *Consider the DS, designed in exercise 10.1, to hit a nail using a point mass robotic system. If* $m = 1$, $k = 25$, $d = 1$, *and the impact is elastic, what would be the force applied to the nail at each cycle? What is the cycle rate?*

10.2 Compliant Motion Generators

Compliant control can be achieved in different ways during the contact phase, depending on the robotic platform and the space in which the task is formulated. Impedance control therefore can be formulated in both Cartesian space and joint space to modify the mechanical impedance of the robot with respect to a desired target in Cartesian or joint space as follows:

$$\Lambda \ddot{\tilde{x}} + D\dot{\tilde{x}} + K\tilde{x} = F_c \tag{10.5a}$$

$$\Lambda \ddot{\tilde{q}} + D\dot{\tilde{q}} + K\tilde{q} = J(q)^T F_c, \tag{10.5b}$$

where $\tilde{q} = q - q_d$, $\tilde{x} = x - x_d$, and $q_d \in \mathbb{R}^D$ and $x_d \in \mathbb{R}^N$ are the desired equilibrium points. Λ, D, and K are the desired inertia, damping, and stiffness matrices, respectively. In order to achieve equation (10.5), the control input τ or $\mathscr{F}$ should be defined as

$$\tau = C(q,\dot{q})\dot{q} + G(q) + M(q)\ddot{q}_d - M(q)\Lambda^{-1}(D\dot{\tilde{q}} + K\tilde{q}) + (M(q)\Lambda^{-1} - I)J(q)^T F_c \tag{10.6a}$$

$$\mathscr{F} = V_x(q,\dot{q})\dot{x} + G_x(q) + M_x(q)\ddot{x}_d - M_x(q)\Lambda^{-1}(D\dot{\tilde{x}} + K\tilde{x}) + (M_x(q)\Lambda^{-1} - I)F_c. \tag{10.6b}$$

It can be easily shown that substituting equations (10.6a) and (10.6b) into the robot's dynamics [see appendix C, equations (C.1) and (C.2)], would result in equation (10.5). Achieving this equation (10.5) requires closing the loop on external force F_c. Measuring the force requires equipping the robot with a force/torque sensor. This is, however, not always feasible. If the desired inertia Λ is identical to the robot inertia $M(q)$ or $M_x(q)$, one can avoid the force-torque sensor by simplifying equation (10.6) to

$$\tau = C(q,\dot{q})\dot{q}_d + G(q) + M(q)\ddot{q}_d - (D\dot{\tilde{q}} + K\tilde{q}) \tag{10.7a}$$

$$\mathscr{F} = V_x(q,\dot{q})\dot{x}_x + G_x(q) + M_x(q)\ddot{x}_d - (D\dot{\tilde{x}} + K\tilde{x}). \tag{10.7b}$$

If the desired trajectory is stationary, one can further simplify the control law [equation (10.7)] to

$$\tau = G(q) - (D\dot{q} + K\tilde{q}) \tag{10.8a}$$

$$\mathscr{F} = G_x(q) - (D\dot{x} + K\tilde{x}). \tag{10.8b}$$

In equation (10.8), gravity is compensated for by the feed-forward term $G(q)$ or $G_x(q)$, and the feedback is used to proportionally act on the configuration error with stiffness matrix K. In addition, velocities are damped through the feedback of velocity acted upon by damping matrix D. Substituting equation (10.8a) into equations (C.1) and (C.2), one can derive the following closed-loop control law:

Substitution of (10.8a) and (10.8b) into (C.1) or (C.2), respectively, yields the following closed-loop behavior:

$$M(q)\ddot{\tilde{q}} + \bar{D}\dot{\tilde{q}} + K\tilde{q} = J(q)^T F_c \tag{10.9a}$$

$$M_x(q)\ddot{\tilde{x}} + \bar{D}_x\dot{\tilde{x}} + K\tilde{x} = F_c, \tag{10.9b}$$

where $\bar{D} = D + C(q,\dot{q})$ and $\bar{D}_x = D + V_x(q,\dot{q})$. Equation (10.8) is the simplest possible implementation of an impedance control architecture and is surprisingly effective, especially in quasi-static cases. Note that, strictly speaking, this controller is valid only for regulations. However, in practice, it can work very well with slowly moving reference configurations.

It can be shown (see exercises 10.3, 10.4, and 10.5) that the motion of the robotic system driven by equation (10.6), (10.7), or (10.8) is asymptotically stable toward the desired trajectory; i.e.:

$$\lim_{t\to\infty} \|q(t) - q_d(t)\| = 0 \qquad \lim_{t\to\infty} \|x(t) - x_d(t)\| = 0$$

$$\lim_{t\to\infty} \|\dot{q}(t) - \dot{q}_d(t)\| = 0 \qquad \lim_{t\to\infty} \|\dot{x}(t) - \dot{x}_d(t)\| = 0 \tag{10.10}$$

$$\lim_{t\to\infty} \|\ddot{q}(t) - \ddot{q}_d(t)\| = 0 \qquad \lim_{t\to\infty} \|\ddot{x}(t) - \ddot{x}_d(t)\| = 0.$$

if

$$0 \prec K, \qquad 0 \prec D, \qquad 0 \prec \Lambda \tag{10.11}$$

in equations (10.6) and (10.7), and

$$0 \prec K, \qquad 0 \prec D \tag{10.12}$$

in (10.8) are met.

Figure 10.4 illustrates the behavior of a simple robotic system with 2 degrees of freedom (DOF), which is driven by equations (10.6), (10.7) or (10.8). As it can be seen, even though the system is stable in all the scenarios, its behavior changes with respect to the control law.

Programming Exercise 10.1   *The aim of this exercise is to help readers to get a better understanding of the impedance control architecture [equations (10.6), (10.7), and (10.8)] and the effect of the open parameters on the generated motion. Open MATLAB and set the directory to the following folder:*

```
1  ch10-DS_compliant/Impedance_controller
```

Study the effects of the following open parameters on the generated motion:

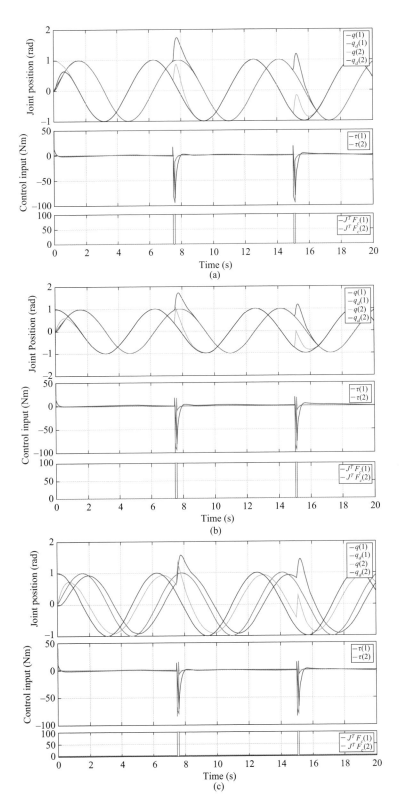

Figure 10.4

The desired trajectory is sinusoidal $q_d = \begin{bmatrix} sin(t) \\ cos(t) \end{bmatrix}$, and the external perturbation torque ($J^T F_c$) is applied to both joints at $t = 7.5s$ and $t = 15s$. The desired inertial, damping, and stiffness matrices are $\Lambda = I_{2\times2}$, $D = 10I_{2\times2}$ and $K = 25I_{2\times2}$, respectively. The resultant behavior of equations (10.6), (10.7), and (10.8) are illustrated in (a), (b), and (c), respectively.

1. *The dynamic specifications of the robot:*
   (a) *Mass of the links*
   (b) *Length of the links*
2. *The initial configuration of the joints:*
   (a) *Position of the joints*
   (b) *Velocity of the joints*[a]
3. *The desired impedance parameters:*
   (a) *The stiffness matrix*
   (b) *The damping matrix*
   (c) *The inertia matrix*
4. *To improve the robustness of the robotic arm to the external perturbations, which parameters would you change while satisfying torque limits? Note that you can change the design of the robot as well.*
5. *What is the effect of the sampling time on the motion of the robots?*

[a] *In the code, the joint acceleration are not initialized. Does the initial value of acceleration have an effect on the generated motion?*

Exercise 10.3   *Consider a robotic system as defined in (C.1) and (C.2) driven by equation (10.6). Prove that the dynamics of the error is asymptotically converging toward zero [i.e., equation (10.11) is satisfied] if $F_c = 066$ and conditions of equation (10.11) are met.*

Exercise 10.4   *Consider a robotic system as defined (C.1) and (C.2) driven by equation (10.7). Prove that the dynamics of the error is asymptotically converging toward zero if $F_c = 0$ and conditions of equation (10.11) are met.*

Exercise 10.5   *Consider a robotic system as defined in (C.1) and (C.2) driven by equation (10.8). Prove that the dynamics of the error is asymptotically converging toward zero [i.e., equation (10.12) is satisfied] if $F_c = 0$, $\ddot{x}_d = \dot{x}_d = 0$, $\ddot{q}_d = \dot{q}_d = 0$ and equation (10.11) are met.*

10.2.1   Variable Impedance Control

In the previous section, we introduced impedance control approaches, where the matrices of inertia Λ, damping D, and stiffness K are constant. This impedance controller is easy to implement, robust toward modeling uncertainties, and computationally inexpensive. Nevertheless, to increase the flexibility and improve the performance of a robotic system, these values can vary as a function of the robot state, time, or external signals like force measurements. While impedance control with constant impedance parameters is an adequate solution in some cases, varying the impedance parameters during the task gives more flexibility and can drastically improve the performance in many tasks.

It can easily happen that one creates impedance profiles that yield an unstable behavior. As an example, let's consider the most common form of impedance control, which is to control the robot in Cartesian space [equation (10.5a)] so that the following dynamic relationship is established between the generalized position error $\tilde{x} \in \mathbb{R}^N$ and generalized force $F_c \in \mathbb{R}^N$:

$$\Lambda \ddot{\tilde{x}} + D(t)\dot{\tilde{x}} + K(t)\tilde{x} = F_c. \tag{10.13}$$

As mentioned in section 10.2, the user-defined virtual inertia Λ, damping D, and stiffness K determine the behavior of the robot when subjected to external torque. If these parameters are constant, the system will be asymptotically stable for any symmetric positive definite matrices Λ, D, and K. If they are varying with time, in order to analyze the stability properties of equation (10.13), one can take as Lyapunov candidate function, the kinetic energy with respect to the velocity error and the virtual potential energy stored by the stiffness:

$$V = \frac{1}{2}\dot{\tilde{x}}^T \Lambda \dot{\tilde{x}} + \frac{1}{2}\tilde{x}^T K(t)\tilde{x}. \tag{10.14}$$

Differentiating V along the trajectories of equation (10.14) with $F_c = 0$ and $\dot{\Lambda} = 0$ yields

$$\begin{aligned}
\dot{V} &= \ddot{\tilde{x}}^T \Lambda \dot{\tilde{x}} + \frac{1}{2}\tilde{x}^T \dot{K}(t)\tilde{x} + \dot{\tilde{x}}^T K(t)\tilde{x} \\
&= -\dot{\tilde{x}}^T D(t)\dot{\tilde{x}} + \frac{1}{2}\tilde{x}^T \dot{K}(t)\tilde{x}.
\end{aligned} \tag{10.15}$$

Equation (10.15) is negative semidefinite if $\dot{K}(t)$ is a negative semidefinite matrix. Hence, we can conclude stability at the origin only if the stiffness profile is constant or decreasing. Assuming $\tilde{x} \neq 0$, increasing the stiffness injects energy, in the form of potential, into the system, and it is intuitively clear that this practice can cause unstable behavior. Hence, it is important to understand to what extent the impedance can be varied without risk of instability.

Different techniques have been proposed to analyze and guarantee the stability of a varying impedance control. One way to stabilize a varying impedance control is online monitoring of the energy of the system and switching to a constant stiffness/damping once the energy is higher than a specific threshold. However this approach yields discontinuities in the stiffness profile, which is usually not desirable for controlling real robots. Another approach is by either injecting damping or by modifying the stiffness profile depending on the energy of the system observed during task execution. The main disadvantage of this approach is the fact that the impedance profile that the robot will use before task execution cannot be determined. Consequently, an engineer who has carefully crafted a task-specific impedance profile could end up watching in frustration as the robot exhibits a totally different behavior than intended as it performs the task.

In this chapter, we introduce three varying impedance control architectures that are inherently stable and can be used in scenarios where the robot's impedance profile needs to be carefully defined before task execution.

10.2.1.1 Independent proportional-derivative LPV controller

By exploiting the properties of matrices based on linear parameter varying (LPV), one can approximate the stiffness and damping matrices[3] as follows:

$$\mathbf{K}(\tilde{x}) = \sum_{k=1}^{K} \gamma_k^K(\tilde{x}) K^k \qquad K^k \in \mathbb{R}^{N \times N} \qquad \gamma_k^K(\tilde{x}) \in \mathbb{R}_{(0,1]}, \forall k$$

$$\mathbf{D}(.) = \sum_{k=1}^{K} \gamma_k^D(.) D^k \qquad D^k \in \mathbb{R}^{N \times N} \qquad \gamma_k^D(.) \in \mathbb{R}_{(0,1]}, \forall k. \tag{10.16}$$

In equation (10.16), stiffness ($\mathbf{K}$) can only be a function of the robot's state. However, damping ($\mathbf{D}$) can be a function of the state of the robot, the desired trajectory, external signals, or even time. Hence, the arguments of the damping matrix are not explicitly specified, which is indicated by (.).

Given the stiffness and damping matrices [equation (10.16)], one can rewrite the impedance control laws [equations (10.6), (10.7), and (10.8)] as

$$\mathscr{F} = V_x(q, \dot{q})\dot{x} + G_x(q) + M_x(q)\ddot{x}_d - M_x(q)\Lambda^{-1}(\mathbf{D}(.)\dot{\tilde{x}} + \mathbf{K}(\tilde{x})\tilde{x})$$

$$+ (M_x(q)\Lambda^{-1} - I)F_c \tag{10.17a}$$

$$\mathscr{F} = V_x(q, \dot{q})\dot{x}_d + G_x(q) + M_x(q)\ddot{x}_d - (\mathbf{D}(.)\dot{\tilde{x}} + \mathbf{K}(\tilde{x})\tilde{x}) \tag{10.17b}$$

$$\mathscr{F} = G_x(q) - (\mathbf{D}(.)\dot{x} + \mathbf{K}(\tilde{x})\tilde{x}), \tag{10.17c}$$

respectively. By substituting equation (10.17) into (C.2), the closed-loop dynamic of a robotic arm is

$$\Lambda\ddot{\tilde{x}} + \mathbf{D}(.)\dot{\tilde{x}} + \mathbf{K}(\tilde{x})\tilde{x} = F_c \tag{10.18a}$$

$$M_x(q)\ddot{\tilde{x}} + \mathbf{D}(.)\dot{\tilde{x}} + \mathbf{K}(\tilde{x})\tilde{x} = F_c \tag{10.18b}$$

$$M_x(q)\ddot{\tilde{x}} + (V_x(q, \dot{q})\dot{x} + \mathbf{D}(.))\dot{\tilde{x}} + \mathbf{K}(\tilde{x})\tilde{x} = F_c. \tag{10.18c}$$

It can be shown (see exercise 10.6) that the closed-loop dynamic of such robotic system [equation (10.18)] is asymptotically stable toward the desired trajectory; i.e.:

$$\lim_{t \to \infty} \|x(t) - x_d(t)\| = 0, \qquad \lim_{t \to \infty} \|\dot{x}(t) - \dot{x}_d(t)\| = 0, \qquad \lim_{t \to \infty} \|\ddot{x}(t) - \ddot{x}_d(t)\| = 0. \tag{10.19}$$

If

$$\begin{cases} K^i + K^{iT} \prec 0, & K^i = K^{iT}, \ \forall i \in \{1, \dots, K\} \\ D^j + D^{jT} \prec 0, & \forall j \in \{1, \dots, K\} \\ 0 < \gamma_k^K(\tilde{x}) \leq 1, & \forall i \in \{1, \dots, K\} \\ 0 < \gamma_k^D(.) \leq 1, & \forall j \in \{1, \dots, K\}. \end{cases} \tag{10.20}$$

As can be seen, the stiffness matrix in the control law [equation (10.18)] can be only a function of the tracking error—nothing else. This might be limiting, however, as it cannot be applied to scenarios where the stiffness matrix needs to vary with respect to the robot's velocity or the measured force rather than the tracking error. Nevertheless, the main advantage of equation (10.16) is the fact that the activation parameters of the stiffness and damping matrices are defined independent of each other.

10.2.1.2 Dependent proportional-derivative-LPV controller

One way to address the shortcoming of approach 1 is to define the stiffness and the damping matrices as follows:

$$\mathbf{K}(.) = \sum_{k=1}^{K} \gamma_k(.) K^k \qquad K^k \in \mathbb{R}^{N \times N}$$

$$\gamma_k(.) \in \mathbb{R}_{(0,1]}. \qquad (10.21)$$

$$\mathbf{D}(.) = \sum_{k=1}^{K} \gamma_k(.) D^k \qquad D^k \in \mathbb{R}^{N \times N}$$

In equation (10.21), stiffness ($\mathbf{K}$) and damping ($\mathbf{D}$) can be a function of the robot's velocity, the desired trajectory, external signals, or even time. Hence, the arguments of the stiffness/damping matrices are not explicitly specified and are indicated by (.). However, the activation parameter of the damping and stiffness matrices must be the same. Based on this definition, similar to equation (10.22), one can rewrite the impedance control laws [equation (10.6)] as

$$\mathscr{F} = V_x(q, \dot{q})\dot{x} + G_x(q) + M_x(q)\ddot{x}_d - M_x(q)\Lambda^{-1}(\mathbf{D}(.)\dot{\tilde{x}} + \mathbf{K}(.)\tilde{x})$$

$$+ (M_x(q)\Lambda^{-1} - I)F_c, \qquad (10.22a)$$

by which, the closed loop dynamic is given by

$$\Lambda\ddot{\tilde{x}} + \mathbf{D}(.)\dot{\tilde{x}} + \mathbf{K}(.)\tilde{x} = F_c. \qquad (10.23a)$$

It can be shown that the closed-loop dynamic of the system [equation (10.23)] is asymptotically stable toward the desired trajectory x_d:

$$\lim_{t \to \infty} \|x(t) - x_d(t)\| = 0, \qquad \lim_{t \to \infty} \|\dot{x}(t) - \dot{x}_d(t)\| = 0, \qquad \lim_{t \to \infty} \|\ddot{x}(t) - \ddot{x}_d(t)\| = 0. \quad (10.24)$$

If there is a P such that

$$\begin{cases} \begin{bmatrix} 0 & I \\ K^k & D^k \end{bmatrix}^T P + P \begin{bmatrix} 0 & I \\ K^k & D^k \end{bmatrix} \prec 0 \\ \\ 0 \prec P, \qquad P^T = P \qquad \forall k \in \{1, \ldots, K\}. \\ \\ 0 < \lambda_k(.) \le 1, \qquad \sum_{k=1}^{K} \lambda_k(.) = 1 \end{cases} \qquad (10.25)$$

10.2.1.3 Energy tank-based variable impedance controller

In this approach,[4] the stability of the impedance-varying controller is studied in a general case (i.e., the damping and stiffness matrices do not need to be expressed as a combination of LPV-based matrices). However, the desired inertia matrix needs to be constant. Hence, we devise the following impedance-varying controller based on equation (10.6):

$$\mathscr{F} = V_x(q, \dot{q})\dot{x} + G_x(q) + M_x(q)\ddot{x}_d - M_x(q)\Lambda^{-1}(D(t)\dot{\tilde{x}} + K(t)\tilde{x})$$

$$+ (M_x(q)\Lambda^{-1} - I)F_c. \qquad (10.26)$$

The damping and stiffness matrices can depend on any external variable. However, for the sake of brevity, we consider solely a dependency on time. By using equation (10.26), the closed-loop dynamic of a robotic arm is

$$\Lambda \ddot{\tilde{x}} + D(t)\dot{\tilde{x}} + K(t)\tilde{x} = F_c. \tag{10.27}$$

Studying the stability of equation (10.27) motivates the search for a less conservative Lyapunov candidate function [equation (10.14)]. In the adaptive control architectures, it is common to construct energy functions of weighted sums of the velocity error and position error. The same approach can be used for varying stiffness/damping controls to establish stability conditions. Consider the following Lyapunov candidate function:

$$V = \frac{1}{2}(\dot{\tilde{x}} + \alpha\tilde{x})^T \Lambda (\dot{\tilde{x}} + \alpha\tilde{x}) + \frac{1}{2}\tilde{x}^T \beta(t)\tilde{x}, \tag{10.28}$$

where

$$\beta(t) = K(t) + \alpha D(t) - \alpha^2 \Lambda. \tag{10.29}$$

with $0 < \alpha$ being some positive constant chosen such that $0 \leq \beta(t)$. This Lyapunov candidate function is a generalized version of a function that is used for the analysis of time-varying scalar systems. Note that $\alpha \to 0 \Rightarrow$ equation (10.28)$\to$ equation (10.14). In contrast to equation (10.14), however, equation (10.28) allows for establishing sufficient constraints for stability that are independent of the state. In other words, the system in equation (10.27), with $F_c = 0$, is globally uniformly converging to the desired trajectory $x_d, \dot{x}_d, \ddot{x}_d$:

$$\lim_{t \to \infty} \|x(t) - x_d(t)\| = 0, \qquad \lim_{t \to \infty} \|\dot{x}(t) - \dot{x}_d(t)\| = 0, \qquad \lim_{t \to \infty} \|\ddot{x}(t) - \ddot{x}_d(t)\| = 0. \tag{10.30}$$

If the following holds:

$$0 \preceq D(t),$$

$$0 \preceq K(t),$$

$$0 < \alpha, \tag{10.31}$$

$$\alpha \Lambda - D(t) \preceq 0,$$

$$\dot{K}(t) + \alpha \dot{D}(t) - 2\alpha K(t) \prec 0.$$

It is perhaps not intuitive why the derivative of the damping appears in equation (10.31), as $\dot{D}$ did not appear in equation (10.15). This means that for a constant stiffness, stability would be ensured by any positive definite D, without any direct constraints on $\dot{D}$. However, increasing the damping too fast can make the system converge to points with $\dot{\tilde{x}} = 0$ and $\tilde{x} = 0$. The presence of $\dot{D}$ prevents this from happening because both $\dot{D}$ and $\dot{K}$ are in effect bounded by this constraint.

Figures 10.5, 10.6, and 10.7 provide illustrative examples of the motion of the 2-DOF robot introduced in figure 10.4, controlled by equations (10.17a), (10.22), and (10.26), respectively. As can be seen, different stiffness and damping profiles can be generated in such a way that the stability of the system is ensured.

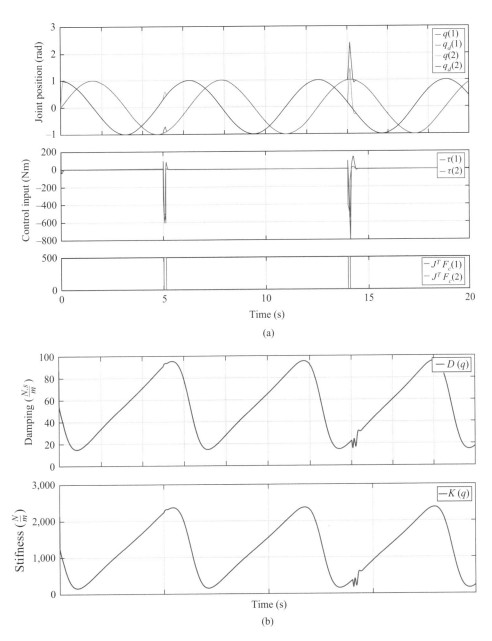

Figure 10.5

The desired trajectory is sinusoidal $q_d = \begin{bmatrix} sin(t) \\ cos(t) \end{bmatrix}$, and the external perturbation torque $(J^T F_c)$ is applied to both joints at $t = 7.5s$ and $t = 15s$. The control law [equation (10.17a)] is used to drive the robot where $\mathbf{K}(\tilde{x}) = \frac{\lambda_1 \mathbf{K}}{\lambda_1 \mathbf{K} + \lambda_2 \mathbf{K}}(\tilde{x})K_1 + \frac{\lambda_2 \mathbf{K}}{\lambda_1 \mathbf{K} + \lambda_2 \mathbf{K}}(\tilde{x})K_2$ and $\mathbf{D}(\tilde{x}) = \frac{\lambda_1 \mathbf{K}}{\lambda_1 \mathbf{K} + \lambda_2 \mathbf{K}}(\tilde{x})D_1 + \frac{\lambda_2 \mathbf{K}}{\lambda_1 \mathbf{K} + \lambda_2 \mathbf{K}}(\tilde{x})D_2$. $\lambda_1 = (q - \mu_1)^T (q - \mu_1)$ and $\lambda_2 = (q - \mu_2)^T (q - \mu_2)$, where $\mu_1 = \begin{bmatrix} -1 \\ 1 \end{bmatrix}$ and $\mu_2 = \begin{bmatrix} 1 \\ 1 \end{bmatrix}$. $D_1 = 10I_{2\times2}$, $D_2 = 100I_{2\times2}$, $K_1 = 25I_{2\times2}$ and $K_2 = 2500I_{2\times2}$. Because the impedance profile is a function of the state of the system, it adjusts with respect to the sensory information. Hence, the perturbations have an effect on the impedance profile.

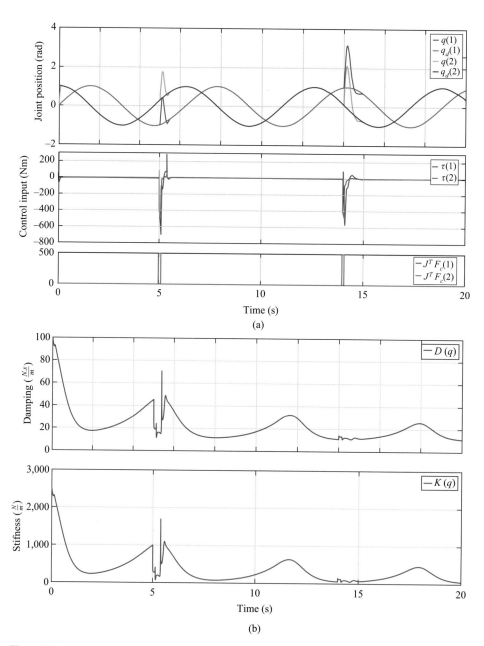

Figure 10.6

The desired trajectory is sinusoidal $q_d = \begin{bmatrix} sin(t) \\ cos(t) \end{bmatrix}$, and the external perturbation torque ($J^T F_c$) is applied to both joints at $t = 7.5$s and $t = 15$s. The control law [equation (10.22)] is used to drive the robot where $\mathbf{K}(\tilde{x}) = \frac{\lambda_1 \mathbf{K}}{\lambda_1 \mathbf{K} + \lambda_2 \mathbf{K}}(\tilde{x})K_1 + \frac{\lambda_2 \mathbf{K}}{\lambda_1 \mathbf{K} + \lambda_2 \mathbf{K}}(\tilde{x})K_2$ and $\mathbf{D}(\tilde{x}) = \frac{\lambda_1 \mathbf{K}}{\lambda_1 \mathbf{K} + \lambda_2 \mathbf{K}}(\tilde{x})D_1 + \frac{\lambda_2 \mathbf{K}}{\lambda_1 \mathbf{K} + \lambda_2 \mathbf{K}}(\tilde{x})D_2$. $\lambda_1 = (q - \mu_1)^T(q - \mu_1) + (Dq - \mu_3)^T(Dq - \mu_3) * t$ and $\lambda_2 = (q - \mu_2)^T(q - \mu_2) + (Dq - \mu_4)^T(Dq - \mu_4)$, where t is time, $\mu_1 = \begin{bmatrix} -1 \\ 1 \end{bmatrix}$, $\mu_2 = \begin{bmatrix} -1 \\ 1 \end{bmatrix}$, $\mu_3 = \begin{bmatrix} 0 \\ 0 \end{bmatrix}$ and $\mu_4 = \begin{bmatrix} -1 \\ -1 \end{bmatrix}$. $D_1 = 10I_{2 \times 2}, D_2 = 100I_{2 \times 2}, K_1 = 25I_{2 \times 2}$ and $K_2 = 2500I_{2 \times 2}$. Because the impedance profile is a function of the state of the system, it adjusts with respect to the sensory information. Hence, the perturbations have an effect on the impedance profile.

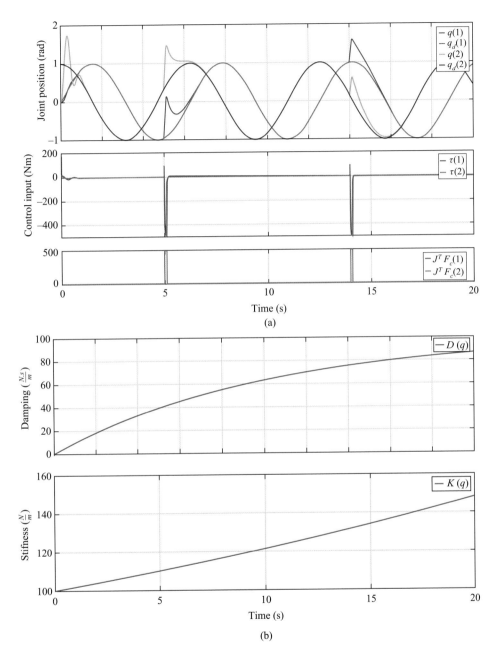

Figure 10.7

The desired trajectory is sinusoidal $q_d = \begin{bmatrix} sin(t) \\ cos(t) \end{bmatrix}$, and the external perturbation torque $(J^T F_c)$ is applied to both joints at $t = 7.5$s and $t = 15$s. The control law [equation (10.26)] is used to drive the robot. The stiffness and damping are increased with respect to the stability constraints [equation (10.31)]. Because the impedance profile is a function of time and not the state of the system, the perturbations do not have an effect on the impedance profile.

Programming Exercise 10.2   *The aim of this exercise is to help readers to get a better understanding of the three variable impedance control approaches and the effect of the open parameters on the generated motion. Open MATLAB and set the directory to the following folder:*

```
1   ch10-DS_compliant/Variable_Impedance_controller
```

For each variable impedance approach, study the effects of the following open parameters on the generated trajectories:

1. *The dynamic specifications of the robot:*
    (a) *Mass of the links*
    (b) *Length of the links*
2. *The initial configuration of the joints:*
    (a) *Position of the joints*
    (b) *Velocity of the joints[a]*
3. *The desired impedance parameters:*
    (a) *The stiffness matrix*
    (b) *The damping matrix*
4. *The activation parameters:*
    (a) *At the position level*
    (b) *At the velocity level*
5. *What is the effect of the sampling time on the generated trajectories?*

Exercise 10.6   *Consider the dynamics of a robotic arm as defined in (C.2) driven by equation (10.17). Prove that the dynamic of the error is asymptotically converging toward zero [equation (10.19) is satisfied] if $F_c = 0$ and conditions of equation (10.20) are met.*

Exercise 10.7   *Consider the task of reaching and contacting a large fixed object presented in section 8.4, with motion dynamics for the robot as defined in (C.2) and driven by equation (10.17). Design the impedance profile such that the robot is stiff if it is far from the surface and compliant if it is close to the surface.*

What would happen if one designed the stiffness profile and kept the damping matrix constant? What role has the inertial matrix?

Exercise 10.8   *Consider two robotic arms driven by the virtual object based DS [equations (7.25) and (7.24)]. If the dynamics of the robotic systems are identical [equation (C.2)], and the low-level control law is equation (10.17), design the impedance profile such that the robots are stiff if (1) they are far from the virtual object, (2) the virtual object is far from the real object, and they are compliant if (1) or (2) is not true.*

What would happen if the robotic systems were not identical (i.e., the dynamic constraints of the robots are different)?

10.3 Learning the Desired Impedance Profiles

Defining the right impedance is very important for accomplishing complex tasks. Often, the right impedance changes as the task unfolds. To this end, significant research efforts have been devoted to the issue of when and how to vary the impedance. Many methods have been developed to determine automatically how to vary the impedance so as to achieve better performance, greater safety, or lower energy consumption. In this section, we introduce two of many approaches for learning/approximating the impedance profile from a set of demonstrations.

10.3.1 Learning VIC from human motions[5]

In this approach, stiffness variations are derived for a compliant controller from a set of position-level demonstrations. A probabilistic model is fit to the demonstrated trajectories, and the stiffness profile is shaped so that the robot adopts a high stiffness in directions of low variance and vice versa. This approach is based on the assumption that the user conveys impedance information by watching how the demonstrated motions vary. High variance would reflect low impedance. The approach is advantageous in that impedance information is derived from kinematic data only. The user, hence, does not need to provide force information as well. One disadvantage, however, is that the assumption that impedance is directly related to kinematic variability in the demonstrations may not be a feature of the task, but rather of poor user skills. The users may end up with a robot using impedance variations that they did not wish to convey. This approach is amenable only to expert demonstrators.

Given a set of demonstration consisting of M trajectories, where each trajectory $m \in \{1, \ldots, M\}$ consists of a set of T^m positions ($x \in \mathbb{R}^N$), velocities ($\dot{x} \in \mathbb{R}^N$), and accelerations ($\ddot{x} \in \mathbb{R}^N$) of the robot's end-effector, as presented in equation (10.21), one can use a controller based on a time-varying mixture of K linear systems:

$$\ddot{\hat{x}} = \sum_{k=1}^{K} \gamma_k(t) \left[K^k(\mu^k - x) - \mathbf{D}\dot{x} \right].$$ (10.32)

In equation (10.32), the stiffness profile is changing with respect to time,[6] but the damping profile is constant. The activation parameters can be approximated by a set of Gaussians $\mathcal{N}(\mu^k, \Sigma^k)$, with centers μ^k equally distributed over time, and variance parameters Σ_i set to a constant value inversely proportional to K:

$$\gamma_k(t) = \frac{\mathcal{N}(t; \mu^k, \Sigma^k)}{\sum_{i=1}^{K} \mathcal{N}(t; \mu^i, \Sigma^i)}.$$ (10.33)

Assuming that K_{min} and K_{max} are the minimum and the maximum of what are dynamically feasible stiffnesses of the robot. These are determined by the user and the limitations of the hardware. By concatenating the demonstrated data points in a matrix $Y = \left[\ddot{x}(\frac{K_{min}+K_{max}}{2})^{-1} + 2\dot{x}(\frac{K_{min}+K_{max}}{2})^{-\frac{1}{2}} + x \right]$, the corresponding activation parameters and the mean of the Gaussians in H and μ, respectively, we can write the linear equation $Y = H\mu$, which can be easily used to estimate the means of the Gaussians:

$$\mu = H^+ Y, \tag{10.34}$$

where H^+ is the pseudoinverse of H. The covariance matrices are estimated to represent the variability and correlation along the movement and among the different demonstrations:

$$\Sigma^k = \frac{1}{N} \sum_{j=1}^{N} (Y_k^j - \bar{Y}^k)(Y_k^j - \bar{Y}^k)^T \qquad \forall k \in \{1, \dots, K\}, \tag{10.35}$$

where $Y_k^j = H_k^j \mathsf{D}(Y_j - \mu_k)$ and $\bar{Y}_k = \frac{1}{N} \sum_{j=1}^{N} Y_k^j$. Furthermore, given the maximum and the minimum eigenvalues of Σ^k, $\forall k \in \{1, \dots, K\}$ indicated by λ_{max} and λ_{min}, respectively, the stiffness matrice K_k is estimated through eigen-components decomposition:

$$K^k = V^k D^k V^{k-1}$$

$$D^k = K_{min} + (K_{max} - K_{min}) \frac{\lambda_k - \lambda_{min}}{\lambda_{max} - \lambda_{min}}, \tag{10.36}$$

where λ_k and V_i are the concatenated eigenvalues and eigenvectors of the inverse covariance matrix (Σ^{k-1}). This defines a stiffness matrix that is proportional to the inverse of the observed covariance (i.e., a high variability results in a lower stiffness and vice versa). The rescale matrix D_k is set so as to ensure that the desired stiffness is executable given the hardware to execute.

In this approach, the stiffness profile is learned offline, and it cannot be changed during the course of the motion execution. In approach 2, we will adapt and vary the stiffness profile online.

10.3.2   Learning VIC from kinesthetic teaching[7]

In this approach, a user interface allows an operator to physically interact with the robotic platform to adjust the desired stiffness variations explicitly. If the users want to reduce the stiffness of the robot in a specific direction, they should perturb/shake the robot in that direction. A higher perturbation amplitude results in less stiffness. Hence, the perturbations can be seen as a means to increase the variability of the demonstrated motions that can be mapped into a reduced stiffness profile. Hence, approaches 1 and 2 are similar, as higher variability will result in a lower stiffness profile. However, in approach 2 the stiffness profile is learned online and does not depend on the robot's position.

Let $x, x^* \in \mathbb{R}^3$ denote the real and the desired positions of the end-effector of the robot, and $\tilde{x} = x - x^*$ is a perturbed data point. Let $\Xi = \{\tilde{x}^i, t^i\}_{i=0}^{N}$ denote the set of observed perturbations with their corresponding time stamps, where N is the number of provided perturbation data points.[8] Consider a sliding temporal window view of length S and the number of the perturbed data point received at the range of $[t - S, t]$ is N_t. Let μ_t and σ_t be the mean and the covariance of the data points in the window $[t - S, t]$:

$$\mu_t = \frac{1}{N_t} \sum_{t-S}^{t} \tilde{x}^t \tag{10.37}$$

$$\Sigma^t = \frac{1}{N_t}\sum_{t-S}^{t} \tilde{x}^t \tilde{x}^{tT} - \mu_t \mu_t^{T}.$$

The defined covariance matrix is symmetric and positive definite. Hence, it can be decomposed into its singular values:

$$\Sigma^t = Q \Lambda Q^T, \tag{10.38}$$

where Λ is a diagonal matrix composed of the singular values $0 < \lambda_t^i$. Given this, the stiffness matrix is defined as follows:

$$K^t = Q \Gamma Q^T$$

$$\Gamma = \begin{bmatrix} \gamma(\sqrt{\lambda_t^1}) & 0 & 0 \\ 0 & \gamma(\sqrt{\lambda_t^2}) & 0 \\ 0 & 0 & \gamma(\sqrt{\lambda_t^3}) \end{bmatrix}, \tag{10.39}$$

where the singular values are set negatively proportional to the square root of the corresponding singular value of the covariance matrix:

$$\gamma(\sqrt{\lambda_t^i}) = \begin{cases} K_{min} & \overline{\sigma} < \sigma_t^i \\ K_{max} - (K_{max} - K_{min})\frac{\sigma_t^i - \underline{\sigma}}{\overline{\sigma} - \underline{\sigma}} & \underline{\sigma} < \sigma_t^i \leq \overline{\sigma}. \\ K_{max} & \sigma_t^i < \underline{\sigma} \end{cases} \qquad \forall i \in \{1,2,3\} \tag{10.40}$$

where K_{min} and K_{max} define the lower and the upper admissible values for the stiffness in any direction, respectively. For instance, the maximum stiffness can be determined to prohibit too high stiffness values in order to ensure the safety of the interaction, or it can be set to the maximum stiffness allowed by the hardware. The minimum stiffness can be set to a low value that ensures that the robot is still capable of unconstrained motion if the stiffness is reduced maximally. The sensitivity of the stiffness as a function of the perturbations is controlled by parameters $\underline{\sigma}$ and $\overline{\sigma}$, which determine the amplitudes required to start reducing the stiffness and to achieve minimum stiffness, respectively. Given this, one can rewrite the Cartesian impedance control law [equation (10.8b)] as follows:

$$\mathscr{F} = G_x(q) - (D\dot{x} + K\tilde{x}), \tag{10.41}$$

where K is defined by equation (10.39). It worth noting that the stability of the impedance control architecture can easily be ensured during the motion execution, as the stiffness and the damping values are constant. However, once the operator interacts/perturbs the robotic arm, the updating phase might cause unstable behavior. To address this problem, one can define D and $\dot{D}$ such that it satisfies the stability condition [equation (10.31)].

10.4 Passive Interaction Control with DSs

In the previous section, we introduced impedance-varying control architectures and showed that how one could learn/estimate the parameters of the controller. In the following section, we introduce approaches to control the compliance of the robotic system *directly* by DSs.

Let $f(x)$ be a continuous DS describing a nominal motion plan with a single equilibrium point x^* such that $f(x^*) = 0$. Furthermore, x^* is a stable equilibrium point. The variable $x \in \mathbb{R}^N$ represents a generalized state variable, which could be robot joint angles or Cartesian positions. Any integral curve of $f(.)$ represents the desired motion of the robot in the absence of perturbations.

Similar to (C.1), consider the dynamics of a robotic system in contact with an environment as follows:

$$M(x)\ddot{x} + C(x, \dot{x})\dot{x} + g(x) = \tau_c + \tau_e. \tag{10.42}$$

The goal of this section is to design a controller τ_c so that equation (10.42) has the following properties:

- Passivity $(\tau_e, \dot{x})$ should be preserved for the controlled system.

- The controller should make the robot move according to $f(.)$ and dissipate kinetic energy in a direction perpendicular to $f(.)$.

- It should be possible to vary task-based impedance of the manipulator, such as how dynamics define how external forces τ_e affect the velocity $\dot{x}$.

Because the goal is to have a passive system, it is not necessary for $f(x)$ to be asymptotically stable. To achieve the aforementioned goals, consider a feedback controller consisting solely of a damping term and a gravity cancellation term:

$$\tau_c = g(x) - D\dot{x}, \tag{10.43}$$

where $D \in \mathbb{R}^{N \times N}$ is some positive semidefinite matrix. It is easy to show that the controller in equation (10.43) renders the system [equation (10.42)] passive with respect to the input τ_e, output $\dot{x}$, with the kinetic energy as storage function. This is true for an arbitrarily varying damping, so long as it remains positive semidefinite. By exploiting this fact, construct a varying damping term that dissipates selectively in directions orthogonal to the desired direction of motion given by $f(x)$. Similar to what is described section 8.4, let $e_1, \ldots, e_N$ be an orthonormal basis for $\mathbb{R}^N$, with e_1 pointing in the desired direction of motion. Hence, let $e_1 = \frac{f(x)}{\|f(x)\|}$, and let $e_2, \ldots e_N$ a be an arbitrary set of mutually orthogonal and normalized vectors. Let the matrix $Q(x) \in \mathbb{R}^{N \times N}$ be a matrix whose columns are $e_1, \ldots, e_N$. This matrix is a function of the state x because the vectors e_1 and all $e_1, \ldots, e_N$ depend on x via $f(x)$. We then define the state-varying damping matrix $D(x)$ as follows:

$$D(x) = Q(x)\Lambda Q(x)^T, \tag{10.44}$$

where Λ is a diagonal matrix with nonnegative values on the diagonal $\lambda_1, \ldots, \lambda_N \geq 0$.

By adjusting these damping values, different dissipation behaviors can be achieved. For example, setting $\lambda_1 = 0$ and $\lambda_2, \ldots, \lambda_N > 0$ results in a system that selectively dissipates energy in directions perpendicular to the desired motion. Hence, external work in irrelevant directions is opposed, while along the integral curves of $f(x)$, the system is free to move. If $\|f(x)\|$ is very small (e.g., near the equilibrium point), finding the basis for the damping becomes ill defined. This can easily be handled by keeping the previous basis if $\|f(x)\| < \eta$, where $\eta > 0$ is some predetermined small threshold value.

While the selective damping in equation (10.43) allowed selective energy dissipation, it cannot drive the robot forward along the integral curves of $f(.)$. Hence, the system would move only if external energy is provided to it, in which case kinetic energy along the desired direction of motion would be accepted while kinetic energy in directions perpendicular to the desired motion would be dissipated. In order to make the robot move along the integral curves of $f(.)$ without external input, we have to add some driving control to equation (10.43). This can be achieved through rather simple means, provided that the nominal task model $f(.)$ is the negative gradient of an associate potential function. This restricted class of DS will be referred to as conservative vector.[9] Given this, the controller [equation (10.43)] is modified with negative velocity error feedback:

$$\tau_c = g(x) - D(x)(\dot{x} - f(x)) = g(x) - D(x)\dot{x} + \lambda_1 f(x). \tag{10.45}$$

The last equality occurs because $f(x)$ is an eigenvector of $D(x)$. If $f(x)$ is a conservative system with an associated potential function $V_f(x)$. Then, the system [equation (10.42)] under control given by equation (10.45) is passive with respect to the input output pair $\tau_e, \dot{x}$. (See exercise 10.9.)

It is important to note that it is never necessary to evaluate the potential function because it is merely its existence that is required. Unfortunately, only very simple tasks can be modeled with conservative DS models, and learned models such as stable estimator of dynamical systems (SEDS), introduced in section 3, or locally modulated dynamical systems (LMDS), introduced in section 8.2, are not generally conservative. The accuracy at which the desired dynamics will be followed in the absence of external wrench depends on the dynamic properties of the manipulator, as well as the desired dynamics.

10.4.1   Extension to Nonconservative Dynamical Systems

Let $f(x)$ be decomposed into a conservative part and a nonconservative part:

$$f(x) = f_c(x) + f_r(x), \tag{10.46}$$

where $f_c(.)$ denotes the conservative part, which has an associated potential function; and $f_r(.)$ denotes the nonconservative part. Any system can be written in this form, although it may not always be trivial to find such a decomposition. Section 10.4.1.1 gives guidelines as to how such a decomposition can be acquired.

We shall consider an additional state variable $s \in \mathbb{R}$, which represents stored energy. It is a virtual state to which we can assign arbitrary dynamics. We shall consider dynamics coupled with the robot state variables $x, \dot{x}$ as follows:

$$\dot{s} = \alpha(s)\dot{x}^T D\dot{x} - \beta_s(z, s)\lambda_1 z, \tag{10.47}$$

where $z = \dot{x}^T f_r(x)$. The scalar functions $\alpha : \mathbb{R} \mapsto \mathbb{R}$ and $\beta : \mathbb{R} \times \mathbb{R} \mapsto \mathbb{R}$ control the flow of energy between the virtual storage s and the robot, and will be defined in the following discussion. It is necessary to put an upper bound on the virtual storage, such that it can store only a finite amount of energy. Let $\bar{s} > 0$ denote this upper bound. Then, $\alpha(s)$ should satisfy

$$\begin{cases} 0 \le \alpha(s) \le 1 & s < \bar{s} \\ \alpha(s) = 0 & s \ge \bar{s} \end{cases} \tag{10.48}$$

Disregarding for the moment the second term in equation (10.47), it is clear that the first term (energy that would otherwise be dissipated) only adds to the virtual storage, so long as the latter remains below its upper bound, $s < \bar{s}$. Now turning to the second term of (10.47), $\beta_s(z, s)$ should satisfy

$$\begin{cases} \beta_s(z, s) = 0 & s \leq 0 \text{ and } z \geq 0 \\ \beta_s(z, s) = 0 & s \geq \bar{s} \text{ and } z \leq 0 \\ 0 \leq \beta(z, s) \leq 1 & \text{elsewhere} . \end{cases} \tag{10.49}$$

Considering the second term in equation (10.47), it is clear that with βs satisfying equation (10.49), transfer to the virtual storage ($z < 0$) is possible only so long as $s < \bar{s}$. Conversely, extraction of energy from the storage ($z > 0$) is possible only so long as $s > 0$. When the storage is depleted, the controller can no longer be allowed to drive the system along f_r if this increases the kinetic energy of the system. Therefore, we introduce the scalar function $\beta_R(z, s)$, whose role is to modify the control signal if the storage is depleted:

$$\tau_c = g(x) - D\dot{x} + \lambda_1 f_c(x) + \beta_R(z, s)\lambda_1 f_r(x), \tag{10.50}$$

where $\beta_R : \mathbb{R} \times \mathbb{R} \mapsto \mathbb{R}$ is a scalar function that should satisfy

$$\begin{cases} \beta_R(z, s) = \beta_s(z, s) & z \geq 0 \\ \beta_R(z, s) \geq \beta_s(z, s) & z < 0. \end{cases} \tag{10.51}$$

Given a nominal task model $f(x)$, which is composed of conservative and nonconservative parts according to equation (10.46) and the robotic system [equation (10.42)] is controlled by equation (10.50), and assuming that the functions α, β_s, β_R satisfy the conditions in equations (10.48), (10.49), and (10.51), respectively. One can prove that $0 < s(0) \leq \bar{s}$ results in a passive closed-loop system with respect to the input-output pair τ_e, x (see exercise 10.10).

10.4.1.1 Decomposition of task DS

The controller described in section 10.4.1 relies on the decomposition of the DS into a conservative and a nonconservative part, as per equation (10.46). As shown in equation (10.45), the conservative part of the DS can always be tracked. The nonconservative part may be scaled to zero if the energy tank is depleted [see equation (10.50)]. Importantly, the passivity of the controller does not rely on the decomposition of equation (10.46) being perfect (in other words, it is not necessary that $f_r(.)$ is purely rotational). For example, any DS can be used with $f_c(.) = 0$ and $f_r(.) = f(.)$. The advantage of extracting a conservative component $f_c(.)$ is that this component can always be followed, even when the energy tank is depleted. Therefore, it is interesting to provide as good a decomposition as possible.

How to extract a conservative component from a DS depends on the method that was used to learn and encode the DS model. As an example, in LMDS, introduced in 8.2, the dynamics $f = G(x)f_O(x)$ that are based on conservative original dynamics $f_O(.)$ have an implicit decomposition into conservative and nonconservative parts:

$$f_c(x) = f_O(x) \tag{10.52a}$$

and

$$f_r(x) = (M(x) - I_{N \times N}) f_O(x), \tag{10.52b}$$

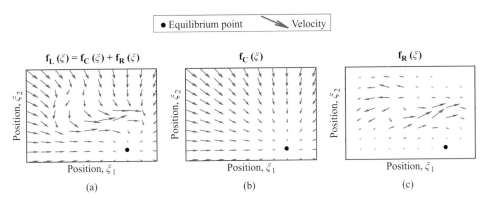

Figure 10.8
The velocity from a DS evaluated over a 2D workspace. (**a**) A task model represented with LMDS. (**b**) The conservative component (linear isotropic system used as the original dynamics in LMDS), as in equation (10.52a). (**c**) Nonconservative component, (10.52b).

where $f_O(x)$ denote the original dynamics, $G(x) \in \mathbb{R}^{N \times N}$ represents the continuous matrix-valued modulation function. This decomposition is illustrated in figure 10.8.

As introduced in section 3, several proposed methods for learning stable DS models in literature use a known Lyapunov function for ensuring the stability of nonlinear DS models during learning. Examples include SEDS. This class of DS has a straightforward decomposition given implicitly by the known Lyapunov function. Let $f_P(x)$ denote such a system, and let $V_P(x)$ denote the associated Lyapunov function. Because $V_P(x)$ is known, a conservative component can be found by taking its gradient:

$$f_c(x) = -\nabla V_P(x) \tag{10.53a}$$

and

$$f_r(x) = f_P(x) - f_c(x) = f_P(x) + \nabla V_P(x). \tag{10.53b}$$

It is possible to combine the LMDS approach with batch-learning methods using a known Lyaupunov function. For example, a stable SEDS model can serve as the original dynamics in LMDS. The conservative part of the SEDS model, equation (10.53a), would then also be the conservative part of the reshaped system, and the nonconservative part is simply the difference between the reshaped dynamics and the conservative part [equation (10.46)].

10.4.1.2 Impedance adjustment

The architecture used in this section differs fundamentally from the classical impedance control framework introduced in section 10.2, in that there is no notion of reference position. Instead, there is only a reference velocity, which is generated online as a function of the robot position. The classical mass-spring-damper model, which is most commonly used in impedance control, has the advantage that a designer can develop an intuition for how the system behavior will change as a result of a modification of the impedance parameters. This section elucidates the link between the proposed controller and this model in order to aid in intuitive understanding of the behavior of the proposed system.

We write the closed-loop dynamics of the system [equation (10.42)] under control $\tau_c = g(x) - D\dot{x} + \lambda_1 f(x)$. Note that the passive controller derived in section 10.4.1 yields the same closed-loop behavior in the ideal case that the virtual storage is never depleted:

$$M(x)\ddot{x} + (D(x) + C(x,\dot{x}))\dot{x} - \lambda_1 f(x) = \tau_e. \tag{10.54}$$

Similar to impedance controllers implemented without force sensing, it is not possible to alter the inertia of the system. We have control of the damping in directions orthogonal to the desired motion via the damping values $\lambda_2 \cdots \lambda_N$, which are allowed to vary with time, state, or any other variable. The stiffness term is replaced by $\lambda_1 f(x)$, which can be interpreted as a nonlinear stiffness term. This interpretation is evident when considering the behavior of equation (10.54) close to the stable equilibrium point of $f(x)$. For simplicity, consider the robot in steady state ($\dot{x} = \ddot{x} = 0$) near the equilibrium point x^*, such that $f(x^*) = 0$. Accounting for steady state and approximating the left side of equation (10.54) with a first-order Taylor expansion around x^* then yields:

$$-\lambda_1 \left.\frac{\partial f}{\partial x}\right|_{x=x^*} (x - x^*) = \tau_e, \tag{10.55}$$

corresponding to a steady-state stiffness equal to the Jacobian of $f(x)$ at the equilibrium point scaled by the value of λ_1. Globally, the term $\lambda_1 f(x)$ can be interpreted as a nonlinear stiffness term *centered on the equilibrium point of* $f(.)$.

While the classical notion of stiffness manifests itself in the vicinity of the equilibrium point of $f(.)$, it is not generally possible to generalize this to stiffness around general points in the workspace. To see this, consider again the steady-state linearization of the left side of equation (10.54), but this time around, there is an arbitrary point x' with $f(x') \neq 0$:

$$-\lambda_1 f(x') - \lambda_1 \left.\frac{\partial f}{\partial x}\right|_{x=x'} (x - x') = \tau_e. \tag{10.56}$$

A key observation is that a stiffness behavior includes symmetry, where perturbations around a point on the desired trajectory are opposed uniformly around the reference trajectory. The DS task model, on the other hand, encodes infinitely many desired trajectories, given by the integral curves of $f(.)$. Hence, if the classical behavior of symmetrically converging toward a fixed trajectory is desired, this should be encoded in the task model $f(.)$. An example of a DS that locally encodes this springlike behavior is given in figure 10.9.

Programming Exercise 10.3   *The aim of this exercise is to help readers to get a better understanding of the control law [equation (10.50)] and how it can be implemented. Open MATLAB and set the directory to the following folder:*

```
ch10-DS_compliant/Passive_Dynamical_system
```

This package provides a graphical user interface (GUI) that allows you to learn a first-order, LPV-based DS and control the robot by equation (10.44) such that it follows the generated motion by the DS.

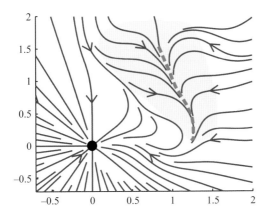

Figure 10.9
A 2D illustration of a task model that locally encodes a stiffness-like behavior. The green dashed path corresponds to the reference trajectory. The shaded area roughly delineates the region in which the behavior of the DS is qualitatively similar to a stiffness attraction toward the reference trajectory. In this example, a perturbation from the nominal trajectory would yield a restoring force within the shaded region. The response to a large perturbation that makes the robot leave the shaded area is instead met with a completely different path to the target point.

Answer the following questions:

1. *Do the eigenvalues of the damping matrix matter? What would happen if the eigenvalues were very small or very big? What if one of them was zero? In which of these situations can the robot reach the target while following the DS?*

2. *How robust is the robot in face of perturbations if the perturbations are (1) aligned and (2) not aligned with the learned DS? You can perturb the robot, in the execution phase, by dragging it in any direction.*

3. *What would happen if the target were defined outside the robot-reachable?*

11 Force Control

Force control is essential to robotics. This is required in all tasks that require manipulating objects dexterously. Typical interactive examples include polishing, assembly, cooperative manipulation, and telemanipulation. Force control is required to stabilize robots and for safety in human-robot interaction. In this chapter, we combine force control with the impedance-based DS introduced in chapter 10 to provide robust control of forces in unforeseen situations and in the face of disturbances.

In chapter 10, we introduced DS-based algorithms for providing compliant behavior in contact phases. This chapter presents control architectures that enhance the interactions by controlling the interaction forces while being compliant. We show how we can simultaneously control the force and the motion of a robotic system when moving on non-planar surfaces. This chapter is also accompanied by hands-on programming exercises to show how one can learn a nonlinear model of the surface on which to apply the force. Codes also show examples of how to learn on the fly an adaptive force profile to compensate for poorly modeled contact forces. Readers are highly encouraged to download the source codes and change the parameters and analyze their effects on the behavior.[1]

11.1   Motion and Force Generation in Contact Tasks with DSs

In this chapter, we introduce a control strategy to perform contact tasks with robustness to large real-time disturbances, which can be either human interactions (e.g., stopping the robot, breaking the contact, and moving the robot arbitrarily) or unexpected changes in the environment (e.g., the position and orientation of the surface/object); see figures 11.1 and 11.2.

As presented in section 10.4 in chapter 10, DSs are suitable for generating impedance control laws that provide a compliant and passive robot behavior. Moreover in section 8.4, we introduced a DS-based approach for achieving stable contact in contact/noncontact scenarios. However, the presented approach was not able to control for the contact forces. By exploiting these two approaches, in this section, we introduce a time-invariant, DS-based framework to control contact forces in contact tasks. This strategy is based on local modulation of the robot's nominal task dynamics in order to generate the desired motion and contact forces when the robot is close to the surface. As a result, this strategy offers stable and accurate motion and contact force generation.

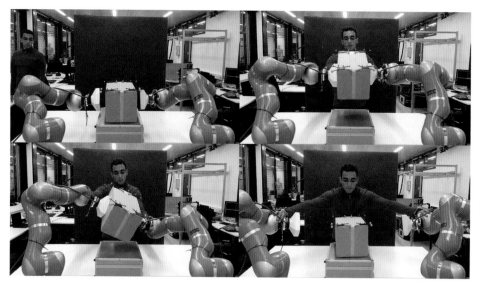

Figure 11.1
Two compliant robot arms reach and grasp a cardboard box (*top left*). A human manipulates the system by changing its pose (*top right and bottom left*) and breaking the grasp (*bottom right*) without endangering safety and stability.

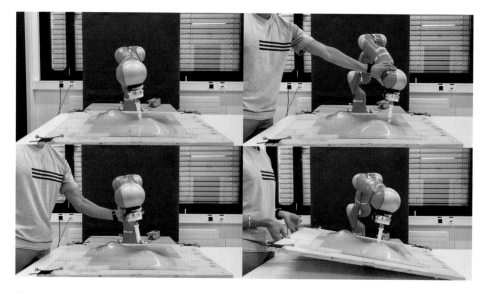

Figure 11.2
A robot arm comes in contact with a surface to perform a circular polishing task (*top left*). Our strategy allows a human to safely interact with the robot while it is moving on the surface (*top right*), break the contact at any moment (*bottom left*), and move the surface (*bottom right*) without compromising the stability of the system.

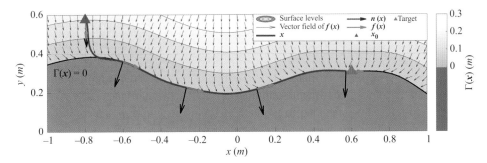

Figure 11.3
Illustration of a robot driven by a nominal DS to come in contact with a surface and move toward a target, starting from an initial position x_0. The normal distance $\Gamma(x)$ and vector $n(x)$ to the surface can be learned using a number learning algorithms, such as support vector regression (SVR) or Gaussian process regression (GPR). Here, we use SVR with a Gaussian kernel ($C = 100$, $\Sigma = 0.01$, $\sigma = 0.20$).

An autonomous DS usually takes as input a state variable (e.g., real position x) and returns the rate of change of that variable [e.g., desired velocity $\dot{x}_d = f(x)$]. It can be seen as a velocity vector field describing the desired behavior for any given position in space.

As illustrated in figure 11.3, in this section, we assume the existence of a nominal DS $f(x)$ that brings the robot into contact with a surface and moves it along the surface. We suppose that the contact surface is nonpenetrable, and that we have an explicit expression for the normal vector $n(x)$ and distance to the surface $\Gamma(x)$ at all points in space. The nominal DS should satisfy the following:

$$\begin{cases} f(x)^T n(x) = 0 & \text{in contact} \\ f(x)^T n(x) > 0 & \text{in free motion.} \end{cases} \tag{11.1}$$

As shown in sections 3 and 8, such dynamics can be learned from human demonstrations and locally modulated to meet these constraints. Once the robot is on the surface, the goal is to apply a state-dependent, desired force profile along the normal to the surface $F_d(x) \in [0, F_{max}]$ with ($F_{max} > 0$). Thereafter, we show how to modulate the nominal DS to generate contact forces in addition to motion.

The generated contact forces are not only the result of the desired motions, but also of the dynamics of the robot. We express the dynamics of N degrees of freedom robotic manipulator in three-dimensional (3D) Cartesian space:

$$M(x)\ddot{x} + C(x, \dot{x})\dot{x} = F_c + F_e, \tag{11.2}$$

where $x \in \mathbb{R}^3$ denotes the robot's position, $M(x) \in \mathbb{R}^{3 \times 3}$ the mass matrix, and $C(x, \dot{x})\dot{x} \in \mathbb{R}^3$ the centrifugal forces, while $F_c \in \mathbb{R}^3$ and $F_e \in \mathbb{R}^3$ represent the control and external forces, respectively. Equation (11.2) presumes that the gravity forces $g(x) \in \mathbb{R}^3$ are already compensated for. The control force F_c allows one to track a desired velocity profile $\dot{x}_d \in \mathbb{R}^3$ and it is obtained from the DS-impedance controller [equation (10.45) in chapter 10], which can be rewritten as

$$F_c = D(x)(\dot{x}_d - \dot{x}) = \lambda_1 \dot{x}_d - D(x)\dot{x}, \tag{11.3}$$

where $D(x) \in \mathbb{R}^{3 \times 3}$ is a state-varying damping matrix constructed in such a way that the first eigenvector is aligned with the desired dynamics $\dot{x}_d$ with positive eigenvalue $\lambda_1 \in \mathbb{R}^+$. The first term in equation (11.3) represents the driving force along the desired dynamics where λ_1 appears as an impedance gain. The last term is the damping force that can be manipulated through the last two eigenvalues of $D(x)$ (λ_2 and $\lambda_3 \in \mathbb{R}^+$) to selectively damp disturbances that are orthogonal to the desired velocity.

In this section, we consider a scenario where the DS is applied only to the translation of the end-effector. The desired end-effector's orientation is tracked using the axis-angle representation. The measured and desired orientation are specified as full rotation matrix by $R \in \mathbb{R}^{3 \times 3}$ and $R_d \in \mathbb{R}^{3 \times 3}$, respectively. The orientation error is computed as $\hat{R} = R_d R^T \in \mathbb{R}^{3 \times 3}$ and the corresponding axis-angle representation $\hat{\zeta}$ extracted to compute a control moment using a proportional derivative–like control law. The control wrench formed by the control moment and force (e.g., F_c) is then converted into joint torques using the robot's Jacobian matrix $J \in \mathbb{R}^{6 \times N}$. Therefore, we assume that the robot has torque-sensing ability and is torque-controlled. Torque-controlled robots allow compliant interaction control (specifically, impedance control, which exhibits satisfactory performance in interaction with stiff environments).

11.1.1 A DS-Based Strategy for Contact Task

To achieve the desired motion and force profile with a single DS, we decompose the system as follows:

$$\dot{x}_d = f(x) + f_n(x), \tag{11.4}$$

with $\dot{x}_d$ being the desired velocity profile and $f_n(x)$ being a modulation term that applies only along the direction normal to the surface. Inserting equation (11.4) into equation (11.3), the control force becomes

$$F_c = \lambda_1 f(x) + \lambda_1 f_n(x) - D(x)\dot{x}. \tag{11.5}$$

The first term represents the driving force along the nominal dynamics, the third term is the damping force, and the second term denotes the modulation force along the normal direction to the surface, which we design as follows:

$$f_n(x) = \frac{F_d(x)}{\lambda_1} n(x). \tag{11.6}$$

As an illustration of the strategy, in figure 11.4, the nominal DS presented in figure 11.3 is modulated to generate a contact force once the robot reaches the surface. Before making contact with the surface, the desired and nominal DS are aligned and identical. Close to contact, the normal modulation component is generated and modulates the nominal DS to produce the desired force. To illustrate the robustness of our approach in face of disturbances, an external force disturbs the robot away from the surface while the robot is moving. The modulated DS reacts to the disturbance by realigning with the nominal one. Once the disturbance disappears, the robot reaches the surface and moves toward the target while applying the desired contact force.

When controlling a robot interacting with unknown environments, one should make sure that the interaction is stable for both performance and safety purposes. A sufficient condition to achieve stability is to ensure the passivity of the whole system. It implies that

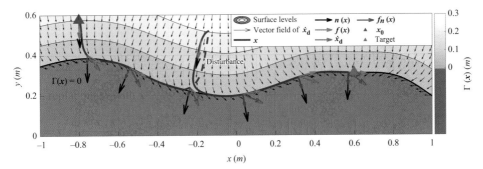

Figure 11.4
Illustration of the modulation approach on the task of reaching and moving on a nonflat surface. The robot is driven by the modulated DS and undergoes a disturbance normal to the surface (dashed line).

the system never generates extra energy, or in other words, that the total energy of the system is bounded by the initial stored energy plus the one injected in the system from the interaction with the environment. However, in order to prove the passivity of the system, equation (11.4) should be modified. For more information, see exercise 11.1.

Programming Exercise 11.1 *The aim of this exercise is to help readers to get a better understanding of the control law [equation (10.50) in chapter 10] and the effect of the open parameters on the generated motion. Open MATLAB and set the directory to the following folder:*

```
1            ch11-Compliance_force/Motion_Force_Generation
```

This package provides a graphical user interface (GUI). You can draw a surface and define an attractor for the DS on the surface. The system will learn the distance function Γ to the surface and generate a linear DS given by equation (11.5) that converges to the surface, moves along the surface while applying a fixed force, and stops at the attractor. Answer the following questions:

1. Does the shape of the surface have an effect on the generated motion/force? Would the robot reach the target in each of the following circumstances?

  (a) The surface is convex.

  (b) The surface is concave.

  (c) The surface has multiple local minima.

2. Observe the distance function. Are all the isolines parallel to the surface? If not, try to improve this by modifying the parameters of the support vector regression technique, in function computeModel.

3. What would happen if the target was not located on the surface? Can the robot reach it?

4. What would happen if the target was located under the surface? Can the robot get close to it?

5. Set the directory to ch11-Compliance_force/Learning-force-adaptation. This code allows you to learn a compensation term for the force profile to adapt to poorly modeled forces at contact. Modify the code to have the RBF functions placed non-uniformly. What is the effect on speed of convergence? Modify the code for a time-varying force error and set the parameters of the learning appropriately so that the adaptation can keep up with the time varying force.

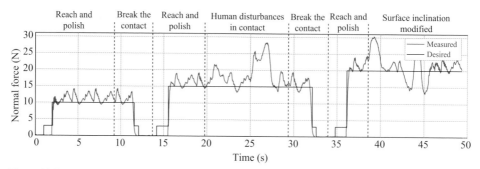

Figure 11.5
Polishing task under various human disturbances: measured versus desired normal force.

11.1.2 Robotic Experiments

In this section, the use case of the presented DS-based strategy is highlighted in two real-world tasks: (1) polishing of a nonflat surface using a single robotic arm robot and (2) reaching, grasping, and manipulating an object with two robotic arms (see exercise 11.2 for more information). In both scenarios, the ability of the approach to generate the desired force profile in the various types of disturbances, such as moving the surface/object unexpectedly prior to and during contact or breaking the contact, is necessary.

In the first task, the DS modulation strategy is tested with a circular polishing task on a nonflat surface, as illustrated in figure 11.2. A robotic arm (KUKA LWR IV+) with 7 degrees of freedom (DOF) is used to perform the task. The robot is equipped with joint torque sensors at the actuators and can be torque-controlled. A six-axis ATI force-torque sensor is also mounted on the end-effector, on which a 3D-printed finger tool is attached. The robot's behavior is evaluated in a simple scenario: the robot comes in contact with the target surface to perform a circular motion on the surface while applying the desired contact force and experiencing disturbances from a human.

Figure 11.5 shows the measured and desired force profiles recorded during the experiment. The robot first reaches the surface to perform the polishing task without experiencing any disturbances. The force generation is relatively accurate, with a root-mean-square (RMS) force error of around 1.9 N (19 percent of the desired force) during this period.

The second experimental scnerio is done with two KUKA LWR IV+ robots to reach and grasp a cardboard box, as shown in figure 11.1. The box has a mass of 0.65 ± 0.05 kg and is tracked by the motion capture system to get its pose. Both robots are equipped with a six-axis ATI force-torque sensor at the end-effector, on which a flat palm is mounted for grasping. The evaluation scenario is designed such that the two arms reach and grasp the object before a human comes and interacts with the system by moving it around, changing its orientation or even breaking the grasp.

Figure 11.6a illustrates the measured and desired contact forces. The RMS force error when the object is grasped, and without human disturbances, are around 1.7 N (11.3 percent of the desired force) for both robots. The noncontact/contact transition in the reaching and grasping phases is smooth, and no instability is observed in the force profiles when the human intentionally breaks the grasp. Similarly, despite disturbances applied to the system

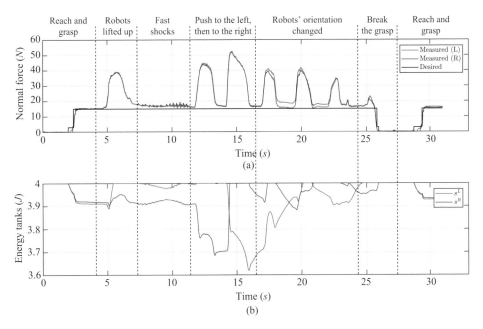

Figure 11.6

Reaching, grasping, and manipulation task under various human disturbances: (a) Measured $(F^{R^T} n^R, F^{L^T} n^L)$ and desired contact force $\left(F_d(x^L, x^R)\right)$. (b) Robots' energy tank s_L and s_R.

after grasping (e.g., fast shocks on the box, changing's system pose), the measured forces remain smooth, guaranteeing the stability and delivering a satisfactory compliant behavior.

Figure 11.6b illustrates the behavior of the energy tanks for both robots (see exercise 11.1). The tanks are initialized at the maximum allowed level, which is set to 4.0 J. When the robots are initially moving toward the object, energy is mainly dissipated. However, this energy cannot be stored in the tanks because they are already full. Close to contact, a desired contact force starts to be generated while the robots are still slightly moving. These nonpassive actions are implemented by extracting energy from the tanks. Once the object is grasped, the tank levels remain constant until the human moves the robots to lift the object. This dissipated energy is stored in the tanks, but in a nonsymmetrical way due to the interaction. When the human applies fast shocks to the object, the tank levels are barely changing, as the robots barely move. Then, moving the system to the left (from the human's point of view) cause the right arm to generate extra energy as it moves in the direction where it applies the force while the left robot dissipates energy. A high amount of energy is extracted from the tank of the right robot to execute this nonpassive action and maintain the grasp. When pushing the system to the right, the opposite behavior happens, with energy being generated by the left robot and dissipated by the right one, leading their associated tanks to be drained and filled, respectively. A similar reasoning can be applied to the other disturbance phases where the human moves the arms to change the object's orientation or break the grasp.

Exercise 11.1   *As mentioned in section 10.4 in chapter 10, the control law [equation (11.3)] is passive for conservative DSs. However, equation (11.4) is not a conservative DS. By using the energy tank approach, modify equation (11.4) such that the system stays passive.*

Exercise 11.2   *In section 11.1.2, the presented control law [equation (11.3)] is used to accomplished a dual-arm scenario. Modify the DS [equation (11.4)] for each robotic arm, such that the robots reach the object, apply a specific amount of force on the object, and move it.*

12 Conclusion and Outlook

We started this book by advocating for robots to be adaptive and to react to disturbances within milliseconds. We took the view that this could be achieved by providing robots with control laws that are inherently adaptive. We chose to use time-invariant dynamical systems (DSs), because they embed multiple paths, all solutions to the problem, and that they do so in closed-form. This enables robots to switch at run time across paths without the need for replanning. Moreover, we showed that a variety of mathematical properties stemming from DS theory could be used to our advantage to provide guarantees on stability, convergence, and boundedness for control laws. We presented a variety of methods that can be used to learn the control laws from data while preserving these theoretical guarantees. Combining machine learning with DS theory draws on the strength of both worlds and allows us to shape robots' trajectories, to avoid obstacles, modulate forces at contact, and move in synchrony with other agents, as required for the task at hand.

Research never stops, and much remains to be done. Among the topics that need immediate attention, we believe the following are most important:

Control in joint space and Cartesian space: While we can control with DSs in any space, we still rely on inverse kinematics to transfer a control deemed stable in Cartesian space in order to generate an adequate posture in joint space. While doing so, we are no longer ensured to have a feasible path in joint space, nor that we will reach the target. To avoid this problem, we rely very much on the fact that the data used to train the system were kinematically feasible. There has been a recent increase in interest in this issue, with different approaches to tackle the problem, from learning a diffeomorphism [113, 108] to determining a joint latent space [131, 121] or enforcing the two systems to contract using contraction theory [122].

Learning region of stability: All the systems presented here are globally stable. The workspace of the robot is not infinite, however. While we say in chapter 9 that one can bound it through a reversal of the obstacle avoidance approach, it may also be interesting to learn automatically the region of attraction. Many approaches have been taken to do so in the literature [73, 13, 154], but these are numerical approaches that depend strongly on the number of hyperparameters of the machine learning approach. Convex approaches are interesting [79, 97], but they generate conservative control laws and do not scale readily to many dimensions [1]. More work is required to enable learning of such regions of attraction for high-DOF robots, such as humanoids, as well as to adapt this modeling to the task at hand, as the region of stability may change over time.

Closed-loop force control: We have presented an initial approach to include explicit control of forces in chapter 11. Recent developments include control with force-feedback with online learning to adapt to the surface in time-dependent systems [46] and for time-invariant DS [4]. This control is done solely at the end point and does not prevent overshoot or instabilities at initial contact when moving fast. For DSs control to be extended to manipulation tasks, there is a need to address these shortcomings.

Inverse dynamics: All the approaches presented here assume that the robot's dynamics was compensated for. This is clearly not the case when controlling some robots, especially humanoid robots. More work is required to ensure accurate tracking and online adaptation of the control law to compensate for poor dynamics. To close, we thank readers for their interest in this book and we hope that they will find these approaches useful for their work. We look forward to hearing feedback from readers.

V APPENDICES

Appendix A: Background on Dynamical Systems Theory

We provide here only a summary of the main concepts and definitions related to dynamical systems (DSs), which would be required to understand the developments in the book. Introductions to the concepts that are core to dynamical systems are made in each chapter as required. Readers who are familiar with dynamical systems are referred to other textbooks, such as [137, 90].

A.1 Dynamical Systems

A DS consists of a set of equations that describe the temporal evolution of a dynamic process. In this book, we are considering deterministic dynamical systems. Let $x \in \mathbb{R}^N$ be the state of the system. The temporal evolution of the system is given by the state time derivative $\dot{x}^* \in \mathbb{R}^N$:

$$\dot{x} = f(x, t), \tag{A.1}$$

where $f : \mathbb{R}^{N+1} \to \mathbb{R}^N$ is a smooth continuous function.

The DS is said to be *autonomous* or *time-invariant* if its evolution does not depend explicitly on time. Then, its temporal evolution simplifies to

$$\dot{x} = f(x). \tag{A.2}$$

In this book, we are restricting our analysis to autonomous dynamical systems. However, many of the results extend to time-dependent dynamical systems. A time-dependent system can always be converted to an autonmous system by expanding the state: $x \in \mathbb{R}^{N+1}$ with $x_{N+1} = t$. As this book treats only of the case of autonomous DS, we refer to these systems broadly as *dynamical systems (DS)*.

A DS can depend on one or more variables. If x and y are two independent variables with $x \in \mathbb{R}^{N_x+1}$ and $y \in \mathbb{R}^{N_y+1}$, we can generate an explicit dependency across the two dynamics:

$$\dot{x} = f(x, y),$$
$$\dot{y} = g(y), \tag{A.3}$$

with $f : \mathbb{R}^{N_x} \to \mathbb{R}^N$, $g : \mathbb{R}^{N_y} \to \mathbb{R}^N$ being two smooth, continuous functions. The dynamics of x and y are then said to be *coupled*.

Dynamical systems are not restricted to first-order differential equations. The system can be of a higher order. For instance, a DS can be characterized by a second-order differential

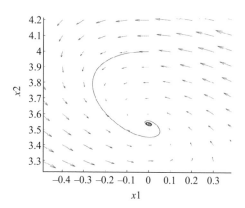

Figure A.1
A visualization of a linear DS. Each arrow represents the DS value at that point [i.e., $\dot{x} = f(x)$], whereas the dark line represents a path integral until the attractor.

equation of the form $\ddot{x} = f(\dot{x}, x)$. However, all dynamical systems can be reduced to a set of first-order different equations by setting

$$\dot{y} = f(x, y),$$
$$y = \dot{x}.$$

(A.4)

A.2   Visualization of Dynamical Systems

The best way to visualize a DS is to plot its vector field. This consists of choosing an array of points in space (which are usually uniform) and plot at each point x an arrow that denotes the vector $\dot{x}$. The direction of the arrow denotes the direction in which the motion moves, whereas the length of the arrow denotes the amplitude.

One can also generate a *path integral* by integrating forward the dynamics from $t = 0$, starting at starting point $x(0)$ and using numerical integration. Many toolboxes would do it for you; however, for information on numerical integration, readers may refer to [141]. The path integral is useful to determine if the system converges or diverges at critical points in the state space. It is used to explore the dynamics of complex dynamical systems whose stability cannot be characterized explicitly. An illustration of the vector field and one path integral is given in figure A.1.

A.3   Linear and Nonlinear Dynamical Systems

A DS is *linear* when it can be written as:

$$\dot{x} = Ax + b,$$

(A.5)

where A is a $N \times N$ matrix and $b \in \mathbb{R}^N$. A sets the shape of the dynamics, whereas b acts as an offset to move the dynamics in space.

A DS is said to be *nonlinear* when f is nonlinear. An example of a nonlinear DS would be

$$\dot{x} = \exp(-\|x\|) \cos(x).$$

(A.6)

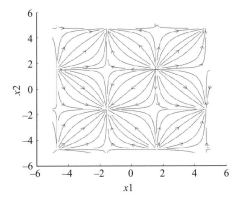

Figure A.2
The nonlinear DS given in equation (A.6), with several locally stable attractors.

As we will see in the next section, the stability of a linear system can be determined analytically. The stability of a nonlinear DS can rarely be analyzed explicitly to determine whether the system, as a whole, is stable. Often, one can characterize the system only locally. For instance, the system in equation (A.6) is locally stable and instable, with a periodicity defined by $\cos(x)$; see figure A.2.

A.4 Stability Definitions

To characterize DSs, one looks for different properties. The most important of all the properties is the notion of *stability*. Characterizing the stability of a DS consists of determining whether the dynamics stabilizes at one or more points in the state space. To determine the stability of a DS, one must first identify the point or points where the dynamics stops. These are called *equilibrium points*:

Definition A.1 *(Equilibrium point). An equilibrium point for a DS is a point* $\mathrm{x} \in \mathbb{R}^N$, *such that* $\mathrm{f}(\mathrm{x}) = 0$.

Finding an equilibrium point is not sufficient to prove stability. One must further study the behavior of the DS around the equilibrium point. An equilibrium would be unstable if the system can be sent away from the equilibrium with a small perturbation. Conversely, it would be stable if the system were always to return to the equilibrium. For instance, the system $x = -x$ has a stable equilibrium at zero, whereas the system $x = -\|x\|$ has an unstable attractor at the origin. A system where the equilibrium is either stable or unstable, depending on the direction of the perturbation, is called a *saddle point*. The system $x = -x$ has a saddle point at zero in two dimensions (2D); for instance, see figure A.3.

Lyapunov stability conditions can be used to assess the stability of an equilibrium point, as described in definition A.2.

Definition A.2 *(Lyapunov stability). An equilibrium point* x^* *is said to be stable in the sense of Lyapunov, or simply stable, if for each* $\epsilon > 0$, *there exists* $\delta(\epsilon) > 0$ *such that*

$$\left\| x\left(t_0\right) - x^* \right\| < \delta \Rightarrow \left\| x(t) - x^* \right\| < \epsilon, \quad \forall t > t_0.$$

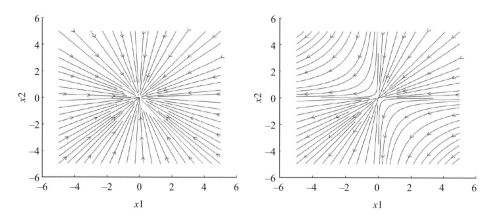

Figure A.3
Linear DS with a stable equilibrium (left) and saddle point (right) at the origin.

One distinguishes between asymptotic and exponential stability. Asymptotic stability ensures that the system will eventually reach the equilibrium, but it does not specify how fast this happens, whereas exponential stability allows one to determine the rate at which the system converges.

Definition A.3   *(asymptotic stability). An equilibrium point x^* is asymptotically stable if it is stable and if there also exists $\delta > 0$ such that*

$$\left\| x\left(t_0\right) - x^* \right\| < \delta \Rightarrow \left\| x(t) - x^* \right\| \to 0, t \to \infty.$$

Definition A.4   *(Exponential stability). An equilibrium point x^* is exponentially stable if it is asymptotically stable and there also exists $\alpha, \beta, \delta > 0$ such that*

$$\left\| x\left(t_0\right) - x^* \right\| < \delta \Rightarrow \left\| x(t) - x^* \right\| \leq \alpha \left\| x(0) - x^* \right\|^{-\beta t}, \forall t.$$

If there is a single stable equilibrium under Lyapunov stability, the system is said to be *globally stable*. When global stability cannot be ensured, one may seek to determine if there is a region around the equilibrium where the dynamics is stable. This region is denoted as the *basin of attraction $B(x^*)$*. The system shown in figure A.2 has one basin of attraction around each of the stable equilibria.

Definition A.5   *(Basin of attraction) A set $\Delta \subseteq \mathbb{R}^N$ is a basin of attraction of an equilibrium point x^* if: $\Delta(x^*) = \{x \in \mathbb{R}^N, \lim_{t \to \infty} f(x) = x^*\}$.*

Another interesting property is boundedness, which allows one to determine if the DS is contained within a region. A basin of attraction is by definition bounded because all paths starting in the basin cannot get out.

Definition A.6   *(Boundedness) A DS is bounded if for each $\delta > 0$, there exists $\epsilon > 0$ such that*

$$\left\| x\left(t_0\right) \right\| < \delta \Rightarrow \left\| x(t) \right\| < \epsilon, \forall t > t_0.$$

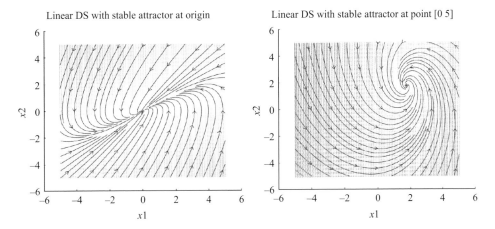

Figure A.4
Linear DS with stable attractor at the origin (*left*) and off-center (*right*).

A.5   Stability Analysis and Lyapunov Stability

Determining the stability of a linear DS is straightforward if matrix A is invertible and accepts,= a real-valued eigendecomposition. The equilibrium is, then, the solution to $Ax + b = 0$, $x^* = -A^{-1}b$. To determine if the equilibrium is stable, one proceeds to an eigenvalue decomposition. $A = V \Lambda V^T$, with V being a matrix composed of the eigenvectors columnwise and Λ being the diagonal matrix of eigenvalues. The following three cases arise:

1. If all eigenvalues are negative, the equilibrium is stable.
2. If all eigenvalues are positive, the equilibrium is unstable.
3. If one eigenvalue is negative and one positive, the equilibrium is a saddle point.

In various chapters of the book, we present methods to determine A and b in such a way as to place the equilibrium at the desired attractor point and to stabilize the system at this point. A can be constructed to ensure that the eigenvalues are strictly negative and b can be chosen in such a way as to shift the equilibrium to the desired attractor point. Figure A.4 top illustrates two linear DSs with a stable equilibrium centered at the origin and off-center.

The stability of nonlinear dynamical systems can be characterized as follows. One first determines the equilibrium points by solving $f(x) = 0$. Note that, unlike linear DSs, the equilibrium point in nonlinear DSs may not be unique or may not be found analytically. The stability of the equilibrium points can then be estimated by linearizing the system locally around the equilibrium through a Taylor expansion and computing the eigenvalue decomposition of the derivative of the function around each equilibrium. The linearization using the first Taylor expansion term holds only if the higher-order terms are negligible. They remain negligible within a small neighborhood of the equilibrium point. To determine this neighborhood (basin of attraction around the equilibrium), one usually proceeds numerically through sampling. To characterize stability analytically, one can revert to the use of

Lyapunov stability theorem and prove that the dynamics described by $f(x)$ is stable at x^*, if there is a Lyapunov function $V(x) : \mathbb{R}^N \to \mathbb{R}^N$ such that

1. $V(x) = 0$ if and only if $x = x^*$.
2. $V(x) \geq 0$, if and only if $x \neq x^*$.
3. $\dot{V}(x) = \nabla V f(x) < 0, \forall x \neq x^*$.

The Lyapunov function acts as an energy function. The minimum of the function is located at the equilibrium (condition 1) and positive elsewhere (condition 2). Condition 3 states that the dynamics moves in such a way that the energy keeps decreasing. The system is hence bound to reach the equilibrium because this is the minimum of the function. A DS that can be characterized by a Lyapunov function is called *conservative*.

Definition A.7 *(Conservative dynamical systems): An autonomous DS f(x) is conservative if and only if there exists a Lyapunov function V_f such that*

$$f(x) = -\nabla V_f(x). \tag{A.7}$$

A.6 Energy Conservation and Passivity

Lyapunov stability specifies that the energy of the system decreases over time before eventually vanishing at the equilibrium. *Passivity* extends this concept to systems that are controlled through an external input u.

As presented in chapter 10, when using a DS to control robots, the output of the DS becomes a trajectory that a low-level controller is tracking. To study the stability of the complete system, one needs to take into account the control input $u \in \mathbb{R}^P$ and verify that the energy injected by this control input does not destabilize the system. In other words, one must verify that the system is *closed-loop passive*. Passivity-based control was introduced in 1989 ([112]).

Consider a closed-loop system whose dynamics is described by

$$\dot{x} = f(x, u). \tag{A.8}$$

Assume the existence of a variable $y = h(x)$, $y \in \mathbb{R}^m$, which keeps track of the changes arising in the state of the system. To ensure that the system remains controllable, one must prove that the system remains *passive*.

Definition A.8 *(Passivity): A system with the form*

$$\dot{x} = f(x, u)$$
$$y = h(x) \tag{A.9}$$

is passive if there is a lower-bounded storage function $V : \mathbb{R}^N \to \mathbb{R}_{0\leq}$ *such that*

$$\underbrace{V(x(t)) - V(x(0))}_{\text{Stored energy}} \leq \underbrace{\int_0^t u(s)^T y(s) ds}_{\text{Supplied energy}} \tag{A.10}$$

is satisfied for all $0 \leq t$, all input functions u, and all initial conditions $x(0) \in \mathbb{R}^N$.

Observe that in the absence of any input (i.e., $u = 0$), we have $V(x(t)) \leq V(x(0))$. With strict inequality, we are back to the Lyapunov stability condition, and the system is guaranteed to go back to equilibrium. This concept is best encapsulated in the following alternative definitions of passivity.

Definition A.9 *(Passivity—definition 2): The system $\dot{x} = f(x, u)$ and $y = h(x)$ is passive if there is a continuously differentiable, positive, semidefinite function $V : \mathbb{R}^N \to \mathbb{R}_{0\leq}$ (the storage function) such that*

$$u^T y \geq \dot{V} = \frac{\delta V}{\delta x} f(x, u), \forall x, u. \tag{A.11}$$

Definition A.10 *(Passivity—definition 3): A system with the form*

$$\dot{x} = f(x, u) \tag{A.12}$$
$$y = h(x)$$

is passive if there is a continuously differentiable, lower-bounded storage function $V : \mathbb{R}^N \to \mathbb{R}_{0\leq}$ such that along the trajectories generated by (A.12)

$$\dot{V}(t) \leq u(t)^T y(t) \tag{A.13}$$

is satisfied for all $0 \leq t$, all input functions u and all initial conditions $x(0) \in \mathbb{R}^N$.

A.7 Limit Cycles

In this chapter, we have discussed only DSs with fixed-point equilibria. However, a DS may also stabilize on a closed path. Such a path is called a *limit cycle*. If the dynamics is initiated on the path, it continues moving along the path indefinitely. Often, the limit cycle is circular or elliptic, and it rotates around a point in space. The trajectories around the limit cycle may be moving toward or away from the path. If they are moving toward it, the limit cycle becomes an attracting surface. If all trajectories lead to the limit cycle, it is referred to as a *stable limit cycle*.

Limit cycles exist only for nonlinear DSs. Determining such limit cycles is difficult in practice. However, creating them can be simpler in some cases. For instance, one can proceed through a change of variable using polar coordinates. In two dimensions, we set ρ, θ as the polar coordinates of x, and we create a DS that converges on a limit cycle by setting a linear DS for ρ, which vanishes at the desired distance from the origin. The rotation is generated by a second DS on θ, which is coupled to ρ:

$$\dot{\rho} = A\rho + b, \tag{A.14}$$
$$\dot{\theta} = f(\theta, \rho).$$

A.8   Bifurcations

In the previous sections, we have seen how to characterize the behavior of the DS through an analysis of global, asymptotic, or exponential stability. Some DSs may change their characteristic behavior due to changes in their input parameters. This is known as a *bifurcation*. After a *bifurcation*, a stable DS can become suddenly unstable, or it may stabilize at a limit cycle in place of stabilizing at a fixed point. In section 4.2 in chapter 4, we present a method by which one can learn a DS with explicit bifurcations. This is useful to embed within a single DS multiple dynamics and control transitions across these dynamics through the bifurcation parameter.

Definition A.11   *(Bifurcation): A time-invariant, autonomous, DS of the form $f(x, \mu)$, $x \in \mathbb{R}^N$, $\mu \in \mathbb{R}^P$, where f is continuous and differentiable. A bifurcation occurs at parameter $\mu = \mu_0$ if there are parameter values μ_1 arbitrarily close to λ_0, with dynamics topologically inequivalent from those at μ_0. For example, the number or stability of equilibria or periodic orbits of f may change with perturbations of μ to μ_0.*

Bifurcation diagrams can be used to divide the μ parameter space into regions of topologically equivalent systems. Bifurcations occur at points that do not lie in the interior of one of these regions.

Hopf bifurcations These correspond to the class of bifurcations that lead to a change from fixed-point equilibrium to the limit cycle. The limit cycle arises when the equilibrium changes stability via a pair of purely imaginary eigenvalues. Here, $f(x, \mu)$ has a set of equilibria $x^*(\mu)$ that depend on the bifurcation parameter μ. If the Jacobian matrix $J(\mu) = \nabla_x f(x^*, \mu)$ has one pair of complex eigenvalues: $\lambda_{1,2}(\mu)$, which becomes purely imaginary when $\mu = \mu_0$. The system changes dynamics as μ passes through μ_0. Its equilibrium changes stability, and a unique limit cycle bifurcates from it.

Appendix B: Background on Machine Learning

B.1 Machine Learning Problems

We define the machine learning problems posed in this book that are used to solve robot learning problems.

B.1.1 Classification

In classification problems, one typically has a data set of inputs/outputs $\{X, Y\} = \{(x^i, y_i)\}_{i=1}^M$, where $x^i \in \mathbb{R}^N$ is the ith N-dimensional input data point (or feature vector) from M samples and $y_i \in \{-1, 1\}$ is the corresponding categorical outcome/output (or class label). The goal of a classification problem is to learn a mapping function $y = f(x) : \mathbb{R}^N \rightarrow \{-1, 1\}$, such that, given a new sample (or query point) of $x' \in \mathbb{R}^N$, we can predict its label; that is, $y' = f(x') \in \{-1, 1\}$ for the binary classification case (see figure B.1). If a data set is linearly separable, then $f(x)$ can be a simple linear function; this, however, is rarely the case for real-world datasets. Hence, one must apply nonlinear classification techniques to find the nonlinear decision boundaries in the data set. In section B.3, we describe how to use Gaussian mixture models (GMMs) to solve a classification problem, while the support vector machine (SVM) technique is described in section B.4.

B.1.2 Clustering

As opposed to classification, in clustering problems we solely have the input data $\{X\} = \{x^i\}_{i=1}^M$. The goal of clustering is to partition a data set into a set of K meaningful subclasses, groups or clusters (see figure B.2). These clusters are denoted by the set $C = \{c_1, \ldots, c_K\}$ and computed based on a defined measure of similarity (or distance) of the input data points. The result of a clustering algorithm is a set of corresponding labels $\{Y\} = \{y_i\}_{i=1}^M$, where $y_i \in \{c_1, \ldots, c_K\}$, which indicate to which group (or cluster) an input data sample belongs to. Throughout this book, we primarily are using GMMs for clustering applications.

B.1.3 Regression

Regression is similar to classification, in the sense that it seeks to learn a mapping function $y = f(x)$ from inputs to outputs. In regression, however, both inputs and outputs are continuous variables; that is, $\{X, Y\} = \{(x^1, y_1), \ldots, (x^M, y_M)\}$ of M samples, where $x^i \in \mathbb{R}^N$ are multidimensional inputs and $y_i \in \mathbb{R}$ is typically a unidimensional output variable (see figure B.3).

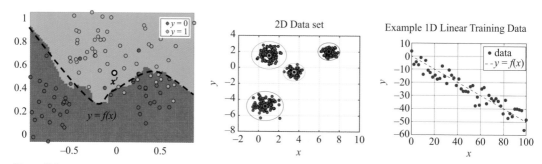

Figure B.1
Classification example on 2D data.

Figure B.2
Clustering example on 2D data.

Figure B.3
Regression example on 1D linear data.

Regression problems are found in a wide spectrum of research, financial, commercial, and industrial areas. For example, one might want to predict the cost of a car based on its attributes, the performance of a CPU given its characteristics, or the exchange rate of a currency, given some relevant economic indicators. Many linear/nonlinear methods exist to obtain this regressive function $f(x)$. In this book, we favor Gaussian mixture regression (GMR), as described in section B.3, as it allows for multivariate outputs, $y \in \mathbb{R}^P$. Yet, we also use SVMs and Gaussian processes (section B.5) to regress unidimensional outputs.

B.2 Metrics

Finally, we end this section by introducing the metrics used to evaluate the learned models, whether it be for model selection, classification, clustering or regression.

B.2.1 Probabilistic Model Selection Metrics

With probabilistic models such as GMM (section B.3) or HMM (see [17]), one can estimate an optimal model size (i.e., the number of K Gaussian components or HMM states) with the following external model metrics:

• **Akaike information criteria (AIC):** The AIC metric is a maximum-likelihood measure that penalizes for model complexity as follows:

$$AIC = -2 \ln \mathcal{L} + 2B, \tag{B.1}$$

where $\mathcal{L}$ the likelihood of the model and B is the total number of model parameters.

• **Bayesian Information Criteria (BIC):** The BIC metric goes even further, penalizing for the number of datapoints as well, with the following equation:

$$BIC = -2 \ln \mathcal{L} + \ln(M)B, \tag{B.2}$$

where M is the total number of datapoints.

B.2.2 Classification Metrics

Many metrics have been proposed to evaluate the performance of a classifier. Most of which, come from a combination of values from the *confusion matrix* (or *error matrix*), as shown in table B.1. In this matrix, the rows represents the real classes of the data and the columns the

Table B.1
Confusion matrix

		Estimated labels	
		Positive	**Negative**
Real labels	**Positive**	TPs	FNs
	Negative	FPs	TNs

estimated classes. The diagonal represents the well-classified examples, while the rest indicate confusion. In the case of a binary classifier (i.e., $y \in \{-1, 1\}$), the following quantities have to be computed:

- **True positives (TPs)**: The number of test samples with a positive estimated label for which the actual label is also positive (good classification)
- **True negatives (TNs)**: The number of test samples with a negative estimated label for which the actual label is also negative (good classification)
- **False positives (FPs)**: The number of test samples with a positive estimated label for which the actual label is negative (classification errors)
- **False negatives (FNs)**: The number of test samples with a negative estimated label for which the actual label is positive (classification errors)

The following metrics can be used to evaluate our classification algorithms.

1. Accuracy: Accuracy represents the percentage of correctly classified data points as follows:

$$ACC = \frac{TP + TN}{P + N}. \tag{B.3}$$

2. $\mathscr{F}$-measure: The $\mathscr{F}$-measure is a well-known classification metric that represents the harmonic mean between Precision ($P = \frac{TP}{TP+FP}$) and Recall ($R = \frac{TP}{TP+FN}$):

$$\mathscr{F} = \frac{2PR}{P + R}. \tag{B.4}$$

It conveys the balance between *exactness* (i.e., precision) and *completeness* (i.e., recall) of the learned classifier.

B.2.3 Clustering Metrics

Clustering is a process of partitioning a set of data (or objects) in a set of meaningful subclasses, called *clusters*. The most popular clustering algorithms, such as K-Means and GMM, represent clusters with K centroids $\mu^k \in \mathbb{R}^N$ (as shown in figure B.2). This is often achieved by minimizing the total squared distance between each point and its closest centroid with the following cost function:

$$J(\mu^1, \ldots, \mu^k) = \sum_{k=1}^{K} \sum_{x^i \in C^k} ||x^i - \mu^k||^2, \tag{B.5}$$

where $C^k \in \{1, \dots, K\}$ is the cluster label. Any centroid-based clustering algorithm then can be evaluated based on equation (B.5). Next, we provide the equations of the clustering metrics that are used in this book.

- **Residual Sum of Squares (RSS):** RSS is, in fact, the cost function that K-means is trying to minimize equation (B.5); therefore,

$$RSS = \sum_{k=1}^{K} \sum_{x^i \in C^k} ||x^i - \mu^k||^2. \tag{B.6}$$

- **AIC:** The AIC metric, introduced for model selection [equation (B.1)], penalizes for model complexity by computing a trade-off between the likelihood of the model and the number of model parameters. With a GMM, equation (B.1) can be used directly to evaluate its clustering performance. Yet, nonprobabilistic clustering approaches, like the K-means algorithm, do not provide a likelihood estimate of the model. Hence, equation (B.1) can be formulated as a metric based on RSS, as follows:

$$AIC_{RSS} = RSS + 2B, \tag{B.7}$$

where $B = (K * N)$ for K clusters and N dimensions.

- **BIC:** As in AIC_{RSS}, we can formulate the BIC metric from equation (B.2) for nonprobabilistic models with the RSS as follows:

$$BIC_{RSS} = RSS + \ln(M)B, \tag{B.8}$$

where B is as before and M is the total number of datapoints.

External Clustering Metrics Comparing the results of different clustering algorithms is often difficult when the labels of the expected clusters (classes) are not given. However, when the labels are available, one can use the $\mathscr{F}$-measure to compare different clustering results.

The $\mathscr{F}$-measure is a well-known classification metric that represents the harmonic mean between Precision ($P = \frac{TP}{TP+FP}$) and Recall ($R = \frac{TP}{TP+FN}$). In the context of clustering, Recall and Precision of the kth cluster with regard to the jth class are $R(s_j, c_k) = \frac{|s_j \cap c_k|}{|s_j|}$ and $P(s_j, c_k) = \frac{|s_j \cap c_k|}{|c_k|}$, respectively, where $S = \{s_1, \dots, s_J\}$ is the set of classes and $C = \{c_1, \dots, c_K\}$ the set of predicted clusters. Further, s_j is the set of data-points in the jth class, whereas c_k is the set of data-points belonging to the kth cluster. The $\mathscr{F}$-measure of the kth cluster with regard to the jth class is

$$\mathscr{F}_{j,k} = \frac{2P(s_j, c_k)R(s_j, c_k)}{P(s_j, c_k) + R(s_j, c_k)}, \tag{B.9}$$

and the $\mathscr{F}$-measure for the overall clustering is then computed as

$$\mathscr{F}(S, C) = \sum_{s_j \in S} \frac{|s_j|}{|S|} \max_k \{\mathscr{F}_{j,k}\}. \tag{B.10}$$

B.2.4 Regression Metrics

In this subsection, we list some of the metrics used throughout the book to evaluate regression results. Given a vector $\hat{Y} = \{\hat{y}_i\}_{i=1}^{M}$ of M predictions and a vector $Y = \{y_i\}_{i=1}^{M}$ of the observed values corresponding to these uni-dimensional predictions, the following metrics can be computed:

- *Mean square error (MSE):*

$$MSE = \frac{1}{M} \sum_{i=1}^{M} \left(\hat{y}_i - y_i \right)^2 \tag{B.11}$$

- *Normalized mean square error (NMSE):* The NMSE is simply the MSE normalized by the variance of the observed values as follows:

$$NMSE = \frac{MSE}{VAR(Y)} = \frac{\frac{1}{M} \sum_{i=1}^{M} \left(\hat{y}_i - y_i \right)^2}{\frac{1}{M-1} \sum_{i=1}^{M} \left(y_i - \mu_Y \right)^2}, \tag{B.12}$$

where μ_Y is the mean of the observed values (i.e., $\mu_Y = \frac{1}{M} \sum_{i=1}^{M} y_i$).

- *Root mean square error (RMSE):* The RMSE yields a measure of the spread of the observed values Y as opposed to the predicted values $\hat{Y}$ as follows:

$$RMSE = \sqrt{\frac{1}{M} \sum_{i=1}^{M} \left(\hat{y}_i - y_i \right)^2}. \tag{B.13}$$

- *Coefficient of determination (R^2):*

$$R^2 = \left(\frac{\sum_{i=1}^{M} \left(y_i - \overline{Y} \right) \left(\hat{y}_i - \overline{\hat{Y}} \right)}{\sqrt{\sum_{i=1}^{M} \left(y_i - \overline{Y} \right)^2} \sqrt{\sum_{i=1}^{M} \left(\hat{y}_i - \overline{\hat{Y}} \right)^2}} \right)^2 = \frac{\left(\sum_{i=1}^{M} \left(y_i - \overline{Y} \right) \left(\hat{y}_i - \overline{\hat{Y}} \right) \right)^2}{\sum_{i=1}^{M} \left(y_i - \overline{Y} \right)^2 \sum_{i=1}^{M} \left(\hat{y}_i - \overline{\hat{Y}} \right)^2}, \tag{B.14}$$

where $\overline{Y}, \overline{\hat{Y}}$ denotes the average of the observed and predicted values of y, respectively.

B.3 Gaussian Mixture Models

A GMM is a parametric probability density function (pdf) represented as a weighted sum of K Gaussian densities. GMMs are commonly used as a parametric model of the probability distribution of a data set $X = \{x^1, \ldots, x^M\}$, where $x^i \in \mathbb{R}^N$. They are popular due to their capability of representing multimodal sample distributions. The pdf of a K-component GMM is of the form

$$p(x|\Theta) = \sum_{k=1}^{K} \gamma_k p(x|\mu^k, \Sigma^k), \tag{B.15}$$

where $p(x|\mu^k, \Sigma^k)$ is the multivariate Gaussian pdf with mean μ^k and covariance Σ^k:

$$p(x|\mu^k, \Sigma^k) = \mathcal{N}(x|\mu^k, \Sigma^k)$$

$$= \frac{1}{(2\pi)^{N/2}|\Sigma^k|^{1/2}} \exp\left\{-\frac{1}{2}(x-\mu^k)^T(\Sigma^k)^{-1}(x-\mu^k)\right\}. \tag{B.16}$$

Here, $\Theta = \{\theta_1, \ldots, \theta_k\}$ is the complete set of parameters $\theta_k = \{\gamma_k, \mu^k, \Sigma^k\}$, where γ_k represents the priors (or mixing weights) of the Gaussian components, which satisfy the constraint $\sum_{k=1}^{K} \gamma_k = 1$.

GMMs are widely used in many areas of engineering; due to their modeling structure and flexibility, they can be used for clustering, classification, and regression purposes.

Θ can be estimated from training data using either a maximum likelihood (ML) parameter estimation approach, through the iterative expectation-maximization (EM) algorithm, or maximum a posteriori (MAP) estimation with fixed values of K. To find the optimal K Gaussians, one must employ model selection techniques or use sampling or variational-based Bayesian nonparametric estimation techniques, which will be introduced next.

B.3.1   Finite Gaussian Mixture Model with EM-Based Parameter Estimation

B.3.1.1   Preliminaries

Finite mixture model: The finite mixture model can be interpreted as a probabilistic hierarchical model, where each kth mixture component is viewed as a cluster represented by an underlying generative distribution $\mathcal{F}$ (e.g., Gaussian, multinomial), parameterized by θ_k (e.g., $\theta_k = \{\mu^k, \Sigma^k\}$ for Gaussian distribution) and its corresponding mixing coefficient π_k. Each data point x^i is then assigned to a cluster k with the clustering assignment indicator variable $Z = \{z_1, \ldots, z_M\}$, where $i : z_i = k$. This process is represented as follows:

$$z_i \in \{1, \ldots, K\}$$

$$p(z_i = k) = \pi_k \tag{B.17}$$

$$x_i|z_i = k \sim \mathcal{F}(\theta_k).$$

Under this hierarchical model, the marginal distribution over Z is defined by the mixing coefficients π_k, viewed as the prior probability of the cluster assignment indicator variable. Through this interpretation, the probability density function of the mixture model is derived by

$$p(x|\Theta, \pi) = \sum_{k=1}^{K} p(z_i = k)f(x|k) = \sum_{k=1}^{K} \pi_k f(x|\theta_k). \tag{B.18}$$

Following equation (B.17), each data point x_i is generated by independently selecting the kth cluster ($z_i = k$) according to the mixing coefficients π_k, and then sampling from that kth distribution, parameterized by θ_k. Further, the posterior probability of a kth component given a data point x_i (i.e., $z_i = k$) is given by

$$p(z_i = k|x^i, \Theta, \pi) = \frac{p(z_i = k, x^i)}{p(x^i|\pi, \Theta)} = \frac{\pi_k f(x^i|\theta_k)}{\sum_{k=1}^{K} \pi_k f(x^i|\theta_k)}, \tag{B.19}$$

which represents the relative importance of each Gaussian component. Equation (B.19) is featured heavily in this book, as it represents the mixing function of the nonlinear DS formulation presented in chapter 3 [i.e., $\gamma_k(x)_i = p(z_i = k|x^i, \Theta, \pi)$].

Parameter estimation: Now, given training data $x \in \mathbb{R}^{N \times M}$ and the structure of the mixture model (namely, the value of K), we seek to estimate the unknown parameters $\Theta = \{\theta_1, \ldots, \theta_K\}$ and $\pi = [\pi_1, \ldots, \pi_K]$ by maximizing the marginal probability of x under equation (B.17):

$$
\begin{aligned}
p(x|\Theta, \pi) &= \sum_Z p(x|Z, \Theta)p(Z|\pi) \\
&= \sum_Z \prod_{i=1}^{M} p(x^i|\theta_{z_i=k})p(z_i = k|\pi) \\
&= \prod_{i=1}^{M} \sum_{k=1}^{K} p(z_i = k|\pi)p(x^i|\theta_k) \\
&= \prod_{i=1}^{M} \sum_{k=1}^{K} \pi_k p(x^i|\theta_k).
\end{aligned}
\tag{B.20}
$$

Equation (B.20) is equivalent to the likelihood of the parameters $\mathscr{L}(\Theta, \pi|x)$. Its first term is the joint probability of x [i.e., $p(x|Z, \Theta) = \prod_{i=1}^{M} f(x^i|\theta_{z_i})$]. Its second term is the probability of the latent variable; $p(Z|\pi) = \prod_{i=1}^{M} p(z_i|\pi) = \prod_{i=1}^{M} \pi_{z_i} = \prod_{k=1}^{K} \pi_k^{M_k}$ [143]. Using these expansions, equation (B.20) yields the well-known likelihood of a finite mixture model [i.e., the last line in equation (B.20)]. Several techniques are available for estimating the parameters of a mixture model. By far, the most popular and well-established method is ML estimation, described next.

B.3.1.2 Finite Gaussian Mixture Model

For a GMM, the hierarchical process in equation (B.17) can be rewritten as follows:

$$z_i \in \{1, \ldots, K\}$$

$$p(z_i = k) = \pi_k \tag{B.21}$$

$$x^i|z_i = k \sim \mathcal{N}(\mu^k, \Sigma^k)$$

for $\theta_k = \{\mu^k, \Sigma^k\}$. A graphical model of this process is shown in figure B.4. The gray-colored node corresponds to observed variables and white nodes correspond to latent variables which must be estimated.

B.3.1.3 ML Parameter Estimation for the Finite Gaussian Mixture Model

The aim of ML estimation is to find the model parameters $\{\Theta, \pi\}$ that maximize the likelihood of the GMM, given the training data set x. For a data set of M training data points, the GMM likelihood $\mathscr{L}(\Theta, \pi|x) = p(x|\Theta, \pi)$ is shown in equation (B.20), assuming the data points are identically and independently distributed. Unfortunately, this equation is a non-linear function of the parameters Θ, and direct maximization is impossible. However, an ML estimate can be obtained iteratively using a special case of the EM algorithm, which

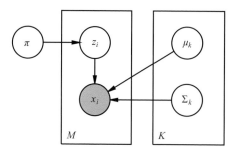

Figure B.4
Graphical model of the finite GMM.

tries to find the optimum of the likelihood. This is equivalent to finding the optimum of the log likelihood as follows:

$$\max_{\Theta,\pi} \quad \log \mathcal{L}(\Theta, \pi|x) = \max_{\Theta,\pi} \quad \log p(x|\Theta, \pi).$$
(B.22)

This can be accomplished via the following steps:

1. **Initialization step:** Initialize the means $\mu = \{\mu^1, \dots, \mu^k\}$, covariance matrices $\Sigma = \{\Sigma^1, \dots, \Sigma^K\}$, and priors $\pi = \{\pi_1, \dots, \pi_k\}$.

2. **Expectation Step:** For each Gaussian $k \in \{1, \dots, K\}$, compute the probability that it is responsible for each point x^i in the data set.

3. **Maximization step:** Reestimate the priors $\gamma = \{\gamma_1, \dots, \gamma_k\}$, means $\mu = \{\mu^1, \dots, \mu^K\}$, and covariance matrices $\Sigma = \{\Sigma^1, \dots, \Sigma^K\}$

4. Go back to step 2 and repeat until $\log \mathcal{L}(\Theta, \pi|x)$ stabilizes.

Next, we describe the computations for each step of the EM algorithm for GMM.

1. Initialization step: Here, we shall initialize for iteration $t = 0$ the set of priors $\pi^{(0)} = \{\pi_1^{(0)}, \dots, \pi_K^{(0)}\}$ to uniform probabilities, the means $\mu^{(0)} = \{\mu^{1(0)}, \dots, \mu^{K(0)}\}$ with the K-means algorithm, and $\Sigma^{(0)} = \{\Sigma^{1(0)}, \dots, \Sigma^{K(0)}\}$, by computing the sample covariances of each assigned cluster from the K-means algorithm.

2. Expectation step (membership probabilities): At each iteration t, estimate, for each Gaussian k, the probability that this Gaussian is responsible for generating each point of the data set. From equation (B.19), the a posteriori probability for the kth component is given by

$$p(z_i = k|x^i, \Theta^{(t)}, \pi^{(t)}) = \frac{\pi_k^{(t)} p(x^i|\mu^k, \Sigma^k)}{\sum_{j=1}^{K} \pi_j^{(t)} p(x^i|\mu_j^{(t)}, \Sigma_j^{(t)})}.$$
(B.23)

These probabilities are the output of the expectation step. They must be computed for each $k \in \{1, \dots, K\}$ for all data-points $i \in \{1, \dots, M\}$.

3. Maximization step (update parameters): To maximize the log-likelihood of the current estimate, we update the priors $\pi^{(t+1)} = \{\pi_1^{(t+1)}, \dots, \pi_K^{(t+1)}\}$ with the following equation:

$$\pi_k^{(t+1)} = \frac{1}{M} \sum_{i=1}^{M} p(z_i = k|x^i, \Theta^{(t)}, \pi^{(t)}),$$
(B.24)

where $p(k|x^i, \Theta^{(t)}, \pi^{(t)})$ is given by equation (B.23). The means are updated by the following equation:

$$\mu^{k(t+1)} = \frac{\sum_{i=1}^{M} p(z_i = k|x_i, \Theta^{(t)}, \pi^{(t)})x^i}{\sum_{i=1}^{M} p(z_i = k|x^i, \Theta^{(t)}, \pi^{(t)})}. \tag{B.25}$$

Finally, the covariance matrices for each kth component are computed with the following equation:

$$\Sigma^{k(t+1)} = \frac{\sum_{i=1}^{M} p(z_i = k|x^i, \Theta^{(t)}, \pi^{(t)})(x^i - \mu^{k(t+1)})(x^i - \mu^{k(t+1)})^T}{\sum_{i=1}^{M} p(z_i = k|x^i, \Theta^{(t)}, \pi^{(t)})}. \tag{B.26}$$

4. Compare $\log \mathcal{L}(\Theta^{(t)}|x)$ to $\log \mathcal{L}(\Theta^{(t-1)}|x)$: To compute the log of equation (B.20) we can reinterpret the log likelihood as

$$\log p(x|\Theta) = \log \left(\prod_{i=1}^{M} p(x^i|\Theta) \right)$$

$$= \sum_{i=1}^{M} \log \left(p(x^i|\Theta) \right) \tag{B.27}$$

$$= \sum_{i=1}^{M} \log \left(\sum_{k=1}^{K} \gamma_k p(x^i|\mu^k, \Sigma^k) \right).$$

B.3.1.4 Model Selection for Finite GMM Parameters

When the value of K is not known, a typical approach is to estimate it through model selection. One begins by estimating the ML parameters of range of $K = [1, \#K]$. Then, for each trained model, we can use the AIC and/or BIC metrics to find the *optimal model*. See section B.2.1, earlier in this appendix, for the *AIC* and *BIC* equations. For a K-component GMM, the computation of the total model parameters B, for equations (B.1) and (B.2) is done by $B = K \times (1 + N + N \times (N-1)/2) - 1$, with -1 corresponding to the priors constraint $\sum_{k=1}^{K} \gamma^k = 1$.

To choose the optimal K that best describes the data with a GMM, we typically estimate the GMM parameters for a range of K values, 10 times each. For each K, we select the best run, which in likelihood terms means the run with the ML. We then plot the values and select the K that yields the best trade-off between likelihood and model complexity. One can also analyze the mean and standard deviation of the 10 estimates' pero kth model. An example of this is shown in figures B.5 and B.6. The data set was sampled from a three-component GMM with almost identical parameters to the estimated ones.

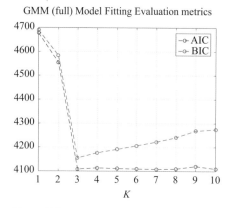

Figure B.5
GMM selection for K_range=1:10

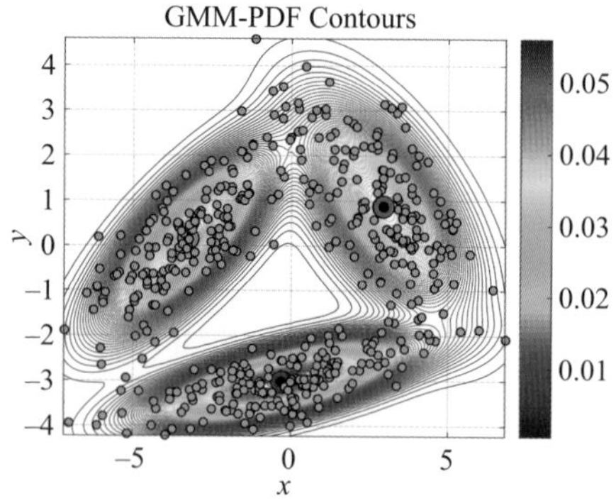

Figure B.6
Optimal fitted GMM K=3 through EM.

B.3.2   Bayesian Gaussian Mixture Model with Sampling-Based Parameter Estimation

B.3.2.1   Preliminaries

Bayesian mixture models: In the Bayesian treatment of mixture models, prior distributions are placed over the parameters, which are now treated as latent variables. Namely, priors are placed over the mixing coefficients $\pi \sim \mathcal{H}_0$ and cluster parameters $\theta_k \sim \mathcal{G}_0$, where $\mathcal{G}_0$ is the *base distribution* (i.e., a distribution on the space of cluster model parameters Θ). To ease computation in Bayesian models, one typically chooses conjugate models for the prior distributions $\mathcal{G}_0$ and $\mathcal{H}_0$. In this Bayesian approach, the vector of mixing coefficients is now considered as a categorical or multinomial distribution, which when sampled gives the probability of $p(z_i = k)$. The conjugate prior distribution over categorical/multinomial distributions is the $\mathcal{D}$irichlet distribution.[1] Thus, the generative process of a Bayesian mixture model with $\mathcal{D}$irichlet prior can then be defined as

$$\pi \sim \mathrm{Dir}(\frac{\alpha}{K}, \ldots, \frac{\alpha}{K})$$

$$\theta_k \sim \mathcal{G}_0(\lambda)$$

$$z_i | \pi = \mathrm{Cat}(\pi) \tag{B.28}$$

$$x_i | z_i = k \sim \mathcal{F}(\theta_k).$$

In this generative model, π is no longer treated as a constant vector of probabilities; rather, it is sampled from a symmetric Dirichlet prior distribution parameterized by the hyperparameter α. Moreover, the cluster model parameters θ_k are also sampled from a base distribution $\mathcal{G}_0$, parameterized by the hyperparameters λ. The data point x^i is then sampled from a distribution parameterized by θ_k. As opposed to the finite mixture model formulation, estimation of the posterior distribution over latent parameters $p(Z, \Theta | x)$ is generally intractable; this is because the joint distribution [50]:

$$p(x, \Theta, Z) = \prod_{k=1}^{K} g_0(\theta_k) \prod_{i=1}^{M} f(x^i | \theta_{z_i}) p(z_i), \tag{B.29}$$

the marginal distribution over Θ:

$$p(\Theta|x) = \int_z p(\Theta|x,Z)p(Z)dZ, \tag{B.30}$$

and the marginal distribution over Z:

$$p(Z|x) = \frac{p(x|Z)p(Z)}{\sum_z p(x|Z)p(Z)} \tag{B.31}$$

require sums over K^M possible cluster assignments Z. However, one can use approximate posterior inference methods, such as sampling or variational-based approaches, to estimate $p(Z,\Theta|x)$. Gibbs sampling[2] is one of the preferred methods to estimate parameters of Bayesian mixture models. To apply Gibbs sampling on any hierarchical model, the conditional posterior distributions of each parameter (i.e., each parameter in $\Phi = \{\Theta, \lambda, \alpha, Z\}$) conditioned on all other parameters must be derived [i.e., $p(\Phi|\Phi_{-1}, x)$]. Next, we provide a run-through of the direct [i.e., samples from $p(Z,\Theta|x)$] and collapsed [i.e., samples from $p(Z|x)$] Gibbs sampling procedures for Bayesian mixture models.

Direct Gibbs sampler: In the case of equation (B.28), assuming that $\mathscr{G}_0$ and F are conjugate and belong to the exponential family, we can sample from $p(Z,\Theta|x)$ by independently sampling two conditional distributions.

The *first* conditional posterior distribution is that of cluster assignments (with mixture weights π integrated out), $p(Z|\Theta, x)$:

$$p(z_i = k|\Theta, Z_{-i}, x, \alpha_0) = p(z_i = k|\theta_k, Z_{-i}, x^i, \alpha)$$
$$\propto p(z_i = k|Z_{-i}, \alpha)p(x^i|\theta_k). \tag{B.32}$$

Here, Z_{-i} denotes all Z except the ith; the first term is the conditional distribution of cluster assignments given by the symmetric Dirichlet distribution [143]:

$$p(z_i = k|Z_{-i}, \alpha) = \frac{M_{(k,-i)} + \alpha/K}{M + \alpha - 1}, \tag{B.33}$$

where M is the number of data-points and $M_{(k,-i)}$ is the number of points belonging to the kth cluster, excluding x_i. The second term of equation (B.32) is simply $p(x_i|\theta_k) = f(x^i|\theta_k)$.

The *second* conditional posterior distribution that one must sample is that of model parameters, $p(\Theta|Z, x)$:

$$p(\theta_k|\theta_{-k}, Z, x, \lambda) = p(\theta_k|x^k, \lambda)$$
$$\propto \mathscr{G}_0(\theta_k|\lambda)p(\theta_k|x^k), \tag{B.34}$$

where θ_{-k} denotes all θ except the kth, the first term is the conjugate prior distribution, and the second is simply the likelihood of data-points assigned to the kth cluster x^k [i.e., $p(\theta_k|x^k) = \mathscr{L}(x^k|\theta_k)$]. The typical Gibbs sampler will sweep through equation (B.32) over all N data-points and K clusters to sample for cluster assignments Z, and then through equation (B.34) over K clusters to sample cluster model parameters Θ; and repeat for T iterations.

Collapsed Gibbs sampler: Often, one can integrate out the model parameters Θ in the conditional distribution of a hierarchical Bayesian model [143]. This is feasible through the use of conjugate priors. For Bayesian mixture models, Θ can be collapsed from equation (B.32) and one needs to only draw samples from the following distribution:

$$p(z_i = k|Z_{-i}, x, \alpha_0, \lambda) \propto p(z_i = k|Z_{-i}, x^{-1}, \alpha_0)p(x_i|z_i = k, Z_{-i}, x^{-1}, \lambda)$$

$$= p(z_i = k|Z_{-i}, \alpha_0)p(x^i|x^{(k,-1)}, \lambda), \tag{B.35}$$

where the first term is the same as in equation (B.32) and the second term is a posterior predictive distribution and can be determined by the following integral:

$$p(x^i|x^{k,-1}, \lambda) = \int_{\theta_k} p(x^i|\theta_k)p(\theta_k|x^{(k,-i)}, \lambda)d\theta_k, \tag{B.36}$$

where $x^{k,-1}$ denotes the data-points belonging to the kth cluster, excluding x^i. Due to conjugacy, and assuming that $\mathcal{G}_0$ and F belong to the exponential family, equation (B.36) can be analytically computed as

$$p(x^i|x_{(k,-i)}, \lambda) = f_k(x^i|S_k, M_k), \tag{B.37}$$

where S_k is the set of sufficient statistics required by the generative distribution $\mathcal{F}$ for the set of points belonging to the kth cluster. Hence, $f_k(x^i, .)$ is defined as the predictive likelihood of x^i, given the observed data-points $x^{(k,-i)}$. Hence, for this sampler, one needs only to compute the updated S_k and M_k values. This method might be slower per iteration, but it converges faster than direct Gibbs sampling [143].

B.3.2.2 Bayesian Gaussian Mixture Model

For Bayesian GMM, we choose the normal-inverse-Wishart (NIW) distribution as the base distribution ($\mathcal{G}_0$) because it is conjugate to our generative distribution ($\mathcal{F}$), the $\mathcal{N}$ ormal distribution. Then, the hierarchical process of a Bayesian Gaussian mixture model as described in equation (B.28) can be re-written as

$$\pi \sim \text{Dir}(\frac{\alpha}{K}, \dots, \frac{\alpha}{K})$$

$$\theta_k \sim \mathcal{N}\text{IW}(\lambda_0)$$

$$z_i|\pi = \text{cat}(\pi) \tag{B.38}$$

$$x_i|z_i = k \sim \mathcal{N}(\theta_k),$$

where $\lambda_0 = \{\mu_0, \kappa_0, \Lambda_0, \nu_0\}$ are the hyperparameters for the $\mathcal{N}$IW distribution, $\theta_k = \{\mu^k, \Sigma^k, \}$ are the parameters for the $\mathcal{N}$ormal distribution, and α is the hyperparameter for the $\mathcal{D}$irichlet distribution. The graphical model of this process is shown in figure B.7. As before, the large gray node corresponds to observed variables, *black nodes* correspond to latent variables, and *small gray nodes* correspond to hyperparameters. Plate notation, as usual, represents repeated variables. The presented model can be described as follows. Given the data set $X \in \mathbb{R}^{N \times M}$ of M samples, we would like to group/cluster them into K clusters. Further, z_i is an indicator variable that takes values of $1, \dots, K$ and stores the cluster assignment of the observed sample x_i, and $\mathcal{N}$ is the generative distribution of x and is parameterized by θ. The distribution of data-points assigned to each kth cluster is thus parameterized by θ_k. θ itself stores multiple parameters (in this case, $\theta = \{\mu, \Sigma\}$),

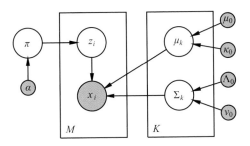

Figure B.7
Graphical model of the Bayesian GMModel.

which parameterize $\mathcal{N}$ and follow the $\mathcal{N}$IW distribution. Finally, π stores the mixing coefficient for every kth cluster and follows a $\mathcal{D}$irichlet distribution with hyperparameters α/K. Also, α is the hyperparameter corresponding to pseudocounts for the $\mathcal{D}$irichlet distribution.

Collapsed Gibbs sampling for Bayesian Gaussian mixture models: The aim of MAP estimation is to find the model parameters $\{\Theta, Z\}$ that maximize the posterior of the Bayesian GMM, given the training data set X.

Bayesian estimates with conjugate $\mathcal{N}$ and $\mathcal{N}$IW distributions Recall that $\mathcal{N}$ is a multivariate Gaussian distribution with mean μ and covariance Σ in the form of equation (B.16). The joint likelihood function of a set of M-independent Gaussian observations is

$$
\begin{aligned}
p(X \mid \mu, \Sigma) &= \prod_{i=1}^{M} \frac{1}{(2\pi)^{d/2}|\Sigma|^{1/2}} \exp\left\{-\frac{1}{2}(x^i - \mu)^T \Sigma^{-1}(x^i - \mu)\right\} \\
&= \frac{1}{(2\pi)^{d/2}|\Sigma|^{1/2}} \exp\left\{-\frac{1}{2}\sum_{i=1}^{M}(x^i - \mu)^T \Sigma^{-1}(x^i - \mu)\right\}, \\
&= \frac{1}{(2\pi)^{d/2}|\Sigma|^{1/2}} \exp\left\{-\frac{1}{2}\text{tr}(\Sigma^{-1}S)\right\}
\end{aligned}
\tag{B.39}
$$

where $S = \sum_{i=1}^{M}(x^i - \mu)(x^i - \mu)^T$ is the matrix of the sum of squares, also known as the *scatter matrix* [105]. The ML estimates of this distribution's parameters given $x_{1:M}$ are the sample mean and covariance:

$$
\hat{\mu} = \frac{1}{M}\sum_{i=1}^{M} x^i, \quad \hat{\Sigma} = \frac{1}{M}\sum_{i=1}^{M}(x^i - \hat{\mu})(x^i - \hat{\mu})^T.
\tag{B.40}
$$

These terms provide sufficient statistics, as they are equivalent to the sums of observations and outer products [143]. On the other hand, the $\mathcal{N}$IW [49] is a four-parameter $\lambda = \{\mu^0, \kappa_0, \Lambda^0, \nu_0\}$ multivariate distribution generated by

$$
\Sigma \sim \text{IW}(\Lambda^0, \nu_0), \quad \mu|\Sigma \sim \mathcal{N}\left(\mu^0, \frac{1}{\kappa_0}\Sigma\right),
\tag{B.41}
$$

where $\kappa_0, \nu_0 \in \mathbb{R}_{>0}$; moreover, $\nu_0 > N - 1$ indicates the degrees of freedom of the N-dimensional scale matrix $\Lambda \in \mathbb{R}^{N \times N}$, which should be $\Lambda \succ 0$. The density of $\mathcal{N}$IW is thus defined by

$$
\begin{aligned}
p(\mu, \Sigma \mid \lambda) &= \mathcal{N}\left(\mu | \mu^0, \frac{1}{\kappa_0} \Sigma\right) \text{IW}(\Sigma \mid \Lambda^0, \nu_0) \\
&= \frac{1}{Z_0} |\Sigma|^{-[(\nu_0 + M)/2 + 1]} \exp\left\{-\frac{1}{2} \text{tr}(\Sigma^{-1}\Lambda^0)\right\}, \\
&\times \exp\left\{-\frac{\kappa_0}{2}(\mu - \mu^0)^T \Sigma^{-1}(\mu - \mu^0)\right\}
\end{aligned}
\tag{B.42}
$$

where $Z_0 = \frac{2^{\nu_0 M/2}\Gamma_N(\nu_0/2)(2\pi/\kappa_0)^{N/2}}{|\Lambda_0|^{\nu_0/2}}$ is the normalization constant. A sample from $\mathcal{N}$IW yields a mean μ and covariance matrix Σ. One first samples a matrix from an Inverse Wishart (IW) distributions parameterized by Λ_0 and ν_0; then μ is sampled from an $\mathcal{N}$ parameterized by μ^0, κ_0, Σ. Because $\mathcal{N}$ and $\mathcal{N}$IW are a conjugate pair, the predictive term in equation (B.36) also follows an $\mathcal{N}$IW [105] with new parameters $\lambda_m = \{\mu^m, \kappa_m, \Lambda^m, \nu_m\}$ computed via the following posterior update equations:

$$
p(\mu, \Sigma | x^{1:M}, \lambda) = \mathcal{N}\text{IW}(\mu, \Sigma \mid \mu^m, \kappa_m, \Lambda^m, \Lambda_m, \nu_m)
$$

$$
\kappa_m = \kappa_0 + M, \qquad \nu_m = \nu_0 + M, \qquad \mu^m = \frac{\kappa_0 \mu^0 + M\bar{x}}{\kappa_m}
\tag{B.43}
$$

$$
\Lambda^m = \Lambda^0 + S + \frac{\kappa_0 M}{\kappa_m}(\bar{x} - \mu^0)(\bar{x} - \mu^0)^T,
$$

where M is the number of samples $x^{1:M}$, whose sample mean is denoted by $\bar{x}$; and S is the scatter matrix, as introduced earlier. The individual marginals are given as follows:

$$
\Sigma | X^{1:M} \sim \text{IW}(\Lambda^m)^{-1}, \nu_m)
$$

$$
\mu | X^{1:M} = t_{\nu_m - N + 1}\left(\mu^m, \frac{\Lambda^m}{\kappa_m(\nu_m - N - 1)}\right),
\tag{B.44}
$$

where $t_{\nu_m - N + 1}$ is a multivariate Student-t distribution with $(\nu_m - N + 1)$ degrees of freedom. By rederiving the posterior while keeping track of the normalization constants (Z) [i.e., $p(\mu, \Sigma | x^{1:M}, \lambda) = \frac{1}{Z_m}\mathcal{N}\text{IW}(\mu, \Sigma | \alpha_m)$], one can find a solution for the marginal likelihood $p(x^{1:M}|\lambda)$, which is given by the following equation [105]:

$$
\begin{aligned}
p(x^{1:M}|\lambda) &= \int\int p(x|\mu, \Sigma)p(\mu, \Sigma|\lambda) d\mu d\theta \\
&= \frac{Z_m}{Z_0} \frac{1}{(2\pi)^{MN/2}} \\
&= \frac{\Gamma_D(\nu_m/2)}{\Gamma_N(\nu_0/2)} \frac{|\Lambda^0|^{\nu_0/2}}{|\Lambda^m|^{\nu_m/2}}\left(\frac{\kappa_0}{\kappa_m}\right)^{N/2} \pi^{-2MN}.
\end{aligned}
\tag{B.45}
$$

Refer to [105] for the complete derivation of equation (B.45).

B.3.3   Bayesian Nonparametric Gaussian Mixture Model with Sampling-Based Parameter Estimation

B.3.3.1   Preliminaries

Bayesian nonparametric mixture models: A Bayesian nonparametric model is none other than a Bayesian mixture model in an infinite-dimensional parameter space Θ; yet it can be evaluated in a finite sample with a subset of parameters that best describes the observed data $X \in \mathbb{R}^{N \times M}$ [111]. Hence, Bayesian nonparametric models learn both the model parameters Θ and the number of necessary parameters K jointly. To convert equation (B.28) into a Bayesian nonparametric mixture model, a Dirichlet process (DP) is placed as a prior $p(Z)$ to the mixing probabilities π, as follows:

$$G \sim \mathrm{DP}(\alpha, \mathscr{G}_0)$$

$$\theta_i \sim G \tag{B.46}$$

$$x_i \sim \mathscr{F}(\theta_i).$$

The DP is an infinite distribution over distributions. Random samples from a DP are in fact discrete and have probability 1, which is done by placing probability masses on an infinite, yet countable collection of points named *atoms* [50]. Such a random sample can be represented as $G = \sum_{k=1}^{\infty} \pi_k \delta_{\theta_k^*}$, where π_k are probabilities assigned to the kth *atom* and θ_k^* are the location of the *atom*, both of which are drawn from $\mathscr{G}_0$ [64]. This process leads to an infinite mixture model, where $K \to +\infty$, with the following density function [143]:

$$p(x|\pi, \theta_1, \dots) = \sum_{k=1}^{\infty} \pi_k f(x|\theta_k). \tag{B.47}$$

Chinese restaurant process: To evaluate equation (B.47) on a finite set of points, the Chinese restaurant process (CRP) is used to represent a DP;[3] and tractably estimate the prior $p(Z)$. The CRP is commonly described by a metaphor of a Chinese restaurant with an infinite number of tables [64]. The process defines a sequence of probabilities for the incoming customers to sit at specific tables. Initially, the first customer sits at the first table. The ith customer then chooses to sit at a table with a probability proportional to the number of customers sitting at that chosen table; otherwise the person sits alone at a new table with a probability proportional to the hyperparameter α (commonly known as the *concentration parameter*). We can summarize this process as follows:

$$p(z_i = k \mid Z_{-i}, \alpha) = \begin{cases} \frac{M_{(k,-i)}}{\alpha + M - 1} & \text{for} \quad k \leq K \\ \frac{\alpha}{\alpha + M - 1} & \text{for} \quad k = K + 1, \end{cases} \tag{B.48}$$

where $M_{(k,-i)}$ is the number of customers sitting at table k, excluding x_i. Intuitively, α defines the probability of a customer's preference to sit alone [64].

B.3.3.2   Chinese restaurant process—Gaussian mixture models

The Gaussian counterpart of the Bayesian nonparametric mixture model in equation (B.46), with $\mathscr{F} = \mathscr{N}(.|\theta)$ and $\mathscr{G}_0 = \mathscr{N}\mathrm{IW}(\lambda)$, can thus be constructed as follows:

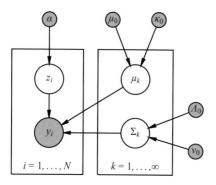

Figure B.8
Graphical representation for the CRP-MM with multivariate Gaussian distribution.

$$z_i \sim \mathrm{CRP}(\alpha)$$

$$\theta_k \sim \mathcal{N}\mathrm{IW}(\beta) \tag{B.49}$$

$$x_i \sim \mathcal{N}(\theta_{z_i}),$$

where $\theta_k = (\mu^k, \Sigma^k)$ represent the mean and covariance matrix of a normal distribution, $\mathscr{F}$. The CRP is used here as a prior on cluster assignments z_i and the NIW [18] prior is used as the base distribution ($\mathscr{G}_0$) with hyperparameters $\lambda = \{\mu^0, \kappa_0, \Lambda^0, \nu_0\}$. In figure B.8, a graphical representation of the Chinese restaurant process–mixture model (CRP-MM) with Gaussian observations is presented. A partition Z is drawn from the CRP, with hyper-parameter α. For each kth cluster, its parameters are drawn from a $\mathcal{N}$IW distribution, with hyperparameter $\{\mu^0, \kappa_0, \Lambda^0, \nu_0\}$. The posterior distribution $p(Z|x, \alpha, \lambda)$ of the CRP-MM can be approximated through the same collapsed Gibbs sampling scheme used for Bayesian Gaussian mixture models, where latent variables Z, are sampled from the following posterior distribution:

$$p(z_i = k|Z_{-i}, x, \alpha, \lambda) \propto p(z_i = k|Z_{-i}, x_{-i}, \alpha)p(x^i|z_i = k, Z_{-i}, x^{-1}, \lambda)$$
$$= p(z_i = k|Z_{-i}, \alpha_0)p(x^i|x^{(k, -1)}, \lambda), \tag{B.50}$$

where the first term corresponds to the prior induced by the CRP in equation (B.48) and the second term is the posterior predictive distribution; in other words, with the predictive likelihood of x^i given the observed data-points $x^{(k, -i)}$, via conjugacy, closed-form solutions for the likelihood exist for the Gaussian distribution [105]. Refer to equation (B.43) for the Bayes estimates.

B.3.4   GMM Applications

After fitting a GMM to a data set via EM-based parameter estimation (section B.3.1) or sampling-based techniques (section B.3.2), one can use this probabilistic model for clustering, classification, and regression, as described next.

B.3.4.1   Clustering with GMM
With a GMM fitted to a data set, one can use the equation of the membership probabilities of each kth Gaussian component (i.e., the probability that this Gaussian is

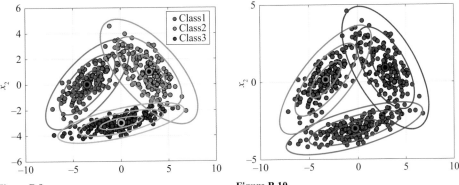

Figure B.9
Data sampled from a GMM ($K = 3$).

Figure B.10
Estimated GMM ($K = 3$) parameters.

responsible for generating each point of the data set), to cluster the data-points. This membership probability is the a posteriori probability for a kth component, given by the following equation:

$$p(k|x^i, \Theta) = \frac{\pi_k p(x^i|\mu^k, \Sigma^k)}{\sum_{j=1}^{K} \pi_j p(x^i|\mu^j, \Sigma^j)}, \tag{B.51}$$

which must be computed for each $k \in \{1, \ldots, K\}$ for all datapoints $i \in \{1, \ldots, M\}$.

Assigning datapoints to clusters: Given the posterior probabilities $p(k|x^i, \Theta^{(t)})$ $\forall i \in M$ and $\forall k \in K$, one can assign a data point to a cluster (one of the kth Gaussian components) in two ways: (1) hard and (2) soft clustering.

Hard clustering with GMM: In hard clustering, the cluster that has the highest probability is assigned to the data point. Namely, for each data point x^i, we shall compute a corresponding cluster label $y_i \in \mathbb{N}$ as

$$y_i = \arg\max_k \{p(k|x^i, \Theta)\} \quad \text{for} \quad y_i \in [1, \ldots, K]. \tag{B.52}$$

Soft clustering with GMM: In soft clustering, a data point is assigned but not labeled (i.e., $y_i = 0$), if the confidence for clustering is low. This confidence is defined by the posterior probability of the kth Gaussian chosen from equation (B.52). For this, we shall define a confidence margin $[t_{min}, t_{max}]$ and use it to ascertain the cluster confidence. That is, if the highest probability is in a range specified by this threshold (given by t_{min} and t_{max}), along with another cluster, it will be deemed as an assignment with *low confidence*, and hence unlabeled. The confidence to label these datapoints is low because more than one cluster has similar probability as specified by the threshold. Otherwise, the data point is assigned to the cluster, as in hard clustering. This is represented in the following equation:

$$y_i = \begin{cases} 0 & \text{if} \quad t_{min} < p(k^*|x^i, \Theta) < t_{max} \quad \text{and} \quad t_{min} < p(k|x^i, \Theta) \quad \exists \quad k \in [1 : -k^* : K] \\ 0 & \text{if} \quad p(k^*|x^i, \Theta) < t_{min} \\ k^* & \text{otherwise} \end{cases}$$

$$\tag{B.53}$$

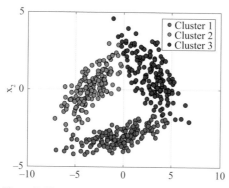

Figure B.11
Output of hard clustering the 2D data set with GMM
($K=3$).

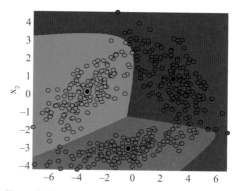

Figure B.12
The cluster boundaries for the GMM ($K=3$) trained
using the 2D data set.

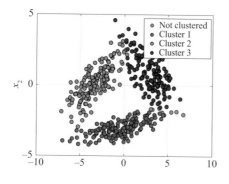

Figure B.13
Output of soft clustering the 2D data set with GMM,
$K=3$, and the soft thresholds set to $[0.3, 0.7]$. Here,
the points in gray do not have confidence to be
clustered completely.

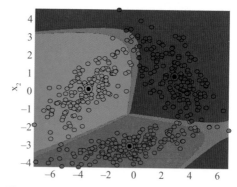

Figure B.14
The cluster boundaries for the GMM trained using the
2D data set with $K=3$ and the soft thresholds set to
$[0.3, 0.7]$. The region in gray is where the confidence
to cluster is low.

for $k^* = \arg\max_{k}\{p(k|x^i, \Theta)\}$. The notation $k \in [1 : -k^* : K]$ refers to any cluster k within the values of $[1, \ldots, K]$ *except* k^*. The cases of equation (B.53) can be interpreted as follows: if the posterior probability of k^* is within the range of $[t_{\min}, t_{\max}]$ and the posterior probability of any other cluster k also falls within this range, then the point is unlabeled. Further, if the posterior probability of k^* is lower than $t_{\min}$, this means that the confidence of labeling this point with any cluster is too low, and hence also should be characterized as unlabeled. Such a case might arise when the number of Gaussians K that is being used is too high.

The difference between clustering with equations (B.52) and (B.53) is illustrated in figures B.11 and B.12, and figures B.13 and B.14 show a two-dimensional (2D) data set sampled from a $K = 3$ GMM. After fitting a GMM with the optimal under BIC parameters (i.e., $K = 3$), one can see that hard clustering tends to label points incorrectly in the boundaries of the clusters, while soft clustering finds the regions in the partition that are ambiguous. This means that the posterior probabilities of more than one cluster are very close, and thus the confidence of labeling those points to any of these clusters is rather low.

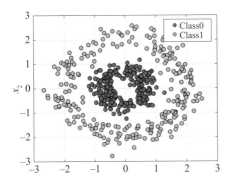

Figure B.15
Concentric circles data set.

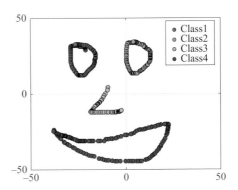

Figure B.16
Self-drawn data set.

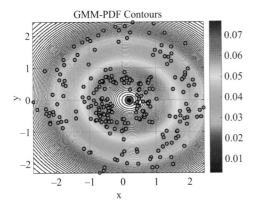

Figure B.17
pdf of GMM ($K = 1$) for class 1.

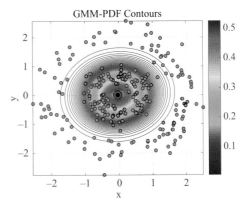

Figure B.18
pdf of GMM ($K = 1$) for class 2.

B.3.4.2 Classification with GMM

To perform classification with GMMs, the first step is to learn a GMM for each class. For example, for the two 2D toy datasets shown in figures B.15 and B.16, one can use GMM with $K = 1$ components to represent each class, as illustrated in figures B.17 and B.18 and figures B.19, B.20, B.21, and B.22, respectively.

Gaussian ML discriminant rule: Given a model for each class, we can use the Gaussian likelihood discriminant rule to classify new test datapoints. An ML classifier chooses the class label that is most likely. For a binary classification problem, a new data point x' belongs to class 1 if the corresponding likelihood is superior to the likelihood of class 2:

$$p(y = 1|x') > p(y = 2|x'). \tag{B.54}$$

Using the Bayes's rule we have

$$p(y = i|x') = \frac{p(x'|y = i)p(y = i)}{p(x')}, \text{ with class } i = 1, 2. \tag{B.55}$$

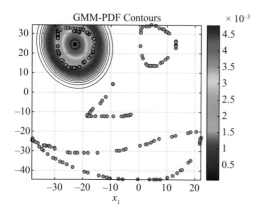

Figure B.19
pdf of GMM ($K = 1$) for class 1.

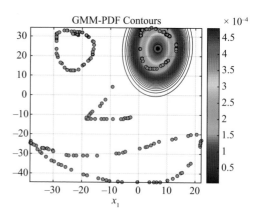

Figure B.20
pdf of GMM ($K = 1$) for class 2.

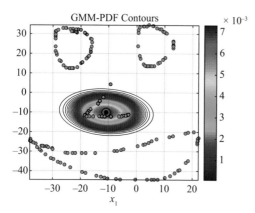

Figure B.21
pdf of GMM ($K = 1$) for class 3.

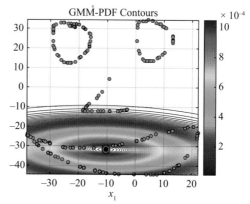

Figure B.22
pdf of GMM ($K = 1$) for class 4.

Replacing this term in the ML discriminant rule for x' in class 1 [equation (B.54)], we obtain

$$p(x'|y=1)p(y=1) > p(x'|y=2)p(y=2). \tag{B.56}$$

In general, one can assume equal class distribution [i.e., $p(y=1)=p(y=2)$]. Given this equality, Eq. B.52 the ML discriminant rule simplifies to:

$$p(x'|y=1) > p(x'|y=2). \tag{B.57}$$

For a GMM composed of K^i multivariate Gaussian functions, the conditional needed densities needed to belong to class i are similar to equation (B.15):

$$p(x'|y=i) = \sum_{k=1}^{K^i} \pi_k p(x'|\mu^k, \Sigma^k), \tag{B.58}$$

with $p(x'|\mu^k, \Sigma^k)$ defined according to equation (B.16).

In the case of a multiclass problem with $i = 1...I$ classes, the ML discriminant rule is the minimum of the log-likelihood (i.e., equivalent to maximizing the likelihood).

Assuming equal class distribution, the ML discriminant rule is

$$c_i(x') = \arg\min_i \left\{ -\log(p(x'|y=i)) \right\}.$$ (B.59)

If one does not make the assumption of equal class distribution, according to equation (B.57), the ML discriminant rule now becomes

$$c_i(x') = \arg\min_i \left\{ -\log(p(x'|y=i)p(y=i)) \right\}.$$ (B.60)

B.3.4.3 Regression with Gaussian Mixture Model

To perform regression with GMMs [i.e., Gaussian mixture regression (GMR), we estimate the joint density of the inputs, $x \in \mathbb{R}^N$, and outputs, $y \in \mathbb{R}^P$, through a K-component Gaussian mixture model as follows:

$$p(x, y|\Theta) = \sum_{k=1}^{K} \pi_k p(x, y|\mu^k, \Sigma^k),$$ (B.61)

where π_k are the priors of each Gaussian component parameterized by $\{\mu^k.\Sigma^k\}$, where

$$\sum_{k=1}^{K} \pi_k = 1, \qquad \mu^k = \begin{bmatrix} \mu_x^k \\ \mu_y^k \end{bmatrix}, \qquad \Sigma^k = \begin{bmatrix} \Sigma_{xx}^k & \Sigma_{xy}^k \\ \Sigma_{yx}^k & \Sigma_{yy}^k \end{bmatrix}.$$ (B.62)

Given this joint density, one can compute the conditional density $p(y|x)$, which will provide the regressive function $y = f(x)$:

$$p(y|x) = \sum_{k=1}^{K} \gamma_k p(y|x; \mu^k, \Sigma^k),$$ (B.63)

where

$$\gamma_k = \frac{\pi_k p(x|\mu_x^k, \Sigma_{xx}^k)}{\sum_{k=1}^{K} \pi_k p(x|\mu_x^k, \Sigma_{xx}^k)}.$$ (B.64)

The γ_k parameters are known as the *mixing weights* (i.e., relative importance) of each Kth regressive model $p(y|x; \mu^k, \Sigma^k)$. The *regressive function* is then obtained by computing the expectation over the conditional density $\mathbb{E}\{p(y|x)\}$ as follows:

$$y = f(x) = \mathbb{E}\{p(y|x)\} = \sum_{k=1}^{K} \gamma_k(x) \tilde{\mu}^k(x),$$ (B.65)

where

$$\tilde{\mu}^k(x) = \mu_y^k + \Sigma_{yx}^k (\Sigma_{xx}^k)^{-1} (x - \mu_x^k)$$ (B.66)

are the local regressive functions, which represent a linear function whose slope is determined by Σ_x^k (the variance of x) and Σ_{xy}^k (the covariance of y and x). The resulting regressive function $y = f(x)$ is thus a linear combination of K regressive models. The advantage of GMR over other nonlinear regressive methods is that one can also compute the uncertainty of the prediction (i.e., the regressed output $\hat{y}$) by computing the variance of the conditional function $Var\{p(y|x)\}$ as follows:

$$Var\{p(y|x)\} = \sum_{i=1}^{K} \gamma_k(x)\left(\tilde{\mu}^k(x)^2 + (\tilde{\Sigma}^k)\right) - \left(\sum_{i=1}^{K} \gamma_k(x)\tilde{\mu}^k(x)\right)^2, \tag{B.67}$$

where $\tilde{\Sigma}^k$ represents the variance of the conditional density for each regressive function and is computed as follows:

$$\tilde{\Sigma}^k = \Sigma_{yy}^k - \Sigma_{yx}^k(\Sigma_{xx}^k)^{-1}\Sigma_{xy}^k. \tag{B.68}$$

$Var\{p(y|x)\}$ represents the uncertainty of the *prediction*, not of the model. To represent the uncertainty of the model, one should compute the likelihood of the joint density [i.e., $\mathscr{L}(\Theta|x,y) = \prod_{i=1}^{M} \sum_{k=1}^{K} \pi_k p(x,y|\mu^k, \Sigma^k)$]. Another advantage of GMR over most regression methods is that it can be used with multidimensional inputs $x \in \mathbb{R}^N$ as well as multidimensional outputs $y \in \mathbb{R}^P$, with the same formulation presented here. For a thorough description of the derivation of this method, we recommend reading Hsi Guang Sung's PhD thesis [145] or the work by Cohn et al. on learning with statistical models [32].

Hence, to learn the regressive function $y = f(x)$ from a data set of inputs $x \in \mathbb{R}^{M \times N}$ and outputs $y \in \mathbb{R}^{M \times P}$, one should estimate the joint density $p(x,y|\Theta)$ with either of the GMM parameter estimation approaches described in sections B.3.1 and B.3.3.2. In figures B.23, B.24, and B.25, we showcase three examples of data sampled from a true regressive function

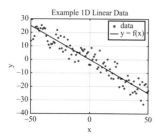

Figure B.23
Regression example on 1D linear data.

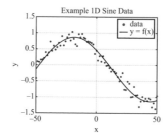

Figure B.24
Regression example on 1D non-linear sinc data.

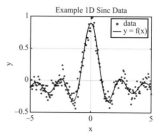

Figure B.25
Regression example on 1D non-linear sinc data.

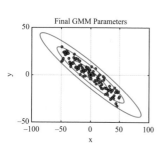

Figure B.26
GMM ($K = 1$) for data from figure B.23.

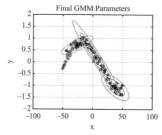

Figure B.27
GMM ($K = 4$) for data from figure B.24.

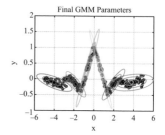

Figure B.28
GMM ($K = 7$) for data from figure B.25.

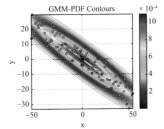

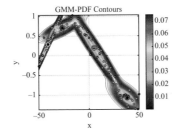

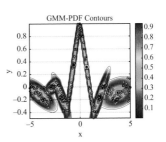

Figure B.29
PDF of the fitted GMM for data from figure B.23.

Figure B.30
PDF of the fitted GMM for data from figure B.24.

Figure B.31
PDF of the fitted GMM for data from figure B.25.

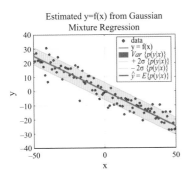

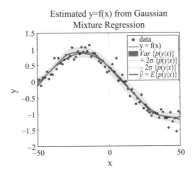

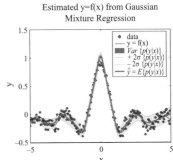

Figure B.32
GMR ($K = 1$) Result of data from figure B.23.

Figure B.33
GMR ($K = 4$) Result of data from figure B.24.

Figure B.34
GMR ($K = 7$) Result of data from figure B.25.

subject to noise. In figures B.26, B.27, and B.28, we illustrate the GMM parameters fitted on the input/output data set. Finally, in figures B.29, B.30, and B.31, we illustrate the joint density of these datasets estimated with the learned GMMs. Once the GMM parameters are estimated for the joint data set $\{x,y\}$, the regressive function can be computed via equation (B.65), as shown in figures B.32, B.33, and B.34, which also illustrate the variance of the learned regressive models.

B.4 Support Vector Machines

SVM is a machine learning technique used to learn about a decision function $y = f(x)$ that maps inputs $x \in \mathbb{R}^N$ to outputs $y \in \mathbb{R}$. For a classification problem, the output is a categorical label $y \in \{\pm 1\}$, while for regression problems, the output is a continuous variable $y \in \Re$. SVMs exploit the so-called kernel trick [128], which lifts nonlinear or nonlinearly separable data to a high-dimensional feature space via a mapping function $\Phi(x) : \mathbb{R}^N \to \mathbb{R}^F$, where F denotes the dimension of the feature space, which is commonly $F > N$. Such mapping is performed with the assumption that in $\mathbb{R}^F$, linear operations can be performed, such as linear classification or linear regression.

Formally, the kernel trick expresses the dot product of two input data samples $x, x' \in \mathbb{R}^N$, lifted to a higher-dimensional Hilbert (feature) space $\mathcal{H}$. It can be represented in terms of a kernel function $k(x, x')$ evaluated on the original input space $\mathbb{R}^N$ as follows:

$$k(x, x') = \langle \Phi(x), \Phi(x') \rangle, \tag{B.69}$$

where $k(x, x') : \mathbb{R}^N \times \mathbb{R}^N \to \mathfrak{R}$ is a positive definite kernel that represents the dot product—that is, $k(x, x') = \langle \Phi(x), \Phi(x') \rangle$, of the data-points lifted to the high-dimensional feature space $\mathbb{R}^{F>N}$. Intuitively, such a kernel function is a measure of similarity between the two datapoints $x, x' \in \mathbb{R}^N$. In nonlinear SVM, this kernel function is used to formulate hyperplane decision functions for nonlinear classification problems (section B.4.1) and ϵ-insensitive loss functions for nonlinear regression problems (section B.4.2). Next, we list a number of well-known kernel examples found in machine learning literature.

Kernel Types

- *Homogeneous polynomial:*

$$k(x, x^i) = \left(\langle x, x^i \rangle \right)^p, \tag{B.70}$$

where $p > 0$ is the hyperparameter and corresponds to the polynomial degree.

- *Inhomogeneous polynomial:*

$$k(x, x^i) = \left(\langle x, x^i \rangle + d \right)^p, \tag{B.71}$$

where $p > 0$ and $d \geq 0$, generally $d = 1$ are the hyperparameters that correspond to the polynomial degree and the offset.

- *Radial basis function (Gaussian):*

$$k(x, x^i) = \exp\left(-\frac{1}{2\sigma^2} ||x - x^i||^2 \right), \tag{B.72}$$

where σ is the hyperparameter and corresponds to the width or scale of the Gaussian kernel centered at x_i .

Throughout this book, the RBF kernel is mainly used. For a more thorough exposition of the kernel trick and SVM theory, see [128].

B.4.1 Classification with SVM (C-SVM)

For a classification problem, an SVM encodes a decision boundary between two classes, $y \in \{\pm 1\}$, by separating them with parallel hyperplanes. The region enclosed by these hyperplanes is the *margin* between the two classes, and the *maximum-margin hyperplane* is the hyperplane that lies equidistant between them.

B.4.1.1 Linear C-SVM

Following [150], the class of hyperplanes in a dot product space $\mathscr{H}$ can be written as

$$\langle \mathbf{w}, x \rangle + b = 0, \tag{B.73}$$

with $\mathbf{w} \in \mathscr{H}$ being the normal vector to the hyperplane and $b \in \mathbb{R}$ indicating the offset of the hyperplane from the origin along $\mathbf{w}$ with $\frac{b}{||\mathbf{w}||}$. Equation (B.73) can thus be used to describe the hyperplanes for each class, with $\langle \mathbf{w}, x \rangle + b = 1$ indicating the boundary of the positive class ($y = +1$). Hence, any data point that lies on or above this boundary belongs to the positive class. On the other hand, the boundary of the negative class can be represented as $\langle \mathbf{w}, x \rangle + b = -1$, and datapoints that lie on or below this boundary will belong to the negative class. These hyperplane equations can be directly used to classify the datapoints; however, we also seek to prevent the datapoints to fall within the margin. To do so, one can define the following constraints for class separation:

$$y_i \langle \mathbf{w}, x^i \rangle + b \geq 1 \quad \forall i = 1, \ldots, M. \tag{B.74}$$

Because equation (B.74) will hold only for strictly linearly separable datasets, slack variables are introduced to allow these constraints to be violated. This results in relaxed class separation constraints that can be interpreted as soft-margin hyperplanes:

$$y_i \langle \mathbf{w}, x^i \rangle + b \geq 1 - \xi_i \quad \forall i = 1, \ldots, M, \tag{B.75}$$

where $\xi_i \in \mathbb{R}$. The aim of the SVM learning algorithm [128] is to estimate $\{\mathbf{w}, b\}$ such that the margin between the two classes is maximized, represented by the distance between the two hyperplanes that separate each class (i.e., $\frac{2}{||\mathbf{w}||}$), while guaranteeing the constraints defined in equation (B.75). This is achieved by minimizing the following constrained optimization problem:

$$\min_{\mathbf{w} \in \mathcal{H}, \xi \in \mathbb{R}^M, b \in \mathbb{R}} \left(\frac{1}{2} ||\mathbf{w}||^2 + \frac{C}{M} \sum_{i=1}^{M} \xi_i \right) \tag{B.76}$$

s.t. $y_i \left(\langle \mathbf{w}, x^i \rangle + b \right) \geq 1 - \xi_i, \quad \forall i = 1, \ldots M,$

where $C \in \mathbb{R}$ is a penalty factor used to trade off between maximizing the margin and minimizing classification errors (i.e., datapoints lying on the wrong side of the hyperplane). As shown in [128], equation (B.76) can be solved via quadratic programming by optimizing the dual of equation (B.76) (as described next). The linear SVM classifier, thus, results in the following decision function:

$$y = f(x) = \mathrm{sgn} \left(\langle \mathbf{w}, x^i \rangle + b \right), \tag{B.77}$$

which is referred to as *C-SVM* or *soft-margin SVM*.

B.4.1.2 Nonlinear C-SVM
Although equation (B.76) finds parameters $\{\mathbf{w}, b\}$ that yield a soft-margin classifier, it is still not suitable for nonlinearly separable datasets. In lieu of this, for nonlinear classification, the datapoints are lifted to the high-dimensional feature space via $\Phi(x) : \mathbb{R}^N \to \mathbb{R}^F$, and the optimization problem defined in equation (B.76) becomes

$$\min_{\mathbf{w} \in \mathcal{H}, \xi \in \mathbb{R}^M, b \in \mathbb{R}} \left(\frac{1}{2} ||\mathbf{w}||^2 + \frac{C}{M} \sum_{i=1}^{M} \xi_i \right) \tag{B.78}$$

s.t. $y_i \left(\langle \mathbf{w}, \Phi(x_i) \rangle + b \right) \geq 1 - \xi^i, \quad \forall i = 1, \ldots M.$

Note that equation (B.78) is identical to equation (B.76), except for the class separation constraints, which are in terms of $\Phi(x^i) \in \mathbb{R}^F$. Furthermore, the normal vector to the hyperplane $\mathbf{w} \in \mathscr{H}$ can be expressed as a linear combination of the input data (feature) vectors:

$$\mathbf{w} = \sum_{i=1}^{M} \alpha_i y_i \Phi(x^i), \tag{B.79}$$

where $\alpha_i > 0$ for the support vectors [i.e., the input data-points that accurately meet the constraint imposed in equation (B.78)]. The expansion of $\mathbf{w}$ (B.79) is then used to represent the optimization constraint as $y_i \left(k(x, x^i) + b \right) \geq 1 - \xi_i$, with $k(\cdot, \cdot)$ being the kernel function in equations (B.70) to (B.72). The nonlinear C-SVM decision function then takes the following form:

$$
\begin{aligned}
y = f(x) &= \mathrm{sgn} \left(\langle \mathbf{w}, \Phi(x^i) \rangle + b \right) \\
&= \mathrm{sgn} \left(\sum_{i=1}^{M} \alpha_i y_i \, \langle \Phi(x), \Phi(x^i) \rangle + b \right) \\
&= \mathrm{sgn} \left(\sum_{i=1}^{M} \alpha_i y_i \, k(x, x^i) + b \right),
\end{aligned}
\tag{B.80}
$$

whose parameters are estimated by maximizing the *Lagrangian dual* of equation (B.78) with regard to α; that is,

$$\underset{\alpha_i \geq 0}{\mathrm{maximize}} \sum_{i=1}^{M} \alpha_i - \frac{1}{2} \sum_{i=1}^{M} \sum_{j=1}^{M} \alpha_i \alpha_j y_i y_j \underbrace{\langle \Phi(x), \Phi(x^i) \rangle}_{k(x^i, x^j)} \tag{B.81}$$

$$\text{under constraint} \quad 0 \geq \alpha_i \geq C \quad \forall i, \quad \sum_{i=1}^{M} \alpha_i y_i = 0.$$

Equation (B.81) is referred to as the *dual* problem, with equation (B.78) being the *primal* one. Equation (B.81) is a quadractic function of the support vector indicator variables α_i subject to linear constraints; hence, it can be solved via quadratic programming, as shown in [128].

Hyperparameters: The efficacy of the nonlinear SVM classifier [equation (B.80)] depends heavily on the parameters that are not optimized in equation (B.78). That would be the kernel hyperparameters and misclassification penalty term C. For example, when using the RBF kernel [equation (B.72)], one needs to find an optimal value for both C and σ. Intuitively, C is a parameter that trades off the misclassification of training examples against the *simplicity* of the decision function. The lower the C, the smoother the decision boundary (with risk of some misclassification). Conversely, the higher the C, the more support vectors are selected far from the margin, yielding a more fitted decision boundary to the data points. Intuitively, σ is the radius of influence of the selected support vectors; if σ is very low, it will be able to encapsulate only those points in the feature space that are very close. On the other hand, if σ is very big, the support vector will influence points farther from them.

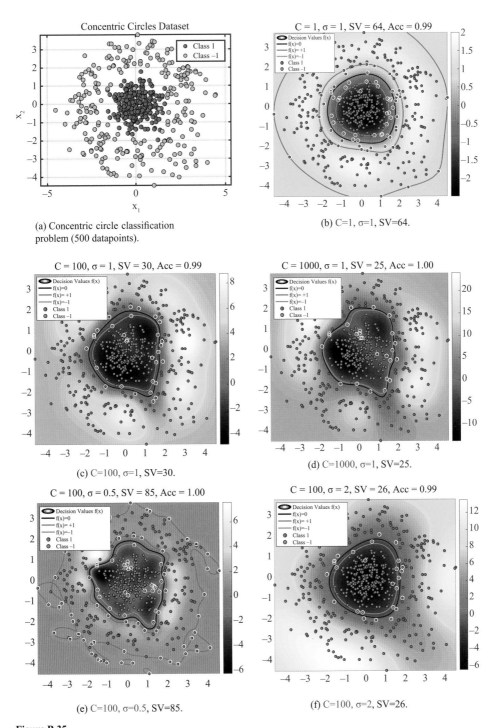

(a) Concentric circle classification problem (500 datapoints).

(b) C=1, σ=1, SV=64.

(c) C=100, σ=1, SV=30.

(d) C=1000, σ=1, SV=25.

(e) C=100, σ=0.5, SV=85.

(f) C=100, σ=2, SV=26.

Figure B.35
Decision boundaries with different hyperparameter values for the circle data set. *Support vector: data points with white edges.*

The effects of the hyperparameters on a nonlinear C-SVM with an RBF kernel are illustrated in figures B.35b–f for the concentric circle data set (figure B.35a). Techniques to find the optimal hyperparameters for SVM are described in section B.4.3, later in this appendix.

B.4.2 Regression with SVM (ϵ-SVR)

In regression, the SVM algorithm (section B.4.1) is modified to approximate the real-valued function $f(x) : \mathbb{R}^N \to \mathbb{R}$, which maps multidimensional inputs $X = \{x^i\}_{i=1}^M$ to unidimensional outputs $Y = \{y_i\}_{i=1}^M$. The goal of SVR is to approximate function $f(x)$ with no more than an ϵ-deviation from the training outputs. Formally, this is defined as maximizing the following ϵ-insensitive loss function [151]:

$$|y - f(x)|_\epsilon = \max\{0, |y - f(x)| - \epsilon\}. \tag{B.82}$$

Intuitively, equation (B.82) indicates that we don't mind having some errors in our regression (predicted outputs), so long as they're within an ϵ-deviation of $f(x)$. Furthermore, it ensures that the regressive function is *flat*.[4] The allowed deviation from the training data is referred to as the ϵ-*insensitive tube*, resulting in the ϵ-SVR algorithm.

B.4.2.1 Linear ϵ-SVR

In ϵ-SVR, $f(x)$ is estimated via the following linear function:

$$f(x) = \langle \mathbf{w}, x \rangle + b \quad \text{with} \quad \mathbf{w} \in \mathcal{H}, b \in \mathbb{R}, \tag{B.83}$$

with the constraint that points lying within the ϵ-tube are not penalized (i.e., $|y - f(x)| \leq \epsilon$). This function can be learned by minimizing the following optimization problem:

$$\min_{\mathbf{w} \in \mathcal{H}, \xi, \xi^* \in \mathbb{R}^M, b \in \Re} \left(\frac{1}{2} ||\mathbf{w}||^2 + C \sum_{i=1}^M (\xi_i + \xi_i^*) \right)$$

$$\text{s.t.} \quad y^i - \langle \mathbf{w}, x^i \rangle - b \leq \epsilon + \xi_i^* \tag{B.84}$$

$$\langle \mathbf{w}, x^i \rangle + b - y^i \leq \epsilon + \xi_i$$

$$\xi_i, \xi_i^* \geq 0, \forall i = 1 \ldots M, \quad \epsilon \geq 0,$$

where $\mathbf{w}$ is the separating hyperplane, ξ_i, ξ_i^* are the slack variables, b is the bias, ϵ is the allowable error, and C is the penalty factor associated with errors larger than ϵ. As in the classification case to optimize for $\mathbf{w}, b$, we maximize the *dual* from equation (B.84) with regard to the Lagrangian multipliers, as described next.

B.4.2.2 Nonlinear ϵ-SVR

To convert equation (B.83) to nonlinear regression, we follow the procedure for nonlinear classification and lift the original data to feature space via the mapping function $\Phi(x) : \mathbb{R}^N \to \mathbb{R}^F$, replace the dot products in the original space with the dot products in the high-dimensional feature space, $\langle \Phi(x), \Phi(x') \rangle$, and subsequently employ the kernel trick [equation (B.69)]. Thus, equation (B.84) becomes

$$\min_{\mathbf{w} \in \mathcal{H}, \xi, \xi^* \in \mathbb{R}^M, b \in \mathbb{R}} \left(\frac{1}{2} ||\mathbf{w}||^2 + C \sum_{i=1}^M (\xi_i + \xi_i^*) \right)$$

s.t. $y^i - \langle \mathbf{w}, \Phi(x_i) \rangle - b \leq \epsilon + \xi_i^*$ (B.85)

$\langle \mathbf{w}, \Phi(x^i) \rangle + b - y^i \leq \epsilon + \xi_i$

$\xi_i, \xi_i^* \geq 0, \forall i = 1 \ldots M, \quad \epsilon \geq 0.$

As in the classification case, equation (B.85) is not solved directly. One must solve the dual optimization problem with regard to the Lagrangian multipliers α; that is,

$$\underset{\alpha_i, \alpha_i^* \geq 0}{\text{maximize}} - \frac{1}{2} \sum_{i=1}^{M} \sum_{j=1}^{M} (\alpha_i^* - \alpha_i)(\alpha_j^* - \alpha_j) \underbrace{\langle \Phi(x^i), \Phi(x^j) \rangle}_{k(x^i, x^j)}$$

$$- \epsilon \sum_{i=1}^{M} (\alpha_i^* + \alpha_i) + \sum_{i=1}^{M} y_i (\alpha_i^* - \alpha_i) \tag{B.86}$$

under constraints: $\sum_{i=1}^{M} (\alpha_i^* - \alpha_i) = 0, \quad \alpha_i^*, \alpha_i \in [0, C/M].$

Furthermore, because $\mathbf{w}$ can be described as a linear combination of a set of training data in the feature space,

$$\mathbf{w} = \sum_{i=1}^{M} (\alpha_i^* - \alpha_i) \Phi(x_i). \tag{B.87}$$

Via the kernel trick, the nonlinear ϵ-SVR regressive function is

$y = f(x) = \langle \mathbf{w}, \Phi(x) \rangle + b$

$$= \sum_{i=1}^{M} (\alpha_i - \alpha_i^*) \langle \Phi(x), \Phi(x^i) \rangle + b \tag{B.88}$$

$$= \sum_{i=1}^{M} (\alpha_i - \alpha_i^*) k(x, x^i) + b,$$

where $\alpha_i, \alpha_i^* > 0$ are the Lagrangian multipliers that define the support vectors.

Hyperparameters: As in nonlinear C-SVM, the predictive accuracy of the approximated $f(x)$ relies heavily on the choice of hyperparameters. Apart from the hyperparameters of the chosen kernel function $k(x, x')$, ϵ-SVR has two open-parameters:

• C: Cost $[0 \rightarrow \infty]$ represents the penalty associated with errors larger than epsilon. Increasing cost value causes a closer fitting to the calibration/training data.

• ϵ: Epsilon represents the minimal required precision.

Next, we discuss/illustrate the effects of each hyperparameter on the resulting $f(x)$.

Effect of ϵ on $f(x)$: When training $f(x)$, there is no penalty associated with points that are predicted within distance ϵ from the y_i. Small values of ϵ force a closer fitting to the

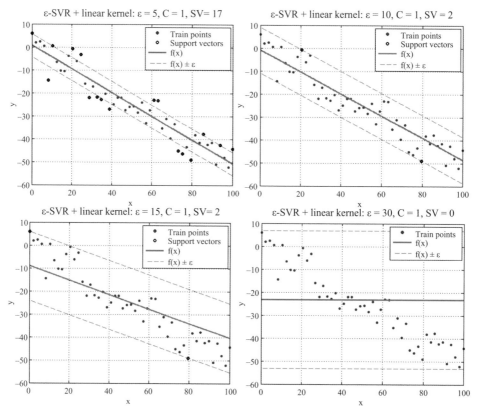

Figure B.36
Effect of ε on regressive function $f(x)$.

training data. Because ϵ controls the width of the insensitive error zone, it can directly affect the number of support vectors. By increasing ϵ, we might get fewer support vectors, but doing so could yield *flatter* estimates (see figure B.36).

Effect of C on f(x): Parameter C determines the trade-off between the model complexity (flatness) and the degree to which deviations larger than ϵ are tolerated in the optimization formulation. For example, if C is too large (infinity), then the objective is to minimize the error at all costs (see figure B.37).

Effect of the kernel hyperparameters on f(x): When using the RBF kernel, we would need to find an optimal value for σ, which is the width in the RBF. Intuitively, σ is the radius of influence of the selected support vectors; if σ is very low, it will be able to encapsulate only those points in the feature space that are very close. On the other hand, if σ is very big, the support vector will influence points farther from them. It can be thought of as a parameter that controls the shape of the separating hyperplane. Thus, the smaller σ is, the more likely one will get more support vectors (for some values of C and ϵ). For more information, see figure B.38.

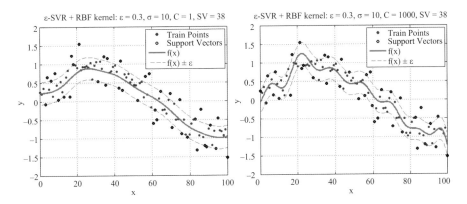

Figure B.37
Effect of C on the regressive function $f(x)$.

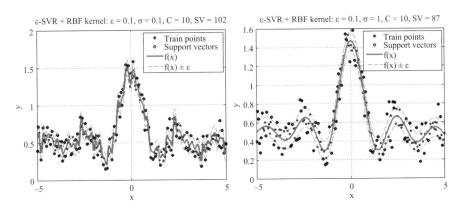

Figure B.38
Effect of σ on the regressive function $f(x)$.

B.4.3   SVM Hyperparameter Optimization

SVM is one of the most powerful nonlinear classification/regression methods to date, because it is able to find a separating hyperplane for nonseparable/nonlinear data on a high-dimensional feature space using the kernel trick. Yet, for it to perform as expected, the best hyperparameters given the data set (and problem) must be found.

Grid search with cross-validation: To tune the optimal hyperparameters in methods such as SVM/SVR and GMM/GMR (from section B.3), we can do an exhaustive search of a grid of the parameter space. This is typically done by learning the decision function for each combination of hyperparameters and computing a metric of performance (from section B.2). However, testing the learned decision function on the data used to train it is a big mistake, due to overfitting. This is done through *cross-validation (CV)*, process which we explain next.

In machine learning, CV consists in *holding out* part of the data to *validate* the performance of the model on unseen samples. The two main CV approaches are *k*-fold and Leave One

Out (LOO). In this book, we generally use the former. In k-fold CV, the training set is split into k smaller sets and the following procedure is repeated k times:

1. Train a model using $k-1$ folds as training data.
2. Validate the model by computing a classification metric on the remaining data.

For each grid search, the previous steps are repeated k times for each combination of hyperparameters in the parameter grid. The overall classification performance can then be analyzed through the statistics (e.g., mean, standard deviation, etc.) of the metrics computed from the k validation sets.

How do we find the optimal parameter choices? Manually seeking the optimal combination hyperparameter values is quite a tedious task. For this reason, we use grid search with cross-validation. First, one must choose admissible ranges for the hyperparameters; in the case of the RBF kernel, it would be for σ, with the additional hyperparameters for classification ($C \in \Re$) and regression ($C \in \Re$ and $\epsilon \in \Re$). Additionally, to get statistically relevant results, one has to cross-validate with a different test/train ratio. A standard way of doing this is applying k-fold CV for each of the parameter combinations and keeping some statistics such as mean and standard deviation of the accuracy for the k runs of that specific combination of parameters; a typical number of fold is $k = 10$.

B.4.3.1 Grid Search with 10-fold Cross Validation

For C-SVM For the circle data set (figure B.35a), we can run C-SVM (an RBF kernel) with 10-fold CV over the following range of parameters: $C\_range = [1:500]$ and $\sigma\_range = [0.25:3]$. The heat maps in figure B.39 represent the mean ACC [equation (B.3)] and F-measure [equation (B.4)] on both training and testing sets. To choose the optimal parameters, one normally ignores the metric on the training test and focus on the performance of the hyperparameters in the testing set. However, when a classifier is showing very poor performance on the testing set, it is useful to analyze the performance on the training set. If the performance on the training set is poor, perhaps the parameter ranges are off, you are not using the appropriate kernel, or your data set is simply not separable using this method.

Choosing the optimal parameters from grid search/CV: One way of choosing the optimal hyperparameter combination is to blindly select the iteration that yields the maximum Test ACC or F-measure. For the CV results presented in figure B.39, a Max. Test ACC of 97 percent with $C = 31.58$ and $\sigma = 3$. The resulting decision boundary for this set of optimal hyperparameters is shown in figure B.40a. As can be seen, the classifier recovers the circular shape of the real boundary from the data set. However, this is not the only optimal value. Analyzing the heat map for Test Mean ACC (figure B.39), the max ACC was chosen from the white area on the right-bottom region of the heat map. Yet, there is another large region with high accuracy at the bottom-center of the heat map where both the C and σ values are smaller. Intuitively, if σ is smaller, we should have more support vectors. The midpoint of this area is the combination of hyperparameters $C = 4, \sigma = 1$, which yields the decision boundary shown in figure B.40b. It yields the same accuracy, and yet the decision boundary is overfitted to the overlapping points in the boundary.

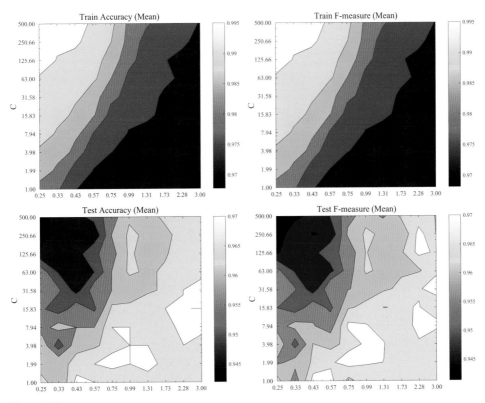

Figure B.39
Heat maps from a grid search on C and σ with 10-fold CV. (top row) Mean *Train* Accuracy and *F*-measure and (right) Mean *Test* Accuracy and *F*-measure.

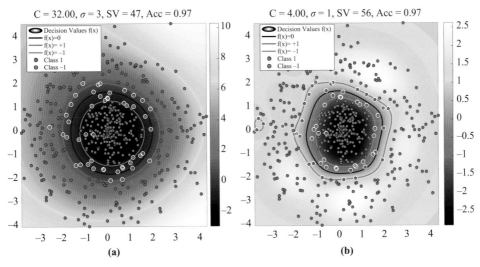

Figure B.40
C-SVM decision boundary with optimal values from grid search with CV.

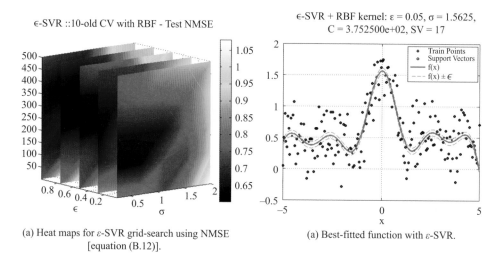

(a) Heat maps for ε-SVR grid-search using NMSE
[equation (B.12)].

(a) Best-fitted function with ε-SVR.

Figure B.41
Grid search result for 1D nonlinear on ε-SVR.

For ϵ-SVR As with C-SVM, for ϵ-SVR, one can run a grid search/CV for the hyperparameters $\{C, \epsilon, \sigma\}$ with the regression metrics introduced in section B.2.4. For the sinc data set (figure B.25), one can run ϵ-SVR (with an RBF kernel) with 10-fold CV over the following range of parameters: $C\_range = [1 : 500]$, $\epsilon\_range = [0.05 : 1]$, and $\sigma\_range = [0.25 : 2]$. By selecting the combination of hyperparameter values that yields the highest performance, according to a performance metric (as described in section B.2.4), on the test data set, we can achieve an approximate regressive function as shown in figure B.41.

B.5   Gaussian Processes Regression

A GPR model is a generalization of the Gaussian pdf (B.16) to functional space. Namely, rather than representing a distributions of points, it represents a distribution of functions, $y = f(x)$, that are consistent with the observed data points. GPR is a probabilistic, nonparametric regression approach. As in any probabilistic approach, it begins with a prior distribution that is then updated as new data are observed; the resulting regressor is then computed as the posterior of this distribution. Following the exposition of ϵ-SVR as in section B.4.2, we begin with the linear case [i.e., Bayesian linear regression (BLR)] and end with the nonlinear GPR formulation.

B.5.1   Bayesian Linear Regression

In BLR, one seeks to find the parameters of a linear regression model; that is,

$$y = f(x) = \mathbf{w}^{\mathsf{T}}x, \tag{B.89}$$

where $y \in \mathbb{R}$ and $\mathbf{w}, x \in \mathbb{R}^{N}$, through a statistical approach. In classical linear regression, the weights $\mathbf{w}$ of equation (B.89) are estimated by minimizing a *loss function* with regard to the input/output data set $\{X, Y\} = \{x^{i}, y_{i}\}_{i=1}^{M}$, where $x^{i} \in \mathbb{R}^{N}$ and $y_{i} \in \mathbb{R}$. This loss function is typically the least-squares error: $||Y - \mathbf{w}^T X||$. Minimizing the least-squares error results in a weight estimate known as ordinary least-squares (OLS):

$$\mathbf{w}_{\text{OLS}} = (X^T X)^{-1} X^T Y. \tag{B.90}$$

In BLR, one assumes that the values of Y contain an additive noise ϵ that follows a zero-mean Gaussian distribution, $\epsilon \sim \mathcal{N}(0, \epsilon_{\sigma^2})$; hence, equation (B.89) takes the following form:

$$y = f(x) + \epsilon \tag{B.91}$$
$$= \mathbf{w}^T x + \epsilon.$$

The assumption that the noise follows $\mathcal{N}(0, \epsilon_{\sigma^2})$ means that we are placing a prior distribution on the noise. Hence, the likelihood of equation (B.91) can be computed as

$$p(Y|X, \mathbf{w}; \epsilon_{\sigma^2}) = \mathcal{N}(Y - \mathbf{w}^T X | I \epsilon_{\sigma^2})$$
$$= \frac{1}{(2\pi \epsilon_{\sigma^2})^{n/2}} \exp\left(-\frac{1}{2\epsilon_{\sigma^2}} ||Y - \mathbf{w}^T X||\right). \tag{B.92}$$

As in the GMM case (section B.3) the parameters of equation (B.92), i.e., the weights $\mathbf{w}$, can be estimated via an ML Estimation (MLE) approach as follows:

$$\nabla_{\mathbf{w}} \log p(Y|X, \mathbf{w}) = -\frac{1}{\epsilon_{\sigma^2}} X^T (Y - \mathbf{w}^T X) \tag{B.93}$$
$$\mathbf{w}_{\text{ML}} = (X^T X)^{-1} X^T Y.$$

Note that equation (B.93) yields the same result as equation (B.90); that is, the MLE does not take into account the prior on the noise. To consider priors on variables, one must instead compute the MAP estimate, as in the Bayesian GMM case (section B.3.2). In BLR, we take advantage of the fact that the likelihood function is Gaussian [equation (B.92)] and add a prior Gaussian distribution over the weights, $\mathbf{w}$. The posterior of equation (B.91), therefore, is

$$\underbrace{p(\mathbf{w}|X, Y)}_{posterior} \propto \underbrace{p(Y|X, \mathbf{w})}_{likelihood} \underbrace{p(\mathbf{w})}_{prior}, \tag{B.94}$$

where $p(\mathbf{w}) = \mathcal{N}(\mathbf{0}, \Sigma_w)$ is the prior distribution on the weights and $p(\mathbf{w}|X, Y)$ is the posterior distribution. Finding the weights with MAP estimation consists of computing the expectation over the posterior; that is,

$$\mathbf{w}_{\text{MAP}} = E\{p(\mathbf{w}|X, Y)\} = \frac{1}{\epsilon_{\sigma^2}} A^{-1} XY \tag{B.95}$$

with $\quad A = \frac{1}{\epsilon_{\sigma^2}} XX^T + \Sigma_w^{-1}.$

Now, to predict an output y^* from a new testing point x^*, we take the expectation of the following *predictive distribution*:

$$p(y^*|x^*, X, Y) = \int p(y^*|x^*, \mathbf{w}) p(\mathbf{w}|X, Y) d\mathbf{w}$$

$$= \mathcal{N}(\frac{1}{\epsilon_{\sigma^2}} x^{*T} A^{-1} XY, \; x^{*T} A^{-1} x^*), \tag{B.96}$$

which yields the following regressor:

$$y^* = f(x^*)$$

$$= E\{p(y^*|x^*, X, Y)\} = \frac{1}{\epsilon_{\sigma^2}} x^{*T} A^{-1} XY. \tag{B.97}$$

The resulting regressor [equation (B.97)] takes into account both the variance in the noise ϵ_{σ^2} and the uncertainty in the weights Σ_w. As with GMM [equation (B.67)], the uncertainty of the prediction can be estimated via the variance of predictive distribution [equation (B.96)]:

$$var\{p(y^*|x^*, X, Y)\} = x^{*T} A^{-1} x^*. \tag{B.98}$$

B.5.2   Estimation of Gaussian Process Regression

GPR is the nonlinear kernelized version of BLR. Rather than modeling a distribution over data-points, GPR models a distribution over functions in a high-dimensional feature space, mapped via $\Phi : \mathbb{R}^N \to \mathbb{R}^F$, as in the nonlinear ϵ-SVR (section B.4.2.2); that is,

$$y = f(x) + \epsilon$$

$$= \mathbf{w}^T \Phi(x) + \epsilon, \tag{B.99}$$

where $y \in \mathbb{R}$, $\mathbf{w}, x \in \mathbb{R}^N$, $\epsilon \sim \mathcal{N}(0, \epsilon_{\sigma^2})$, and $\mathbf{w} \sim \mathcal{N}(0, \Sigma_w)$. Because a GP is a collection of random variables that have a joint Gaussian distribution, it is parameterized by a mean function, $\mu(x) = E\{f(x)\}$, and covariance (kernel) function, $k(x, x') = E\{(f(x) - \mu(x))(f(x') - \mu(x'))\}$, which represents the covariance between a set of random variables for a real process $f(x)$, as follows:

$$f(x) \sim GP\left(\mu(x), k(x, x')\right). \tag{B.100}$$

One can then interpret equation (B.99) with a GP [equation (B.100)] as follows:

$$\mu(x) = E\{f(x)\} = \Phi(x)^T E\{x\} = 0 \tag{B.101}$$

$$k(x, x') = \text{cov}(f(x), f(x')) = E\{f(x)f(x')\}$$

$$= \Phi(x)^T E\{\mathbf{w}\mathbf{w}^T\} \Phi(x') = \Phi(x)^T \Sigma_w \Phi(x). \tag{B.102}$$

As in equation (B.101), one normally assumes that $\mu(x) = 0$. Hence, assuming noisy observations, one can model the joint distribution of a set of input/output training pairs $\{X, Y\} = \{x^i, y_i\}_{i=1}^{M}$ and a set of testing input/output pairs $\{X^*, Y^*\} = \{(x^i)^*, y_i^*\}_{i=1}^{M^*}$ as follows:

$$\begin{bmatrix} Y \\ Y^* \end{bmatrix} \sim \mathcal{N}\left(0, \begin{bmatrix} \overbrace{K(X,X) + \sigma_\epsilon^2 I}^{\text{cov}(Y)} & \overbrace{K(X,X^*)}^{\text{cov}(Y,Y^*)} \\ \underbrace{K(X^*,X)}_{\text{cov}(Y,Y^*)} & \underbrace{K(X^*,X^*)}_{\text{cov}(Y*)} \end{bmatrix}\right). \tag{B.103}$$

The term $\epsilon_{\sigma^2} I$ represents the variance of the assumed additive, identically and independently distributed Gaussian noise ϵ. Further, as denoted in equation (B.103), the matrix $K(X, X) \in \mathbb{R}^{M \times M}$ represents the covariances between all M training points, with each i, jth element given by $[K(X, X)]_{ij} = k(x^i, x^j)$, where $k(\cdot, \cdot)$ is the covariance (kernel) function. The remaining covariance matrices $K(X, X^*) \in \mathbb{R}^{M \times M^*}, K(X^*, X) \in \mathbb{R}^{M^* \times M}$ represent the covariance between the training and test points, while $K(X^*, X^*) \in \mathbb{R}^{M^* \times M^*}$ indicates the covariances between the test points.

Then, as in BLR, to predict the outputs $Y^* \in \mathbb{R}^{M^*}$ of a set of testing points $X^* \in \mathbb{R}^{N \times M}$, we take the expectation of the following predictive distribution:

$$p(Y^*|X^*, X, Y) = \mathcal{N} \Big(\underbrace{K(X^*, X)[K(X, X) + \epsilon_{\sigma^2} \mathbb{I}_M]^{-1} Y}_{\mathbb{E}\{p(Y^*|X,Y)\}},$$

$$\underbrace{K(X^*, X^*) - K(X^*, X)[K(X, X) + \epsilon_{\sigma^2} \mathbb{I}_M]^{-1} K(X, X^*)}_{\text{cov}(Y^*)} \Big), \tag{B.104}$$

The GPR regressive function for a set of testing points X^* is, then,

$$Y^* = f(X^*)$$

$$= E\{p(Y^*|X^*, X, Y)\} \tag{B.105}$$

$$= K(X^*, X)[K(X, X) + \epsilon_{\sigma^2} \mathbb{I}_M]^{-1} Y,$$

with the uncertainty of the predictions being computed as

$$\text{cov}(Y^*) = K(X^*, X^*) - K(X^*, X)[K(X, X) + \epsilon_{\sigma^2} \mathbb{I}_M]^{-1} K(X, X^*). \tag{B.106}$$

One can rewrite equations (B.105) and (B.106) in a compact form for a single test pair $\{x^*, y^*\}$, with $\mathbf{k}(x^*) = \mathbf{k}(x^*, X) \in \mathbb{R}^M$ denoting the vector of covariances between the testing point and the training data set and $K = K(X, X) \in \mathbb{R}^{M \times M}$, as follows:

$$y^* = f(x^*)$$

$$= E\{p(y^*|x^*, X, Y)\} \tag{B.107}$$

$$= \mathbf{k}(x^*)^T [K + \epsilon_{\sigma^2} \mathbb{I}_M]^{-1} Y,$$

and the uncertainty of the predicted output as the variance of the predictive distribution:

$$\text{var}\{y^*\} = k(x^*, x^*) - \mathbf{k}(x^*)^T [K + \epsilon_{\sigma^2} \mathbb{I}_M]^{-1} \mathbf{k}(x^*). \tag{B.108}$$

The resulting regressive function [equation (B.107)] can be further interpreted as a linear combination of weighted kernel functions evaluated on the test point x^*, as follows:

$$y^* = \mathbf{k}(x^*)^T [K + \epsilon_{\sigma^2} \mathbb{I}_M]^{-1} Y$$

$$= \sum_{i=1}^{M} \alpha_i k(x^i, x^*) \quad \text{with} \quad \alpha = [K + \epsilon_{\sigma^2} \mathbb{I}_M]^{-1} Y. \tag{B.109}$$

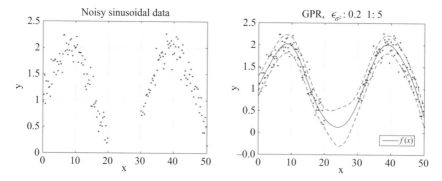

Figure B.42
Sinusoidal data with a hole.

Notice how equation (B.109) has a similar structure to SVR [equation (B.88)]; however, in GPR *all the points* are used in the computation of the predicted signal (i.e., $\alpha_i > 0 \quad \forall i$). For a more thorough description of how these equations are computed, see [118].

Finally, as with any other kernel method, one can use many types of kernel functions. Throughout this book, we primarily use the RBF, also known as the Gaussian or squared exponential (SE) kernel, for GPR:

$$k(x,x') = \sigma_y^2 \exp\left(-\frac{1}{2l^2} \sum_{i=1}^{N} (x^i - x^{i'})^2\right), \tag{B.110}$$

where $l \in \mathbb{R}_+$ is the length scale that modifies the width of the SE function; σ_y^2 is the output variance, which is a scaling factor that indicates the average distance of the estimated regressive function from its mean; and N is the dimensionality of the input data-points. Often σ_y^2 is assumed to be $\sigma_y^2 = 1$; hence, the only hyperparameter to estimate for this kernel is the length scale, l. This assumption is taken throughout this book. Further, with this assumption, equation (B.110) reduces to equation (B.72) for SVMs.

Hyperparameters of GPR with RBF kernel: Using equation (B.110) as the kernel function, we have two hyperparameters to tune for the GPR regressive function:

- ϵ_{σ^2} : Variance of the signal noise.
- l: Variance of the kernel, also known as *kernel width* or *length scale*.

An example of a regressive function learned with GPR is illustrated in figure B.42. In this case, the data set is a sinuosidal with a hole. As can be seen, with an SE kernel using $l = 0.5$ and $\epsilon_{\sigma^2} = 0.2$, the shape of the sinusoidal function can be recovered even with missing data. Next, we discuss and illustrate the effects of each hyperparameter on the resulting $f(x)$.

Effect of ϵ_{σ^2} on $f(x)$: Intuitively, ϵ_{σ^2} can be thought of in terms of how much trust we place in our training data. If you set the value of ϵ_{σ^2} high, it means that your training values y are possibly very corrupted and do not accurately reflect the underlying function. If instead you set ϵ_{σ^2} as a small value, then y accurately reflects the underlying function that you are trying to regress. In figures B.43 and B.44, we illustrate the effect that ϵ_{σ^2} has on the regressor prediction.

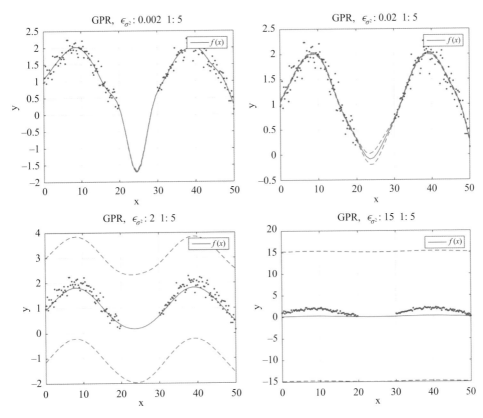

Figure B.43

Effect of noise variance, ε_{σ^2}, on the regressor: The kernel width is kept constant at $l = 5$ and the noise parameter ε_{σ^2} varies from 0.002 to 15. The shape of the regression line seems to be slightly affected. As the noise level increases to infinity, the value of the regressor function $\rightarrow 0$.

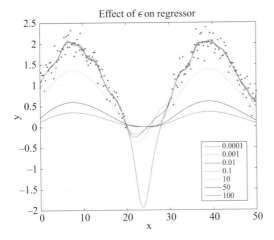

Figure B.44

Effect of noise variance, ε_{σ^2}, on the regressor: A more systematic evaluation of the effect of noise on the regressor.

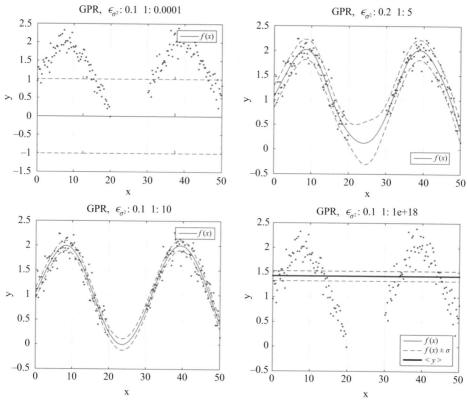

Figure B.45
Effect of kernel variance, l on the regressor: (a) Very small variances causes the regressor function to be zero everywhere except on test points. (b)–(c) Variance is within the scale of the input space, which results in normal behavior. (c) The variance (kernel width) is extremely large. We can see that the regressor value lies exactly on top of the mean of the output data, y.

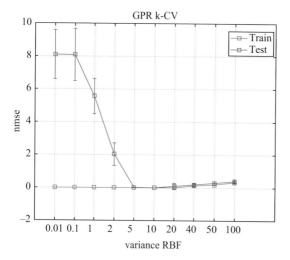

Figure B.46
Grid search through CV for the sinusoidal with hole data set.

Effect of l on f(x): As the variance of the kernel function goes to zero, the kernel function $k(x', x') = 1$ will equal 1 only when the test point x' is equal to training point x; it will be zero otherwise $[k(x, x) = 0]$. As a result, the regressor function will return a nonzero value only when a test point is within a very small distance from a training point (determined by l). When the variance of the kernel tends to infinity, the Gram matrix will equal 1 $[K(X, X) = 1]$, and the regressor will end up giving the mean value of the data.

B.5.3 GPR Hyperparameter Optimization

The optimal hyperparameter values can be found by maximizing the marginal likelihood via Bayesian estimation techniques [118]. Yet they can also be found by performing k-fold CV, as described in section B.4.3 earlier in this appendix. By exploring the range of hyperparameters of $l = [0.01 - 100]$ and $\epsilon_{\sigma^2} = [0 - 8]$, one can see that the best value for the kernel variance is $l = [5 - 10]$.

Appendix C: Background on Robot Control

C.1 Multi-rigid Body Dynamics

In this section, we describe the Cartesian and the joint space controllers that are commonly used for controlling the motion of a robotic arm. For the sake of brevity, we assumed that readers are familiar with the dynamic of a robotic system, both in the task and joint spaces.[1]

The dynamic equation of a robotic manipulator with D revolute joints can be expressed by the following second-order, nonlinear differential equation:

$$M(q)\ddot{q} + C(q,\dot{q})\dot{q} + G(q) = \tau + J(q)^T F_c, \tag{C.1}$$

where $M(q) \in \mathbb{R}^{D \times D}$, $C(q,\dot{q}) \in \mathbb{R}^{D \times D}$, and $G(q) \in \mathbb{R}^{D \times 1}$ are the Inertia, Coriolis, Centrifugal, and Gravitational force matrices, respectively. $J(q) \in \mathbb{R}^{N \times D}$ is the task Jacobian matrices, $\tau \in \mathbb{R}^{D \times 1}$ is the control signal, and $F_c \in \mathbb{R}^{N \times 1}$ is interaction forces between the robot and the environment. Here, $q \in \mathbb{R}^{D \times 1}$ is the joints' position. The robot dynamic [equation (C.1)] in the task space can be written as

$$M_x(q)\ddot{x} + V_x(q,\dot{q})\dot{x} + G_x(q) = \mathscr{F} + F_c, \tag{C.2}$$

where

$$M_x(q) = J^{-T}M(q)J^{-1}, \quad V_x(q,\dot{q}) = J^{-T}C(q,\dot{q})J^{-1} - J^{-T}M(q)J^{-1}\dot{J}J^{-1}$$

$$G_x(q) = J^{-T}g(q), \quad \mathscr{F} = J^{-T}\tau, \quad \dot{x} = J(q)\dot{q}, \quad \ddot{x} = J(q)\ddot{q} + \dot{J}(q)\dot{q}. \tag{C.3}$$

Here, $x \in \mathbb{R}^N$ is the end-effector state. When $N \neq D$, the Jacobian is not invertible. In this case, one can use the Moore-Penrose pseudoinverse; that is, $J^{-1} = J(JJ^T)^{-1}$.[2]

Once the robot is in contact with the environment, the external force is no longer zero, and we have $F_c \neq 0$. When the environment is static, nondeformable, and nonpenetrable, the robot is constrained by the environment to move in the direction specified as follows: $J_c(q)\dot{q} = 0$, $J_c(q)\ddot{q} + \dot{J}_c(q)\dot{q} = 0$, where $J_c(q) \in \mathbb{R}^{D \times N}$ is the constrained Jacobian.

C.2 Motion Control

C.2.1 Preliminaries

Robotic systems can be controlled in several modalities; they can be position-, velocity, or torque-controlled. In all these cases, a reference position, velocity, or torque/force (respectively) is sent to a low-level robot controller to generate the desired motion. Hence, to move a robot with regard to a desired trajectory, one should generate these control signals.

Assuming a torque-controlled robot, the most common control law for motion generation is the linear proportional-derivative (PD) controller, as follows:

$$F_c = -k(x - x_d) - d(\dot{x} - \dot{x}_d), \tag{C.4}$$

where $x_d, \dot{x}_d \in \mathbb{R}^N$ are the desired reference position and velocity that a robot should track, while $x, \dot{x} \in \mathbb{R}^N$ are the current position and velocity of the robot, $F_c \in \mathbb{R}^N$ is the control signal sent to the robot in task space, and $k, d \in \mathbb{R}_+$ are positive gain matrices. The proportional gain k acts as a virtual spring that reduces the position error $(x - x_d)$, while the derivative gain d acts as a damper that reduces the velocity error $(\dot{x} - \dot{x}_d)$ [96]. Other variants of equation (C.4) include setting $k = 0$, which yields the D-controller or setting $d = 0$, which yields a P-controller. More comprehensive control architectures includes the dynamic of the robot, the desired acceleration, or interaction forces. In depth analysis of such control-laws is presented in chapter 10.

C.2.2   Motion Control with Dynamical Systems (DSs)

A DS-based motion generator will yield a desired velocity, but not a desired position. To generate the desired references for control laws derived from equation (C.4) with a DS $f(x)$, we can use the following forward integration law. Given an initial condition $x_d(0)$, $x_d(t), \dot{x}_d(t)$ are computed by solving the following initial value problem:

$$\begin{cases} x_d(t) = x_d(0) + \int_0^t f(x_d(\tau))d\tau \\ \dot{x}_d(t) = f(x_d(t)). \end{cases} \tag{C.5}$$

The numerical solution of the initial value problem is fed directly to equation (C.4). For velocity-controlled robots, one can send $\dot{x}_d$ directly as the reference, while for position-controlled robots, one can send x_d [equation (C.5)].

C.2.3   Inverse Kinematic

To translate the desired end-effector motion $(x_d \in \mathbb{R}^N)$ of the robot to the desired joints' motion $(q_d \in \mathbb{R}^D)$, one needs to solve the Inverse Kinematic (IK) problem. The IK problem can be solved on various levels (namely, position, velocity, and acceleration levels). The position-level solution is preferable over the other solutions, in that the exact desired end-effector location can be achieved. However, the analytical/closed-formed solution is not always available in dexterous robotic platforms, as one end-effector location might represent several joint configurations (i.e., redundant robots where $N < D$). Moreover, the solution might not be continuous or smooth enough for a robot to follow. To address these drawbacks, velocity-level inverse kinematic solutions have been proposed. The damped least-squares method is one of the most popular velocity-level IK solutions. Given the Jacobin matrix $J(q) \in \mathbb{R}^{N \times D}$, the desired joint velocity with respect to the desired end-effector velocity is

$$\dot{q}_d = J^T(q)(J(q)J^T(q) + \lambda^2 I)^{-1}\dot{x}_d, \tag{C.6}$$

where $\lambda \in \Re$ is a nonzero damping constant, and λ should be large enough to avoid unstable behaviors near singularities.[3] But it should not be too large, as it makes it less precise. Apart from the computational simplicity, the main advantage of this method is the smoothness of the generated joint commands, as the solution is continuous at the position level by definition. To increase the continuity of the motion, the following acceleration-level IK solution are proposed:

$$\ddot{q}_d = J^T(q)(J(q)J^T(q))^{-1}(\ddot{x}_d + \dot{J}(q)\dot{q}). \tag{C.7}$$

These methods have not been very popular, as the time derivative of the Jacobin matrix is needed.

Appendix D: Proofs and Derivations

D.1 Proofs and Derivations for Chapter 3

D.1.1 Collapsed Gibbs Sampler and Sampling Equations

Likelihood of Partitions for $Z = \mathbf{Z}(C)$ The likelihood of a partition $Z = \mathbf{Z}(C)$ is computed as the product of the probabilities of the customers $\mathbf{X}$ sitting at their assigned tables Z:

$$p(\mathbf{X}|\mathbf{Z}(C), \lambda) = \prod_{k=1}^{|\mathbf{Z}(C)|} p(\mathbf{X}_{\mathbf{Z}(C)=k}|\lambda), \tag{D.1}$$

where $|\mathbf{Z}(C)|$ denotes the number of unique tables emerging from $\mathbf{Z}(C)$; that is, K in a finite mixture model, and $\mathbf{Z}(C) = k$ is the set of customers assigned to the kth table. Further, each marginal likelihood in equation (D.1) takes the following form:

$$p(\mathbf{X}_{\mathbf{Z}(C)=k}|\lambda) = \int_\theta \left(\prod_{i \in \mathbf{Z}(C)=k} p\left(x^i \mid \theta\right) \right) p\left(\theta \mid \lambda\right) d\theta. \tag{D.2}$$

Because $p(x^i \mid \theta) = \mathcal{N}\left(x^i \mid \mu, \Sigma\right)$ and $p(\theta \mid \lambda) = NIW(\mu, \Sigma \mid \lambda)$, equation (D.2) has an analytical solution that can be derived from the posterior $p(\mu, \Sigma|\mathbf{X})$, as presented in section B.3.2.2 in a appendix B.

Collapsed Gibbs Sampler for *PC*-GMM Equation (3.45) in chapter 3 is sampled via a two-step procedure:

Step 1. The ith customer assignment is removed from the current partition $\mathbf{Z}(C)$. If this causes a change in the partition [i.e., $\mathbf{Z}(C_{-i}) \neq \mathbf{Z}(C)$], the customers previously sitting at $\mathbf{Z}(c_i)$ are split into two (or more) groups and the likelihood must be updated via equation (D.1).

Step 2. A new customer assignment c_i must be sampled; by doing so, a new partition $\mathbf{Z}(c_i = j \cup C_{-i})$ is generated. This new customer assignment might change the current partition $\mathbf{Z}(C_{-i})$. If $\mathbf{Z}(C_{-i}) = \mathbf{Z}(c_i = j \cup C_{-i})$, the partition is unchanged and the ith customer either joined an existing table or sat alone. If $\mathbf{Z}(C_{-i}) \neq \mathbf{Z}(c_i = j \cup C_{-i})$, the partition was changed; specifically, $c_i = j$ caused two tables to merge: table l, which is where the ith customer was sitting prior to step 1, and table m, which is the new table assignment emerging from the new sample $\mathbf{Z}(c_i = j)$. Due to these effects on the partition, instead of explicitly sampling from equation (3.45), [19] proposed sampling from the following distribution:

$$p(c_i = j \mid C_{-i}, \mathbf{X}, \mathbf{S}, \alpha, \lambda) \propto \begin{cases} p(c_i = j | \mathbf{S}, \alpha) \Lambda(\mathbf{X}, C, \lambda) & \text{if } \texttt{cond} \\ p(c_i = j | \mathbf{S}, \alpha) & \text{otherwise,} \end{cases} \qquad (D.3)$$

where $\texttt{cond}$ is the condition of $c_i = j$ merging tables m and l and $\Lambda(\mathbf{X}, C, \lambda)$ is equivalent to

$$\Lambda(\mathbf{X}, C, \lambda) = \frac{p(\mathbf{X}_{(\mathbf{Z}(C)=m \,\cup\, \mathbf{Z}(C)=l)} | \lambda)}{p(\mathbf{X}_{\mathbf{Z}(C)=m} | \lambda) p(\mathbf{X}_{\mathbf{Z}(C)=l} | \lambda)}. \qquad (D.4)$$

This procedure is iterated T times for a predefined number of iterations. The entire procedure is summarized in Algorithm D.1.

Algorithm D.6 Collapsed Gibbs Sampler for $\dot{x}$-SD-CRP

Input: $\mathbf{X}, \dot{\mathbf{X}}$ ▷ Data
          $\alpha, \lambda = \{\mu_0, \kappa_0, \Lambda_0, \nu_0\}$ ▷ Hyperparameters
Output: $\Psi = \{K, C, Z, \Theta\}$ ▷ Inferred clusters and cluster indicators
          Compute pairwise $\dot{x}$-similarity values [equation (3.42) in chapter 3]
1: **procedure** Gibbs-Sampler($\mathbf{X}, \mathbf{S}, \alpha, \lambda$)
2:       Set $\Psi^{t-1} = \{C, K, Z\}$, where $c_i = i$ for $C = \{c_1, \ldots, c_N\}$
3:       **for** `iter t = 1 to T` **do**
4:             Sample a random perm. $\tau(\cdot)$ of integers $\{1, \ldots, N\}$.
5:             **for** `obs i = `$\tau(1)$` to `$\tau(N)$ **do**
6:                   **Remove** customer assignment c_i from the partition
7:                   **if** $\mathbf{Z}(C_{-i}) \neq \mathbf{Z}(C)$, **then**
8:                         **Update** likelihoods according to equation (D.1)
9:                   **end if**
10:                  **Sample** new cluster assignment
11:                  $c_i^{(i)} \sim p(c_i = j | C_{-i}, \mathbf{X}_{-i}, \mathbf{S}, \alpha)$ [equation (D.3)]
12:                  **if** $\mathbf{Z}(C_{-i}) \neq \mathbf{Z}(c_i = j \cup C_{-i})$ **then**
13:                        **Update** table assignments Z.
14:                  **end if**
15:            **end for**
          Sample table parameters Θ_γ from $\mathcal{NIW}$ posterior
16:            update equation (B.43) in appendix B.
17:      **end for**
18: **end procedure**

D.2 Proofs and Derivations for Chapter 4

D.2.1 Expansions for RBF Kernel

The formulation given previously is generic and can be applied to any kernel. Here, we give the RBF kernel–specific expressions for the block matrices in the dual of equation 4.7:

$$(\mathbf{K})_{ij} = y_i y_j k(\mathbf{x}_i, \mathbf{x}_j) = y_i y_j e^{-d\|\mathbf{x}_i - \mathbf{x}_j\|^2}$$

$$(\mathbf{G})_{ij} = y_i \left(\frac{\partial k(\mathbf{x}_i, \mathbf{x}_j)}{\partial \mathbf{x}_j} \right)^T \hat{\mathbf{x}}_j = -2dy_i e^{-d\|\mathbf{x}_i - \mathbf{x}_j\|^2} (\mathbf{x}_j - \mathbf{x}_i)^T \hat{\mathbf{x}}_j.$$

Replacing $\mathbf{x}_j$ with $\mathbf{x}^*$ in this equation, we get

$$(\mathbf{G}_*)_{ij} = y_i \left(\frac{\partial k(\mathbf{x}_i, \mathbf{x}^*)}{\partial \mathbf{x}^*} \right)^T \mathbf{e}_j = -2dy_i e^{-d\|\mathbf{x}_i - \mathbf{x}^*\|^2} (\mathbf{x}^* - \mathbf{x}_i)^T \mathbf{e}_j$$

$$(H)_{ij} = \hat{\mathbf{x}}_i^T \frac{\partial^2 k(\mathbf{x}_i, \mathbf{x}_j)}{\partial \mathbf{x}_i \partial \mathbf{x}_j} \hat{\mathbf{x}}_j = \hat{\mathbf{x}}_i^T \left[\frac{\partial}{\partial \mathbf{x}_i} \left\{ -2de^{-d\|\mathbf{x}_i - \mathbf{x}_j\|^2} (\mathbf{x}_j - \mathbf{x}_i) \right\} \right] \hat{\mathbf{x}}_j$$

$$= 2de^{d\|\mathbf{x}_i - \mathbf{x}_j\|^2} \left[\hat{\mathbf{x}}_i^T \hat{\mathbf{x}}_j - 2d \left\{ \hat{\mathbf{x}}_i^T (\mathbf{x}_i - \mathbf{x}_j) \right\} \left\{ (\mathbf{x}_i - \mathbf{x}_j)^T \hat{\mathbf{x}}_j \right\} \right].$$

Again, replacing $\mathbf{x}_j$ with $\mathbf{x}^*$, we get

$$(H_*)_{ij} = \hat{\mathbf{x}}_i^T \frac{\partial^2 k(\mathbf{x}_i, \mathbf{x}^*)}{\partial \mathbf{x}_i \partial \mathbf{x}^*} \mathbf{e}_j = 2de^{-d\|\mathbf{x}_i - \mathbf{x}^*\|^2} \left[\hat{\mathbf{x}}_i^T \mathbf{e}_j - 2d \left\{ \hat{\mathbf{x}}_i^T (\mathbf{x}_i - \mathbf{x}^*) \right\} \left\{ (\mathbf{x}_i - \mathbf{x}^*)^T \mathbf{e}_j \right\} \right].$$

Replacing $\mathbf{x}_i$ also with $\mathbf{x}^*$, we get

$$(H_{**})_{ij} = \mathbf{e}_i^T \frac{\partial^2 k(\mathbf{x}^*, \mathbf{x}^*)}{\partial \mathbf{x}^* \partial \mathbf{x}^*} \mathbf{e}_j = 2d \left(\mathbf{e}_i^T \mathbf{e}_j \right).$$

D.3 Proofs and Derivations for Chapter 5

D.3.1 Preliminaries for Stability Proofs

Here, we state some mathematical preliminaries that are essential for our proofs.

D.3.1.1 Inequalities for quadratic forms

Let $A \in \mathbb{R}^{M \times M}$ be $A = A^T \prec 0$, and $A = V \Lambda V^T$ be its eigenvalue decomposition with eigenvalues sorted as $\lambda_1 \leq \cdots \leq \lambda_M < 0$. Given a quadratic form $x^T A x$ with $x \in \mathbb{R}^M$, its upper bound can be derived as follows:

$$x^T A x = x^T V \Lambda V^T x = (V^T x)^T \Lambda (V^T x)$$

$$= \sum_{i=1}^M \lambda_i (v_i^T x)(v_i^T x) = \sum_{i=1}^M \lambda_i x^T (v_i v_i^T) x \tag{D.5}$$

$$\leq \lambda_M \sum_{i=1}^M x^T (v_i v_i^T) x = \lambda_M x^T x.$$

Denoting $\lambda_{\max}(A) = \lambda_M$, we have $x^T A x \leq \lambda_{\max}(A)\|x\|^2$. By denoting $\lambda_{\min}(A) = \lambda_1$ and following the same derivation as equation (D.5), the quadratic form has the following upper/lower bounds:

$$\lambda_{\min}(A)\|x\|^2 \leq x^T A x \leq \lambda_{\max}(A)\|x\|^2 < 0. \tag{D.6}$$

This shows the fundamental result that given $A = A^T \prec 0$, a quadratic function $f(x) = x^T A x < 0$. Now assume the quadratic form $x^T A y$ with $x, y \in \mathbb{R}^M$. Following equation (D.5), its upper/lower bounds are

$$\lambda_{\min}(A) x^T y \leq x^T A y \leq \lambda_{\max}(A) x^T y. \tag{D.7}$$

Hence, ensuring that a quadratic function $f(x, y) = x^T A y < 0$ requires $x^T y > 0$. These derivations hold for $A = A^T \succ 0$, yet with eigenvalues sorted as $0 < \lambda_1 \leq \cdots \leq \lambda_M$.

D.3.1.2   Special Matrix Properties

Outer Product: For a matrix $A \in \mathbb{R}^{M \times N}$, where $M \leq N$ and rank$(A) = M$, then $AA^T \succ 0$.

Further, for a square matrix $A \in \mathbb{R}^{M \times M}$ that is $A \succ 0$ (or $\prec 0$) and a matrix $B \in \mathbb{R}^{M \times N}$ with rank$(B) = M$, then $BAB^T \succ 0$ (or $\prec 0$).

Schur Complement: Let $\mathscr{A} = \mathscr{A}^T \in \mathbb{R}^{MN \times MN}$ be a square matrix partitioned into four submatrix blocks as follows:

$$\mathscr{A} = \begin{bmatrix} A & B \\ B^T & C \end{bmatrix} \text{ with } A \in \mathbb{R}^{M \times M} B \in \mathbb{R}^{M \times N}, C \in \mathbb{R}^{N \times N}, \tag{D.8}$$

with rank$(B) = M$ and $N \leq M$. If $A = A^T$ and det$(A) \neq 0$, then the Schur complement of A with regard to $\mathscr{A}$ is defined as

$$S = C - B^T A^{-1} B. \tag{D.9}$$

Via equation (D.9), the block-diagonalization of equation (D.8) yields

$$\mathscr{A} = \begin{bmatrix} A & B \\ B^T & C \end{bmatrix} \begin{bmatrix} \mathbb{I}_M & \emptyset \\ B^T A^{-1} & \mathbb{I}_M \end{bmatrix} \begin{bmatrix} A & \emptyset \\ \emptyset & S \end{bmatrix} \begin{bmatrix} \mathbb{I}_M & A^{-1}B \\ \emptyset & \mathbb{I}_M \end{bmatrix}, \tag{D.10}$$

and det$(\mathscr{A}) = $ det(A) det(S), known as *Schur's formula* [33]. Hence, the definiteness properties of equation (D.8) are defined as follows:

- $\mathscr{A} \succ 0$ if and only if $A \succ 0$ and $S \succ 0$.
- $\mathscr{A} \prec 0$ if and only if $A \prec 0$ and $S \prec 0$.

Special Case Saddle-Point Matrices: A square matrix $\mathscr{A} = \mathscr{A}^T \in \mathbb{R}^{MN \times MN}$ partitioned as shown in equation (D.8) is said to be a *saddle-point matrix* if $A \prec 0$, $C \succeq 0$ and its Schur complement $S = C - B^T A^{-1} B \succ 0$. The saddle-point matrix $\mathscr{A} \nsucc 0$ is a symmetric indefinite matrix with M negative and N positive eigenvalues.

Given $A = A^T \prec 0$, $C = C^T \succeq 0$ and $x, f \in \mathbb{R}^M$, and $y, g \in \mathbb{R}^N$ a saddle-point linear system [12] is defined as

$$\begin{bmatrix} A & B \\ B^T & C \end{bmatrix} \begin{bmatrix} x \\ y \end{bmatrix} = \begin{bmatrix} f \\ g \end{bmatrix} \quad \text{or} \quad \mathscr{A} u = b. \tag{D.11}$$

To solve for equation (D.11), [12] noted that $\mathscr{A}$ is a special case of a saddle-point matrix that can be transformed into a matrix $\hat{\mathscr{A}}$ whose spectrum is entirely contained in the half-plane $\Re(\lambda) < 0$ as follows:

$$\hat{\mathscr{A}} = \mathscr{J} \mathscr{A} = \begin{bmatrix} \mathbb{I}_M & \emptyset \\ \emptyset & -\mathbb{I}_M \end{bmatrix} \begin{bmatrix} A & B \\ B^T & C \end{bmatrix} = \begin{bmatrix} A & B \\ -B^T & -C \end{bmatrix}. \tag{D.12}$$

Using equation (D.12), a linear system, equivalent to equation (D.11), can be constructed as follows:

$$\begin{bmatrix} A & B \\ -B^T & -C \end{bmatrix} \begin{bmatrix} x \\ y \end{bmatrix} = \begin{bmatrix} f \\ -g \end{bmatrix} \quad \text{or} \quad \hat{\mathscr{A}} u = \hat{b}. \tag{D.13}$$

$\hat{\mathscr{A}} \prec 0$ is proved to be negative definite, as shown in theorem 3.6 from the seminal work in [12]; that is, its symmetric part is $\frac{1}{2}(\hat{\mathscr{A}} + \hat{\mathscr{A}}^T) \prec 0$.

Next, we show how equation (D.12) can be used for indefinite quadratic forms. Let $\mathscr{A} \in \mathbb{R}^{2M \times 2M}$ be a saddle-point matrix partitioned as in equation (D.8), with $A = A^T \prec 0$, $C = \emptyset$, $M = N$, and $S = -B^T A^{-1} B \succ 0$. Given $x, y \in \mathbb{R}^M$, an indefinite quadratic form is

$$f(x,y) = \begin{bmatrix} x \\ y \end{bmatrix}^T \underbrace{\begin{bmatrix} A & B \\ B^T & \emptyset \end{bmatrix}}_{\mathscr{A} \not\succ 0} \begin{bmatrix} x \\ y \end{bmatrix} \not\succ 0, \tag{D.14}$$

with $\mathscr{A} \not\succ 0$ having M negative and M positive eigenvalues. This means that $f(x,y) < 0$ for some values of $[x; y] \in \mathbb{R}^{2M}$ and $f(x,y) > 0$ for others. Via equation (D.12), a quadratic form, equivalent to equation (D.14), can be defined as

$$\hat{f}(x,y) = \begin{bmatrix} x \\ y \end{bmatrix}^T \underbrace{\begin{bmatrix} A & B \\ -B^T & \emptyset \end{bmatrix}}_{\hat{\mathscr{A}} \prec 0} \begin{bmatrix} x \\ y \end{bmatrix} + y^T 2B^T x. \tag{D.15}$$

From equation (D.15), we can see that the first term will always be negative $\forall x, y \in \mathbb{R}^M$, and the indefiniteness of equation (D.14) emerges from the residual term $y^T 2B^T$. Hence, to transform equation (D.15) into a negative-definite quadratic form (i.e., $\hat{f}(x,y) < 0$), it would suffice to ensure the following condition:

$$\hat{f}(x,y) < 0 \text{ if and only if } \begin{bmatrix} x \\ y \end{bmatrix}^T \hat{\mathscr{A}} \begin{bmatrix} x \\ y \end{bmatrix} < -y^T 2B^T x. \tag{D.16}$$

Following equation (D.6), and defining $\mathscr{H} = \frac{1}{2}(\hat{\mathscr{A}} + \hat{\mathscr{A}}^T) \prec 0$ as the symmetric part of $\hat{\mathscr{A}}$, we can compute the bounds of the

$$\lambda_{\min}(\mathscr{H})||[x;y]||^2 \leq \begin{bmatrix} x \\ y \end{bmatrix}^T \hat{\mathscr{A}} \begin{bmatrix} x \\ y \end{bmatrix} \leq \lambda_{\max}(\mathscr{H})||[x;y]||^2 < 0. \tag{D.17}$$

By expanding $\mathscr{H}$, we find that $\mathscr{H} = A$ and equation (D.17) is reduced to

$$\lambda_{\min}(A)||[x]||^2 \leq \begin{bmatrix} x \\ y \end{bmatrix}^T \hat{\mathscr{A}} \begin{bmatrix} x \\ y \end{bmatrix} \leq \lambda_{\max}(A)||[x]||^2 < 0. \tag{D.18}$$

Thus, equation (D.16) becomes

$$\hat{f}(x,y) < 0 \text{ if and only if } \lambda_{\max}(A)||[x]||^2 < -y^T 2B^T x. \tag{D.19}$$

D.3.2 Stability of Linear Locally Active Globally Stable Dynamical Systems

Here, we wish to prove theorem 5.1. That is, that the combined DS in equation (5.1) in chapter 5 with the activation function [equation (5.3)] is globally asymptotically stable at the point x_g^* for the Lyapunov function proposed in equation (5.5). This can be proved if the following conditions hold: (I) $V(x_g^*) = 0$, (II) $V(x) > 0 \; \forall \; x \neq x_g^*$, (III) $\dot{V}(x_g^*) = 0$, and (IV) $\dot{V}(x) < 0 \; \forall \; x \neq x_g^*$. From equation (5.5) it is straightforward to see that conditions (I–II) are met. To prove (III–IV), we compute the time derivative of $V(x)$ as

$$\dot{V}(x) = \nabla_x V(x)^T \frac{d}{dt} x(t) = \nabla_x V(x)^T f(x)$$

$$= \nabla_x V(x)^T \Big[\underbrace{\alpha(x) f_g(x) + \overline{\alpha}(x) f_l(\boldsymbol{h}(x), x)}_{(5.1)} \Big]$$

$$= \nabla_x V(x)^T \Big[\alpha(x)(A_g x + \boldsymbol{b}_g) + \overline{\alpha}(x)\Big(\boldsymbol{h}(x)(A_{l,a} x + \boldsymbol{b}_{l,a}) \tag{D.20}$$

$$+ (1 - \boldsymbol{h}(x))(A_{l,d} x + \boldsymbol{b}_{l,d} - \lambda(x)\nabla_x \boldsymbol{h}(x))\Big)\Big]$$

$$= \nabla_x V(x)^T \Big[\alpha(x) \underbrace{A_g \tilde{x}_g}_{(5.16)} + \overline{\alpha}(x)\Big(\underbrace{\mathbf{A}_l(\boldsymbol{h}(x))\tilde{x}_l}_{(5.18)} - \lambda(x)\nabla_x \boldsymbol{h}(x)\Big)\Big].$$

The gradient of $V(x)$ is

$$\nabla_x V(x) = (P_g^T + P_g)\tilde{x}_g + \beta_l^2(x)\Big[P_l \tilde{x}_l + P_l^T \tilde{x}_g \Big], \tag{D.21}$$

with

$$\beta_l^2(x) = 2\beta(x)\tilde{x}_g^T P_l \tilde{x}_l. \tag{D.22}$$

Following section D.3.1.1, the upper/lower bounds of equation (D.22) are defined as

$$0 \le \lambda_{\min}(P_l)2\beta(x)\tilde{x}_g^T \tilde{x}_l < \beta_l^2(x) \le \lambda_{\max}(P_l)2\beta(x)\tilde{x}_g^T \tilde{x}_l. \tag{D.23}$$

By evaluating equation (D.20) at x_g^*, we see that $\dot{V}(x_g^*) = 0$, ensuring condition (III). For ease of readability in the following derivations, we simplify the notations to $\mathbf{A}_l(\boldsymbol{h}(x)) = \mathbf{A}_l$, $\nabla_x \boldsymbol{h}(x) = \nabla_x \boldsymbol{h}$, $\nabla_x V(x) = \nabla_x V$, $\alpha(x) = \alpha$, $\overline{\alpha}(x) = \overline{\alpha}$, $\lambda(x) = \lambda$, and $\beta_l^2(x) = \beta_l^2$. To ensure condition (IV) [i.e., $\dot{V}(x) < 0$], we begin by grouping the terms in equation (D.20) as follows:

$$\dot{V}(x) = \alpha \underbrace{\tilde{x}_g^T A_g^T \nabla_x V}_{\dot{V}_g(x): \text{ Global Component}} + \overline{\alpha} \underbrace{\Big(\tilde{x}_g^T \mathbf{A}_l^T - \lambda \nabla_x \boldsymbol{h}^T\Big)\nabla_x V}_{\dot{V}_l(x): \text{ Local Component}} \tag{D.24}$$

$$= \alpha(x)\dot{V}_g(x) + \overline{\alpha}(x)\dot{V}_l(x).$$

According to equation (5.17), the activation function takes the values $0 < \alpha(x) \le 1$. We now consider two cases: $\alpha = 1$ and $0 < \alpha < 1$. When $\alpha(x) = 1$, equation (D.24) becomes $\dot{V}_g(x)$; hence, we must ensure that

$$\dot{V}_g(x) < 0 \ \forall \ x \ne x_g^*. \tag{D.25}$$

Negative definiteness of global component $\dot{V}_g(x)$: Expanding $\dot{V}_g(x)$ from equation (D.24) and grouping the similar terms yields

$$\dot{V}_g(x) = \tilde{x}_g^T A_g^T \nabla_x V(x)$$

$$= \tilde{x}_g^T (P_g A_g)\tilde{x}_g + \tilde{x}_g^T (A_g^T P_g)\tilde{x}_g + \beta_l^2 \Big(\tilde{x}_g^T (A_g^T P_l)\tilde{x}_g + \tilde{x}_g^T (A_g^T P_l)\tilde{x}_l\Big)$$

$$= \tilde{x}_g^T \bigg(\underbrace{A_g^T P_g + P_g A_g}_{Q_g \prec 0 \ (5.16)} + \underbrace{\beta_l^2}_{\geq 0} \underbrace{A_g^T P_l}_{Q_g^l \prec 0 \ (5.16)} \bigg) \tilde{x}_g + \underbrace{\beta_l^2}_{\geq 0} \tilde{x}_g^T \underbrace{A_g^T P_l}_{Q_g^l \prec 0 \ (5.16)} \tilde{x}_l$$

$$= \tilde{x}_g^T \underbrace{(Q_g + \beta_l^2 Q_g^l)}_{\prec 0} \tilde{x}_g + \underbrace{\tilde{x}_g^T \beta_l^2 Q_g^l \tilde{x}_l}_{\leq 0 \ \text{Proof below}} < 0. \tag{D.26}$$

By defining $Q_{gg}^l = Q_g + \beta_l^2 Q_g^l$, which is $Q_{gg}^l = (Q_{gg}^l)^T \prec 0$, and following section D.3.1.1, the first term in equation (D.26) has the following bounds:

$$\underbrace{\lambda_{\min}(Q_{gg}^l)}_{<0} ||\tilde{x}_g||^2 \leq \tilde{x}_g^T Q_{gg}^l \tilde{x}_g \leq \underbrace{\lambda_{\max}(Q_{gg}^l)}_{<0} ||\tilde{x}_g||^2 < 0. \tag{D.27}$$

This proves that $\tilde{x}_g^T Q_{gg}^l \tilde{x}_g < 0 \ \forall \, x \neq x_g^*$. Further, via section D.3.1.1 and equation (D.23), the second term in equation (D.26) has the following upper/lower bounds:

$$\underbrace{\lambda_{\min}(Q_g^l)}_{<0} \underbrace{\beta_l^2 \tilde{x}_g^T \tilde{x}_l}_{\geq 0} \leq \tilde{x}_g^T \beta_l^2 Q_g^l \tilde{x}_l \leq \underbrace{\lambda_{\max}(Q_g^l)}_{<0} \underbrace{\beta_l^2 \tilde{x}_g^T \tilde{x}_l}_{\geq 0} \leq 0. \tag{D.28}$$

This proves that $\tilde{x}_g^T \beta_l^2 Q_g^l \tilde{x}_l \leq 0 \forall \, x \neq x_g^*$. Hence, if the conditions stated in equation (5.16) hold, then equation (D.25) holds.

Boundedness of local component $\dot{V}_l(x)$: For $0 < \alpha(x) < 1$, we must ensure that $\alpha(x)\dot{V}_g(x) + \bar{\alpha}(x)\dot{V}_l(x) < 0$. Expanding $\dot{V}_l(x)$ from equation (D.24) and grouping similar terms yields

$$\dot{V}_l(x) = \left(\tilde{x}_l^T \mathbf{A}_l^T - \lambda \nabla_x \boldsymbol{h}^T \right) \nabla_x V$$

$$= \tilde{x}_l^T \left(\mathbf{A}_l^T P_g^T + \mathbf{A}_l^T P_g \right) \tilde{x}_g + \tilde{x}_l^T \left(\beta_l^2 \mathbf{A}_l^T P_l^T \right) \tilde{x}_l$$

$$\quad + \tilde{x}_l^T \left(\beta_l^2 \mathbf{A}_l^T P_l^T \right) \tilde{x}_g - \lambda \nabla_x \boldsymbol{h}^T \nabla_x V$$

$$= \tilde{x}_l^T \bigg(\underbrace{2 \mathbf{A}_l^T P_g}_{Q_l^g} + \beta_l^2 \underbrace{\mathbf{A}_l^T P_l}_{Q_l} \bigg) \tilde{x}_g + \tilde{x}_l^T \bigg(\beta_l^2 \underbrace{\mathbf{A}_l^T P_l}_{Q_l} \bigg) \tilde{x}_l - \lambda \nabla_x \boldsymbol{h}^T \nabla_x V \tag{D.29}$$

$$= \tilde{x}_l^T \bigg(\underbrace{Q_l^g}_{Q_{l+}^g \nprec 0 \ [\text{equation (5.18)}]} + \beta_l^2 \underbrace{Q_l}_{Q_{l+} \nprec 0 \ [\text{equation (5.18)}]} \bigg) \tilde{x}_g$$

$$\quad + \tilde{x}_l^T \beta_l^2 \underbrace{Q_l}_{Q_{l+} \nprec 0 \ [\text{equation (5.18)}]} \tilde{x}_l - \underbrace{\lambda(x) \nabla_x \boldsymbol{h}^T \nabla_x V}_{\geq 0 \ \text{via} \ [\text{equation (5.15)}]},$$

with Q_+ indicating the symmetric part of a matrix. Plugging equations (D.26) and (D.29) into equation (D.24) and grouping similar terms yields

$$\dot{V}(x) = \tilde{x}_g^T \underbrace{\alpha \bigg(\overbrace{Q_g}^{\prec 0} + \beta_l^2 \overbrace{Q_g^l}^{\prec 0} \bigg)}_{Q_G(x) \prec 0 \ \text{via equations (5.16)(5.17)}} \tilde{x}_g + \tilde{x}_l^T \underbrace{\bigg(\alpha \beta_l^2 \overbrace{Q_g^l}^{\prec 0} + \bar{\alpha} \bigg(\overbrace{Q_l^g}^{\nprec 0} + \beta_l^2 \overbrace{Q_l}^{\nprec 0} \bigg) \bigg)}_{Q_{LG}(x) \nprec 0 \ \text{via equations (5.16)(5.17)(5.18)}} \tilde{x}_g$$

$$+ \tilde{x}_l^T \underbrace{\left(\overline{\alpha}\beta_l^2 \overset{\not\prec 0}{\overbrace{Q_l}} \right)}_{Q_L(x) \not\prec 0 \text{ via equations (5.17)(5.18)}} \tilde{x}_l - \underbrace{\overline{\alpha}\lambda(x)\nabla_x \boldsymbol{h}^T \nabla_x V}_{\geq 0 \text{ via equation (5.15)}} . \tag{D.30}$$

While the first and last term of equation (D.30) are negative definite, the inner terms can take on both positive and negative values. This is the result of allowing a locally attractive behavior around a point in the state space that is not the global attractor x_g^*. Let $\tilde{x}_{gl} = [\tilde{x}_g; \tilde{x}_l] \in \mathbb{R}^{2M}$ be an augmented state vector and $\boldsymbol{Q}(x) \in \mathbb{R}^{2M \times 2M}$ a state-dependent matrix. Neglecting the modulation term in equation (D.30) (which is always negative and active only around the local virtual attractor $\boldsymbol{x}_l^*$), we can transform equation (D.30) into a conservative Q-matrix-dependent form, as follows:

$$\dot{V}_Q(x) = \tilde{x}_{gl}^T \underbrace{\begin{bmatrix} Q_G(x) & \emptyset \\ Q_{LG}(x) & Q_L(x) \end{bmatrix}}_{\boldsymbol{Q}(x)} \tilde{x}_{gl}$$

$$= \tilde{x}_{gl}^T \underbrace{\begin{bmatrix} Q_G(x) & \frac{1}{2}Q_{LG}(x)^T \\ \frac{1}{2}Q_{LG}(x) & 0 \end{bmatrix}}_{\boldsymbol{Q}_+(x) \not\prec 0} \tilde{x}_{gl} + \tilde{x}_l^T \underbrace{Q_{L+}(x)}_{\not\prec 0} \tilde{x}_l. \tag{D.31}$$

It is straightforward to see that if $\dot{V}_Q(x) < 0 \Rightarrow \dot{V}(x) < 0 \; \forall x = x_g^*$. To ensure such conditions, it would suffice to prove that $\boldsymbol{Q}_+(x) \prec 0$ and to derive a lower bound for the remaining indefinite term. This is not possible, however, as $\boldsymbol{Q}_+(x)$ is a saddle-point matrix (see section D.3.1.2). Namely, the Schur complement of $Q_G(x)$ with regard to $\boldsymbol{Q}_+(x)$ is

$$S_Q(x) = -(1/4)Q_{LG}(x) \underbrace{Q_G^{-1}(x)}_{\prec 0} Q_{LG}(x)^T \succ 0, \tag{D.32}$$

regardless of the definiteness of $Q_{LG}(x)$; see section D.3.1.2. Hence, $\boldsymbol{Q}_+(x)$ will always be indefinite with M (-) and M (+) eigenvalues. Yet, due to the spectral properties of saddle-point matrices, we can employ the $\mathscr{J}$-transform approach described in section D.3.1.2 and transform $\boldsymbol{Q}_+(x)$ into $\hat{\boldsymbol{Q}}_+(x) = \mathscr{J}\boldsymbol{Q}_+(x) \prec 0$ via equation (D.12), yielding a $\dot{V}_Q(x)$ with the form

$$\dot{V}_Q(x) = \tilde{x}_{gl}^T \underbrace{\begin{bmatrix} Q_G(x) & \frac{1}{2}Q_{LG}(x)^T \\ -\frac{1}{2}Q_{LG}(x) & 0 \end{bmatrix}}_{\hat{\boldsymbol{Q}}_+(x) \prec 0} \tilde{x}_{gl} + \tilde{x}_l^T Q_{LG}(x)\tilde{x}_g + \tilde{x}_l^T Q_{L+}(x)\tilde{x}_l, \tag{D.33}$$

which is equivalent to equation (D.31). Hence, in order to ensure $\dot{V}_Q(x) < 0$, we must impose an upper bound on the residual indefinite terms. This bound is derived by applying equation (D.19) to equation (D.33); thus, $\dot{V}_Q(x) < 0$ can be ensured if

$$\lambda_{\max}(Q_G(x))\|\tilde{x}_g\|^2 + \lambda_{\max}(Q_{L+}(x))\|\tilde{x}_l\|^2 < -\tilde{x}_l^T Q_{LG}(x)\tilde{x}_g, \tag{D.34}$$

which is the condition stated in equation (5.19). Intuitively, equation (D.34) defines the upper bound of the dissipation rate of the locally active dynamics. The local DS is thus allowed to exhibit converging behaviors in a local region of the state space around a local virtual attractor x_l^*, while still ensuring global asymptotic stability in x_g^*. Hence, although $\dot{V}(x)$ has an indefinite structure, if equations (5.16), (5.17), (5.18), and (5.19) are met, $\dot{V}(x) < 0 \ \forall x \in \mathbb{R}^M$; that is, equation (5.1) is globally asymptotically stable (GAS) with regard to the global attractor x_g^*.   □

D.3.3 Stability of Nonlinear Locally Active Globally Stable Dynamical Systems

Here, we wish to prove theorem 5.2, that is, that the nonlinear DS [equation (5.22)] is *globally asymptotically* stable at the attractor $x_g^* \in \mathbb{R}^M$, for the Lyapunov function proposed in equation (5.27). This can be proved if the following conditions hold (I) $V(x_g^*) = 0$, (II) $V(x) > 0 \ \forall \ x \neq x_g^*$, (III) $\dot{V}(x_g^*) = 0$, and (IV) $\dot{V}(x) < 0 \ \forall \ x \neq x_g^*$. From equation (5.27), it is straightforward to see that conditions (I–II) are met. To prove (III–IV), we compute the time derivative of $V(x)$ as

$$\dot{V}(x) = \nabla_x V(x)^T \frac{d}{dt} x(t) = \nabla_x^T f(x)$$

$$= \nabla_x V^T \underbrace{\left[\sum_{k=1}^{K} \gamma_k(x) \Big(\alpha(x)(A_g^k x + \boldsymbol{b}_g^k) + \overline{\alpha}(x) f_l^k(\boldsymbol{h}_k(x), x) \Big) \right]}_{\text{via equation (5.22)}}$$

$$\dot{V}(x) = \nabla_x V^T \left[\sum_{k=1}^{K} \gamma_k(x) \Big(\alpha(x)(A_g^k x + \boldsymbol{b}_g^k) + \overline{\alpha}(x) \Big(\boldsymbol{h}_k(x) f_{l,a}^k(x) \right.$$

$$\left. + \overline{\boldsymbol{h}}_k(x)) f_{l,d}^k(x) - \lambda(x) \nabla_x \boldsymbol{h}(x) \Big) \Big) \right]$$

$$= \nabla_x V^T \left[\sum_{k=1}^{K} \gamma_k(x) \Big(\alpha(x)(A_g^k x + \boldsymbol{b}_g^k) + \overline{\alpha}(x) \cdot \Big(\boldsymbol{h}_k(x)(A_{l,a}^k x + \boldsymbol{b}_{l,a}^k) \right. \tag{D.35}$$

$$\left. + \overline{\boldsymbol{h}}_k(x)(A_{l,d}^k x + \boldsymbol{b}_{l,d}^k - \lambda_k(x) \nabla_x \boldsymbol{h}_k(x)) \Big) \Big) \right]$$

$$= \nabla_x V^T \left[\sum_{k=1}^{K} \gamma_k(x) \Big(\alpha(x) \underbrace{A_g^k \tilde{x}_g}_{(5.29)} + \overline{\alpha}(x) \Big(\underbrace{\mathbf{A}_l^k(\boldsymbol{h}_k(x)) \tilde{x}_k}_{(5.31)} - \lambda_k(x) \nabla_x \boldsymbol{h}_k(x) \Big) \Big) \right].$$

The gradient of $V(x)$ is defined as

$$\nabla_x V(x) = (P_g^T + P_g) \tilde{x}_g + \sum_{j=1}^{K} \beta_j^2(x) \left[P_l^j \tilde{x}_j + (P_l^j)^T \tilde{x}_g \right], \tag{D.36}$$

with

$$\beta_j^2(x) = 2\beta_j(x) \tilde{x}_g^T P_l^j \tilde{x}_j. \tag{D.37}$$

To avoid confusion in the following derivations, we define the index of the kth local P-QLF's with j and preserve the index k for the locally active DS parameters. Following section D.3.1.1, the bounds of equation (D.37) are defined as

$$0 \le \lambda_{\min}(P_l^j) 2\beta_j(x) \tilde{x}_g^T \tilde{x}_j < \beta_j^2(x) \le \lambda_{\max}(P_l^j) 2\beta_j(x) \tilde{x}_g^T \tilde{x}_j. \tag{D.38}$$

By evaluating equation (D.35) at x_g^*, we see that $\dot{V}(x^*) = 0$, ensuring condition (III). For ease of readability in the following derivations, we simplify the notations to $\mathbf{A}_l^k(\boldsymbol{h}_k(x)) = \mathbf{A}_l^k$, $\nabla_x h_k(x) = \nabla_x \boldsymbol{h}_k$, $\nabla_x V(x) = \nabla_x V$, $\alpha(x) = \alpha$, $\overline{\alpha}(x) = \overline{\alpha}$, $\lambda_k(x) = \lambda_k$, $\gamma_k(x) = \gamma_k$, and $\beta_j^2(x) = \beta_j^2$. To ensure condition (IV) [i.e., $\dot{V}(x) < 0$], we begin by grouping the terms in equation (D.35) as

$$\dot{V}(x) = \sum_{k=1}^K \gamma_k \Big[\alpha \underbrace{\tilde{x}_g^T (A_g^k)^T \nabla_x V}_{k\text{th Global Comp.:}\dot{V}_g^k(x)} + \overline{\alpha} \underbrace{\Big(\tilde{x}_k^T (\mathbf{A}_l^k)^T - \lambda_k \nabla_x \boldsymbol{h}_k^T \Big) \Big) \nabla_x V}_{k\text{th Local Comp.:}\dot{V}_l^k(x)} \Big]$$

$$= \sum_{k=1}^K \underbrace{\gamma_k}_{\substack{>0 \\ \text{equation (5.30)}}} \underbrace{\Big(\alpha \dot{V}_g^k(x) + \overline{\alpha} \dot{V}_l^k(x) \Big)}_{\dot{V}^k(x)}. \tag{D.39}$$

To ensure $\dot{V}(x) < 0$, it suffices to ensure that $\dot{V}^k(x) < 0 \; \forall k = 1, \ldots, K$. Because $\dot{V}^k(x)$ is equivalent to equation (D.24), we follow the same derivation as in section D.3.2 to ensure its negative definiteness; that is, we begin by ensuring that

$$\dot{V}_g^k(x) < 0 \; \forall \; x \ne x_g^*. \tag{D.40}$$

Negative definiteness of the kth global component $\dot{V}_g^k(x)$: Expanding $\dot{V}_g^k(x)$ from equation (D.39) and grouping the similar terms together yields

$$\dot{V}_g^k(x) = \tilde{x}_g^T (A_g^k)^T \nabla_x V(x)$$

$$= \tilde{x}_g^T (A_g^k)^T (P_g + P_g^T) \tilde{x}_g + \tilde{x}_g^T (A_g^k)^T \underbrace{\Big(\sum_{j=1}^K \beta_j^2 P_l^j \Big) \tilde{x}_g}_{\mathbf{P}_l(x)} + \tilde{x}_g^T (A_g^k)^T \Big(\sum_{j=1}^K \beta_j^2 P_l^j \tilde{x}_j \Big)$$

$$= \tilde{x}_g^T \Big(\underbrace{P_g A_g^k + (A_g^k)^T P_g}_{Q_g^k \prec 0 \text{ equation (5.29)}} + \underbrace{(A_g^k)^T \mathbf{P}_l(x)}_{(Q_g^{l,k})_+ \prec 0 \text{ equation (5.29)}} \Big) \tilde{x}_g$$

$$+ \sum_{j=1}^K \underbrace{\beta_j^2}_{\ge 0} \Big(\tilde{x}_g^T \underbrace{(A_g^k)^T}_{(\cdot)_+ \prec 0 \text{ equation (5.29)}} P_l^j \tilde{x}_j \Big) \tag{D.41}$$

$$= \underbrace{\tilde{x}_g^T \Big(Q_g^k + Q_g^{l,k} \Big) \tilde{x}_g}_{<0 \text{ Proof below.}} + \sum_{j=1}^K \overset{\ge 0}{\beta_j^2} \underbrace{\tilde{x}_g^T \Big((A_g^k)^T P_l^j \Big) \tilde{x}_j}_{\le 0 \text{ Proof below.}} < 0$$

Next, we prove that $\tilde{x}_g^T \Big(Q_g^k + Q_g^{l,k} \Big) \tilde{x}_g < 0$. Note that

$$\tilde{x}_g^T \left(Q_g^k + Q_g^{l,k} \right) \tilde{x}_g = \tilde{x}_g^T \underbrace{Q_g^k}_{\prec 0 \ (5.29)} \tilde{x}_g + \tilde{x}_g^T \underbrace{(Q_g^{l,k})_+}_{\prec 0 \ (5.29)} \tilde{x}_g < 0, \tag{D.42}$$

with $(Q_g^{l,k})_+$ signifying the symmetric part of $Q_g^{l,k}$. Each of the terms have the following upper/lower bounds:

$$\underbrace{\lambda_{\min}(Q_g^k)}_{<0} ||\tilde{x}_g||^2 \leq \tilde{x}_g^T Q_g^k \tilde{x}_g \leq \underbrace{\lambda_{\max}(Q_g^k)}_{<0} ||\tilde{x}_g||^2 < 0 \tag{D.43}$$

$$\underbrace{\lambda_{\min}(Q_g^{l,k})_+}_{<0} ||\tilde{x}_g||^2 \leq \tilde{x}_g^T (Q_g^{l,k})_+ \tilde{x}_g \leq \underbrace{\lambda_{\max}(Q_g^{l,k})_+}_{<0} ||\tilde{x}_g||^2 < 0, \tag{D.44}$$

showing explicitly that the first term in equation (D.41) $< 0 \ \forall x \in \mathbb{R}^M \setminus x = x_g^*$.

Next, we show that $\beta_j^2 \tilde{x}_g^T \left((A_g^k)^T P_l^j \tilde{x}_j \right) \leq 0$. Following section D.3.1.1 and defining $(A_g^k)^T = V \Lambda V^T$, we can find the upper bound of this term:

$$\beta_j^2 \tilde{x}_g^T \left((A_g^k)^T P_l^j \right) \tilde{x}_j = \tilde{x}_g^T \left(V \Lambda V^T \beta_j^2 P_l^j \right) \tilde{\xi}_j = (V)^T \tilde{x}_g)^T \Lambda (V^T \beta_j^2 P_l^j \tilde{x}_j)$$

$$= \sum_{i=1}^{M} \lambda_i (V_i^T \tilde{x}_g^T)(V_i^T \beta_j^2 P_l^j \tilde{x}_j) = \sum_{i=1}^{M} \lambda_i \tilde{x}_g^T (V_i V_i^T) \beta_j^2 P_l^j \tilde{x}_j$$

$$\leq \lambda_{\max}((A_g^k)_+) \sum_{i=1}^{M} \tilde{x}_g^T (V_i V_i^T) \beta_j^2 P_l^j \tilde{x}_j \tag{D.45}$$

$$= \lambda_{\max}((A_g^k)_+) \tilde{x}_g^T \beta_j^2 P_l^j \tilde{x}_j$$

$$= \underbrace{\lambda_{\max}((A_g^k)_+)}_{\prec 0 \text{equation (5.29)}} 2 \underbrace{\beta_j}_{\geq 0} \underbrace{(\tilde{x}_g^T P_l^j \tilde{x}_j)^2}_{\geq 0} \leq 0,$$

proving that $\beta_j^2 \tilde{x}_g^T \left((A_g^k)^T P_l^j \tilde{x}_j \right) \leq 0 \ \forall \ x \neq x_g^*$. Hence, if the conditions stated in equation (5.29) in chapter 5 hold, then equation (D.40) holds.

Boundedness of kth local component $\dot{V}_l^k(x)$: For $0 < \alpha(x) < 1$, we must ensure that $\dot{V}^k(x) = \alpha(x)\dot{V}_g^k(x) + \overline{\alpha}(x)\dot{V}_l^k(x) < 0$. This condition is equivalent to ensuring boundedness on the local component in the linear LAGS-DS case derived in section D.3.2. As in equation (D.29), we expand $\dot{V}_l^k(x)$ from equation (D.39) and group similar terms:

$$\dot{V}_l^k(x) = \left(\tilde{x}_k^T (\mathbf{A}_l^k)^T - \lambda_k \nabla_x h_k^T \right) \nabla_x V$$

$$= \tilde{x}_k^T \left((\mathbf{A}_l^k)^T P_g^T +^T (A_g^k)^T P_g + (\mathbf{A}_l^k)^T \underbrace{\left(\sum_{j=1}^{K} \beta_j^2 P_l^j \right)}_{\mathbf{P}_l(x)} \right) \tilde{x}_g$$

$$+ \tilde{x}_k^T (\mathbf{A}_l^k)^T \left(\sum_{j=1}^{K} \beta_j^2 P_l^j \tilde{x}_j \right) - \lambda_k \nabla_x \boldsymbol{h}_k^T \nabla_x V$$

$$= \tilde{x}_k^T \Big(\underbrace{2(\mathbf{A}_l^k)^T P_g}_{(Q_l^{g,k})_+ \not\prec 0 \text{ equation (5.31)}} + \underbrace{(\mathbf{A}_l^k)^T \mathbf{P}_l(x)}_{(Q_l^k)_+ \not\prec 0 \text{ equation (5.31)}} \Big) \tilde{x}_g \tag{D.46}$$

$$+ \sum_{j=1}^{K} \underbrace{\overset{\geq 0}{\overbrace{\beta_j^2}} \tilde{x}_k^T \Big(\overset{\prec 0 \text{ equation (5.31)}}{\overbrace{(\mathbf{A}_l^k)^T}} P_l^j \Big) \tilde{x}_j}_{\leq 0 \text{ via equation (D.45)}} - \underbrace{\lambda(x) \nabla_x \boldsymbol{h}^T \nabla_x V}_{\geq 0 \text{ via equation (5.15)}}$$

Grouping equations (D.46) and (D.41) together yields the following kth Lyapunov derivative component $\dot{V}^k(x)$:

$$\dot{V}^k(x) = \tilde{x}_g^T \underbrace{\alpha \Big(\overset{\prec 0}{\overbrace{Q_g^k}} + \overset{+ \prec 0}{\overbrace{Q_g^{l,k}}} \Big)}_{Q_{G+}^k(x) \prec 0 \text{ via equations (5.29) and (5.30)}} \tilde{x}_g$$

$$+ \tilde{x}_k^T \underbrace{\overline{\alpha} \Big(\overset{\not\prec 0}{\overbrace{Q_l^{g,k}}} + \overset{\not\prec 0}{\overbrace{Q_l^k}} \Big)}_{Q_{LG}^k(x) \not\geq 0 \text{ via equations (5.16)(5.17), and (5.18)}} \tilde{x}_g \tag{D.47}$$

$$+ \sum_{j=1}^{K} \underbrace{\beta_j^2 \Big(\tilde{x}_g^T \alpha \big((A_g^k)^T P_l^j \big) \tilde{x}_j + \tilde{x}_k^T \big((\mathbf{A}_l^k)^T P_l^j \big) \tilde{x}_j \Big)}_{\leq 0 \text{ via equations (5.29),(5.30),(5.31), and (D.45)}} - \underbrace{\overline{\alpha} \lambda(x) \nabla_x \boldsymbol{h}^T \nabla_x V}_{\geq 0 \text{ via equation (5.15)}}.$$

As in the linear case, equation (D.47) is an indefinite quadratic form; hence, each kth component of our DS induces a kth saddle-point Lyapunov derivative, which can be bounded following the transformation derived for equation (D.33). Following section D.3.2, equation (D.47) yields a $\mathscr{J}$-transformed Q-dependent term as follows:

$$\dot{V}_Q^k(x) = \overline{x}_{gk}^T \underbrace{\begin{bmatrix} Q_{G+}^k(x) & \frac{1}{2} Q_{LG}^k(x)^T \\ -\frac{1}{2} Q_{LG}^k(x) & 0 \end{bmatrix}}_{\hat{Q}_+^k(x) \prec 0} \overline{x}_{gk} + \tilde{x}_k^T Q_{LG}^k(x) \overline{x}_g$$

$$+ \sum_{j=1}^{K} \underbrace{\beta_j^2 \Big(\tilde{x}_g^T \alpha \big((A_g^k)^T P_l^j \big) \tilde{x}_j + \tilde{x}_k^T \overline{\alpha} \big((\mathbf{A}_l^k)^T P_l^j \big) \tilde{x}_j \Big)}_{\leq 0 \text{ via equations (5.29),(5.30),(5.31), and (D.45)}} \tag{D.48}$$

for $\tilde{x}_{gk} = [\tilde{x}_g; \tilde{x}_k] \in \mathbb{R}^{2M}$. If we ensure that $\dot{V}_Q^k(x) < 0$, then $\dot{V}^k(x) < 0$ follows, and consequently $\dot{V}(x) < 0$. This can be achieved by imposing a bound on the residual indefinite term in equation (D.48). Following equation (D.19), we see that the first term in equation (D.48) has an upper bound of

$$\tilde{x}_{gk}^T \hat{Q}_+^k(x) \tilde{x}_{gk} < \lambda_{\max}(Q_{G+}^k(x)) ||\tilde{x}_g||^2 < 0. \tag{D.49}$$

Following equation (D.45), the jth component of the last term in equation (D.48) has the following upper bound:

$$\beta_j^2 \left(\tilde{x}_g^T \alpha \left((A_g^k)^T P_l^j \right) \tilde{x}_j + \tilde{x}_k^T \overline{\alpha} \left((\mathbf{A}_l^k)^T P_l^j \right) \tilde{x}_j \right) <$$
$$2\beta_j \left(\alpha \lambda_{\max}((A_g^k)_+)(\tilde{x}_g^T P_l^j \tilde{x}_j)^2 + \overline{\alpha} \lambda_{\max}(\mathbf{A}_l^k)(\tilde{x}_k^T P_l^j \tilde{x}_j)^2 \right) \leq 0. \tag{D.50}$$

Thus, $\dot{V}_Q^k(x) < 0$ can be ensured if the indefinite term is bounded by

$$\lambda_{\max}(Q_{G+}^k(x))||\tilde{x}_g||^2 < -\tilde{x}_k^T Q_{LG}^k(x)\tilde{x}_g$$
$$- \sum_{j=1}^{K} 2\beta_j \left(\alpha \lambda_{\max}((A_g^k)_+)(\tilde{x}_g^T P_l^j \tilde{x}_j)^2 + \overline{\alpha} \lambda_{\max}(\mathbf{A}_l^k)(\tilde{x}_k^T P_l^j \tilde{x}_j)^2 \right), \tag{D.51}$$

which is the condition stated in equation (5.32). As in the linear case, equation (D.51) defines the upper bound of the dissipation rate of the kth locally active dynamics. Hence, although $\dot{V}(x)$ has an indefinite structure, if equations (5.29), (5.30), (5.31), and (5.32) are met, the DS in equation (5.22) is GAS at the attractor x_g^* [i.e., $\lim_{t \to \infty} ||x(t) - x_g^*||$]. $\square$

D.4   Proofs and Derivations for Chapter 9

D.4.1   Proof of Theorem 9.1

Consider a hypersurface $X^b \subset \mathbb{R}^d$ corresponding to the boundary points of a hypersphere obstacle in $\mathbb{R}^d$ with a center x^o and a radius r^o. Impenetrability of the obstacle's boundaries is ensured if the normal velocity at boundary points $x^b \in X^b$ vanishes:

$$n\left(x^b\right)^T \dot{x}^b = 0 \quad \forall x^b \in X^b, \tag{D.52}$$

where $n\left(x^b\right)$ is the unit normal vector at a boundary point x^b:

$$n\left(x^b\right) = \frac{x^b - x^o}{\|x^b - x^o\|} \xrightarrow{\tilde{x}^b = x^b - x^o} n\left(x^b\right) = \frac{\tilde{x}^b}{r} \quad \forall x^b \in X^b. \tag{D.53}$$

We have

$$n\left(x^b\right)^T \dot{x}^b = n\left(x^b\right) E\left(\tilde{x}^b, r^o\right) D\left(\tilde{x}^b, r^o\right) E\left(\tilde{x}^b, r^o\right)^{(-1)} f(.) \tag{D.54}$$

Considering the fact that $n\left(x^b\right)$ is equal to the first eigenvector of $E\left(\tilde{x}^b, r^o\right)$ and the first eigenvalue is zero for all points on the obstacle boundary, yields:

$$n\left(x^b\right)^T \dot{x}^b = \left[\begin{array}{c} 1 \\ [\mathbf{0}]_{d-1} \end{array} \right]^T D\left(\tilde{x}^b, r^o\right) E\left(\tilde{x}^b, r^o\right)^{(-1)} f(.)$$
$$= [\mathbf{0}]_d^T E\left(\tilde{x}^b, r^o\right)^{(-1)} f(.) = 0. \tag{D.55}$$

D.4.2   Proof of Theorem 9.2

Here, we first prove impenetrability, and then convergence.

The initial DS can be written as a linear combination of two vectors: one contained in the tangent hyperplane $\mathbf{f}_e(x)$ and the other parallel to the reference direction $\mathbf{f}_r(x) \parallel \mathbf{r}(x)$:

$$\mathbf{f}(x) = \mathbf{f}_r(x) + \mathbf{f}_e(x) = \|\mathbf{f}_r(x)\| \mathbf{r}(x) + \|\mathbf{f}_e(x)\| \mathbf{e}(x), \tag{D.56}$$

where $\mathbf{e}(x)$ is a linear combination of all tangent vectors $\mathbf{e}_i(x)$, $i = 1..d - 1$, which are described in equation (9.4) in chapter 9.

For any point on the boundary ($\Gamma(x) = 1$), the modulated DS follows from equation (9.3) and the condition in equation (9.7):

$$\dot{x} = \lambda_r(x)\mathbf{f}_r(x) + \lambda_t(x)\mathbf{f}_e(x) = \lambda_t(x)\|\mathbf{f}_e(x)\| \mathbf{e}(x) \quad \forall x \in \mathscr{X}^b.$$

According to the *von Neuman* boundary condition, impenetrability is ensured if there is no velocity in the normal direction on the surface of the obstacle:

$$\mathbf{n}(x)^T \dot{x} = \mathbf{n}(x)^T \mathbf{e}(x)\|\dot{x}\| = 0 \quad \forall x \in \mathscr{X}^b, \tag{D.57}$$

as per the definition, the tangent hyperplane is orthogonal to $\mathbf{n}(x) = d\Gamma(x)/dx$. $\square$

The proof of convergence to the attractor consists of four steps:

1. Reduce the d-dimensional problem to two dimensions (2D) (section D.4.2.1).

2. Prove the existence of cone-shaped invariant sets, including the attractor shown by the blue and red cones in figure D.1 (sections D.4.2.2 and D.4.2.3).

3. Prove that all trajectories except the saddle-point line reach this invariant set (sections D.4.2.3 and D.4.2.4).

4. Prove convergence to the attractor through contraction in this invariant set (section D.4.2.5).

D.4.2.1   Reduction to two dimensional problem

Let us consider a linear, two dimensional plane spanned by the current position x, the attractor x^*, and the reference point inside the obstacle x^r. By definition, the reference direction $\mathbf{r}(x)$ from equation (9.4) and the initial DS $\mathbf{f}(x)$ from equation (9.1) are also contained in the reference plane $\mathscr{S}^r(x)$:

$$\{x, x^*, x^r\} \in \mathscr{S}^r(x) \quad \Rightarrow \quad \mathbf{r}(x), \mathbf{f}(x) \in \mathscr{S}^r(x). \tag{D.58}$$

Using equation (D.56), we can write

$$\dot{x} = \mathbf{M}(x)\mathbf{f}(x) \quad = \lambda_r(x)\mathbf{f}_r(x) + \lambda_e(x)\mathbf{f}_e(x). \tag{D.59}$$

Hence, the following can be concluded:

$$\mathbf{f}_r(x) \in \mathscr{S}^r(x) \Rightarrow \mathbf{f}_e(x) \in \mathscr{S}^r(x) \Rightarrow \dot{x} \in \mathscr{S}^r(x). \tag{D.60}$$

In other words, the modulated DS is parallel to the plane. Hence, any motion starting in the plane $\{x\}_0 \in \mathscr{S}(\{x\}_0)$, remains in this plane for all time [i.e., $\{x\}_t \in \mathscr{S}(\{x\}_0) t = 0..\infty$].

Further proofs of convergence can be conducted in two dimensions, and they are applicable to the d-dimensional case. Without loss of generality, for further proof, the attractor is

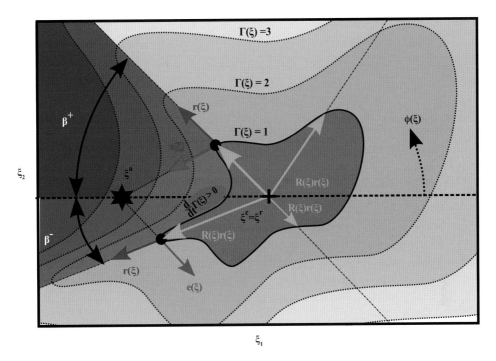

Figure D.1
Any trajectory starting outside the obstacle in the positive or negative half-plane ends up in the invariant cone-regions red, blue or violet.

placed at the origin $x^* = \mathbf{0}$, and the reference point of the obstacle is $x^r = [d_1 \ 0]$, with $d_1 > 0$ (figure D.1).

D.4.2.2 Region with movement away from the obstacle

We want to show that there exists a region exists where the DS is moving away from the obstacle [i.e., $d\Gamma(x) \ dt > 0$]. Close to the obstacle with $\Gamma(x) - 1 << 1$, following equality for any level function $\Gamma(x)$ given in equation (9.2) holds that $\Gamma(x)R(x)\mathbf{r} = x - x^r$, with $R(x) \in \mathbb{R}_{\geq 0}$ being the distance to the center to the surface of the obstacle in the reference direction (figure D.1). We define $\tilde{\mathbf{E}}(x) = [R(x)\mathbf{r} \ \mathbf{e}]$, where $\mathbf{e}(x)$ from equation (9.4) is perpendicular to $d\Gamma(x)/dx$, to express the change of the level function close to the surface:

$$\left[\tfrac{d\Gamma}{dt} \ 0\right]^T = \frac{d}{dt}\left(\tilde{\mathbf{E}}(x)^{-1}(x - x^r)\right)$$

$$= -\tilde{\mathbf{E}}(x)^{-1}\frac{d}{dt}\tilde{\mathbf{E}}(x)\tilde{\mathbf{E}}(x)^{-1}\left(x - x^r\right) + \tilde{\mathbf{E}}(x)^{-1}\mathbf{M}(x)x$$

$$= -\begin{bmatrix} 0 & (\cdot) \\ (\cdot) & (\cdot) \end{bmatrix}\begin{bmatrix} (\cdot) \\ 0 \end{bmatrix} - \mathbf{diag}\left(1/R, \ 1\right)\mathbf{D}\,\mathbf{E}^{-1}x.$$

The first diagonal element of $\tilde{\mathbf{E}}(x)^{-1}\frac{d}{dt}\tilde{\mathbf{E}}(x)$ is zero because the derivative of the first row, the vector to the surface, is parallel to the tangent [i.e., $\frac{d}{dt}R(x)\mathbf{r}(x) \parallel \mathbf{e}(x)$]. Furthermore, the second element of $\tilde{\mathbf{E}}(x)^{-1}(x - x^r)$ is zero, since $x - x^r \parallel \mathbf{r}(x)$. If the position of the robot expressed in the **r-e**-basis is negative, then $\left(\mathbf{E}^{-1}x\right)_r = x_r < 0 \Rightarrow \frac{d}{dt}\Gamma(x) = -R(x)\lambda_r(x)x_r >$

0, and the DS moves away from the obstacle (the purple region in figure D.1). This is intuitively the region where the initial DS is already moving away from the robot: $\mathbf{f}(x)^T \cdot \frac{d}{dx}\Gamma(x) < 0$.

D.4.2.3 Convergence to Cone region

In order for any trajectory except for the saddle-point trajectory $\mathcal{X}^S$ given in equation (9.11) to converge to the cone region, the reference angle $\phi(x) = \arctan(x_2/(x_1 - d_1))$ must decrease, where $d_1 > 0$ corresponds to the position of the reference point on the x_1-axis. This time derivative is evaluated with equation (9.3) as

$$\dot{\phi}(x) = \tilde{x} \times \mathbf{f}(x)/\|\tilde{x}\|^2 = \lambda_e(x)d_1 x_2/\|\tilde{x}\|^2 \tag{D.61}$$

and $\tilde{x} = x - [d_1 \ 0]$.

As the eigenvalue $\lambda_e(x)$ is larger than zero, equation (D.61) is positive for $x_2 > 0$, and negative on the other half-plane. This leads to two deductions. First, any point starting outside the obstacle and not on the saddle point reaches a cone region with $|\pi - \phi(x)| \leq \beta \leq \min(\beta^+, \beta^-)$, with $\beta > 0$. Second, any region outside the obstacle that contains the attractor x^* and has its center at x^r is invariant.

D.4.2.4 Staying outside the obstacle for a finite time

While any trajectory would end up within the cone boundary in finite time, the invariant set is only strictly outside the obstacle $x \in \mathcal{X}^e$. We therefore want to show that the robot stays outside the obstacle for a finite time.

Components in the direction of the $\mathbf{n}(x) = d\Gamma/dx$ are referred to as $(\cdot)_n$. Moreover, the velocity is bounded in a real system, with the maximum velocity in the normal direction v_n. The time to transit between arbitrary levels Γ_0 and Γ_1 can be evaluated with equation (9.6) as

$$t^b(x) = \int_{\Gamma_0}^{\Gamma_1} \frac{d\Gamma}{\dot{x}_n} = \int_{\Gamma_0}^{\Gamma_1} \frac{d\Gamma}{\lambda_r(x)f_n(x)} = \int_{\Gamma_0}^{\Gamma_1} \frac{1}{f_n(x)} \frac{\Gamma}{\Gamma - 1} d\Gamma$$

$$\geq \frac{1}{v_n}\left[\Gamma_0 + \log(\Gamma_0 - 1) - \Gamma_1 - \log(\Gamma_1 - 1)\right]. \tag{D.62}$$

Starting outside the obstacle $\Gamma_0 > 1$, and having the goal on the surface $\Gamma_1 = 1$, the time results are $\lim_{\Gamma_1 \to 1} t^b(x) \to \infty$.

From this, it follows that any point starting outside the obstacle does not reach the surface in a finite time.

D.4.2.5 Contraction analysis

The modulated DS from equation (9.3) restated as a function of $\mathbf{g}(\cdot)$:

$$\dot{\mathbf{x}} = \mathbf{g}(x, x) = \mathbf{M}(\mathbf{x})\mathbf{f}(x) = \mathbf{M}(x)x. \tag{D.63}$$

Using partial contraction theory, the virtual system is chosen as a function of the new variable $\gamma \in \mathbb{R}^d$:

$$\dot{\gamma} = \mathbf{g}(\gamma, \mathbf{x}) = -\mathbf{M}(\mathbf{x})\gamma. \tag{D.64}$$

If the system $\mathbf{g}(\cdot)$ is contracting with respect to γ, and it has the two particular solutions, $\gamma = x$ and $\gamma = x^* = \mathbf{0}$. It follows from [155] that the x-system exponentially tends to $x^* = \mathbf{0}$.

Hence, we need to show that the system is contracting with respect to γ. A possible contraction metric is $\mathbf{P}(x) = \Theta(x)^T \Theta(x)$, with

$$\Theta(x) = \hat{\mathbf{E}}(x)^{-1} = \left[w_r(x)\mathbf{r}(x) \quad w_e(x)\mathbf{e}(x) \right]^{-1}, \tag{D.65}$$

where $\Theta(x)$ is a uniformly invertible square matrix. The system is contracting with respect to the metric $\mathbf{P}(x)$ if the symmetric part of the generalized Jacobian $\mathbf{F}_{sym} = \mathbf{F} + \mathbf{F}^T$ is negative definite, with

$$\mathbf{F} = \frac{d}{dt}\Theta\,\Theta^{-1} + \Theta\frac{\partial \mathbf{g}(\gamma, x)}{\partial \gamma}\Theta^{-1} \tag{D.66}$$

$$= -\hat{\mathbf{E}}^{-1}(x)\frac{d}{dt}\hat{\mathbf{E}}(x) - \hat{\mathbf{E}}(x)^{-1}\mathbf{E}(x)\mathbf{D}(x)\mathbf{E}(x)^{-1}\hat{\mathbf{E}}(x),$$

where $\hat{\mathbf{E}}(x)^{-1}\mathbf{E}(x) = \mathbf{diag}\left(w_r(x),\, w_e(x)\right)$. Hence, the second term is $-\mathbf{diag}(\lambda_r(x),\, \lambda_e(x))$. The rows of the first term $\dot{\mathbf{r}}$ and $\dot{\mathbf{e}}$ are evaluated as

$$\hat{\mathbf{E}}^{-1}\frac{d}{dt}\hat{\mathbf{E}} = \hat{\mathbf{E}}^{-1}\left[\dot{w}_r\mathbf{r} + w_r\frac{d}{dt}\mathbf{r} \quad \dot{w}_e\mathbf{e} + w_e\frac{d}{dt}\mathbf{e} \right] \tag{D.67}$$

$$= \begin{bmatrix} \frac{\dot{w}_r}{w_r} - \|\dot{\mathbf{r}}\|\frac{1}{\tan\epsilon} & \frac{w_e}{w_r}\|\dot{\mathbf{r}}\|\mathrm{sign}(\mathbf{e}^T \cdot \dot{\mathbf{r}})\frac{1}{\sin\epsilon} \\ \frac{w_r}{w_e}\|\dot{\mathbf{e}}\|\mathrm{sign}(\mathbf{r}^T \cdot \dot{\mathbf{e}})\frac{1}{\sin\epsilon} & \frac{\dot{w}_e}{w_e} - \|\dot{\mathbf{e}}\|\frac{1}{\tan\epsilon} \end{bmatrix},$$

with $\cos\epsilon(x) = \mathbf{r}(x)^T \cdot \mathbf{e}(x)$, $\dot{\mathbf{r}} = \frac{d}{dt}\mathbf{r}(x)$, and $\dot{\mathbf{e}} = \frac{d}{dt}\mathbf{e}(x)$. Furthermore, under the condition that $\hat{\mathbf{E}}(x)$ has full rank, we have $\epsilon \in\,]0, \pi[$.

Notes

Chapter 1

1. Many commercial robots, such as the KUKA lightweight series or the Franka robot arm, come with a fairly accurate inverse dynamics controller, provided by the manufacturer. However, if you find out that the model is not precise enough, it is recommended to proceed to an estimation of the parameters through system identification.

Chapter 2

1. http://www.icub.org/bazaar.php

2. http://www.simlab.co.kr/Allegro-Hand.htm

3. https://www.shadowrobot.com/products/dexterous-hand/

4. Backdrivability can also be achieved through active compliance.

5. This simplistic rule is not always true. You may sometimes need more samples, and in other cases, fewer samples may be sufficient.

6. Part of this section's material was originally published in [87].

7. The flow incurs a discontinuity at the ball due to division by zero when computing $g(x)$.

8. This assumes a controller that does not fully compensate for the robot's dynamics, which is often the case even for very sophisticated platforms, like those presented in these examples. Of course, one option is to learn a better model of the dynamics.

Chapter 3

1. An $N \times N$ real symmetric matrix A is positive definite if $x^T A x > 0$ for all nonzero vectors $x \in \mathbb{R}^N$, where x^T denotes the transpose of x. Conversely, A is negative definite if $x^T A x < 0$. For a nonsymmetric matrix, A is positive (negative) definite if and only if its symmetric part $\tilde{A} = \left(A + A^T \right) / 2$ is positive (negative) definite.

2. This approach was originally published in [43].

3. This approach was originally introduced in [100].

Chapter 4

1. This section is an adaptation of the original publication in [134].

2. The SVM classifier function is bounded if the radial basis function (RBF) is used as a kernel.

3. Bold fonts represent vectors; x^i denotes the ith vector and x^i denotes the ith element of vector x.

4. Source code for learning is available at http://asvm.epfl.ch

5. This section is adapted from [69].

6. As $\mu = (x_1^2 + x_2^2)$ for $\mu > 0$ and $x_1, x_2 \in \Omega$, the ω-limit set of the DS (i.e., an invariant set of points such that for time $t_i \rightarrow \infty, f(x, \mu, t_i) \in \Omega$ [52]).

Chapter 5

1. This approach was originally introduced in [42].

2. This approach was originally published in [99].

Chapter 6

1. This work was published first in [135].

2. This assumes that there is a single grasp configuration for a given object. As the reach-and-grasp dynamics may vary depending on the object to be grasped, it may be necessary to relearn a system for each object's class.

3. It is also worth mentioning that the two trajectories are fairly similar when initialized close to the demonstration envelope.

4. This work was published first in [94].

Chapter 7

1. It is expected that chapter 6 be read before this chapter.

2. The material presented in this chapter was published originally in a number of papers [102, 101, 104, 103, 100].

3. We assume that the DS [equation (7.25)] is fast enough to converge to an acceptable neighborhood around the desired trajectory before t^*.

Chapter 8

1. The material presented in this chapter is adapted from [102, 138, 85].

2. For the modulation to be locally active, because $\gamma(x, x^o)$ is strictly positive, the function should be truncated to 0, as will be described in section 8.2.2.2.

3. This section is adapted from [85].

4. Adapted from [138].

5. For the sake of simplicity, in this chapter, we call the velocity normal to the surface the *normal velocity*.

6. It is important to note that $y = \sum_{i=1}^{d} e^i(x)e^i(x)^T y, \ \forall y \in \mathbb{R}^N$.

7. Second-order, linear parameter varying (LPV)–based dynamical systems are elaborated on in more detail in section 7.2.

Chapter 9

1. This chapter is adapted from [57, 71, 104].

2. The element-wise (of each Cartesian vector element) weighted mean of the modulated DS $\dot{x}^o$ for each obstacle may create stationary points.

3. In the two-dimensional (2D) case, this hypersphere is a line that represents the angle between the initial DS $f(x)$ and the modulated DS $\dot{x}_k$. It has a magnitude strictly smaller than π.

4. As an example, $N_R = 2, \mathbf{q} = q^{12}$. Due to hardware limitations, it is not possible to construct a data set of collision boundaries for more than two 7-DOF arms at once; for example, for three arms, the approximate size of the data set is $(3 * 7 * 3) \times 1,000^3$, while it is $(3 * 7 * 2) \times 1,000^2$ for two arms.

5. Due to redundancy, q_7^i has no effect on the joint configurations.

6. This run time includes the construction of $F(q^{ij})$, $J(q^{ij})$ and multiplication by equation (9.23). The Eigen Library is used for such operations, which has underlying dynamic allocation strategies, inducing the standard seen in figure 9.12.

Chapter 10

1. The materials presented in this chapter is an adaptation of the following works [102, 101, 104, 103, 100, 84, 87, 27].

2. From now on, we mainly refer to the impedance concept. This discussion, however, can easily be extended to the admittance concept.

3. For simplicity, we just present the Cartesian control laws here. However, one can easily extend this to the joint space controllers.

4. This approach is previously proposed in [83].

5. Adapted from [27]

6. The activation parameters can be a function of time, state, or external variables. For the sake of brevity, we will just write it as a function of time.

7. Adapted from [84]

8. In order to distinguish the sensory noises from the perturbation coming from the operator, one can incorporate a low-pass and a high-pass filter before computing the desired stiffness. The role of these filters is twofold. First, the low-pass filter removes high-frequency sensory noises and interactions coming from other sources than the operator (e.g., contact with the environment). While separating the interaction signal in the frequency domain does not guarantee that such effects are avoided, it does make them less probable. Second, the high-pass filter allows the operator to comfortably perform slow perturbations to feel the stiffness of the robot.

9. The definition of conservative dynamical systems is provided in appendix A.

Chapter 11

1. The material presented in this chapter is an adaptation of the original publication in [5].

Appendix B

1. The $\mathscr{D}$irichlet distributions are the best suited for representing distributions over the simplex (i.e., the set of N-vectors whose components add up to 1). Hence, it's a useful prior distribution on discrete probability distributions over categorical variables.

2. Gibbs sampling is a Markov chain Monte Carlo (MCMC) sampling method, which is commonly used to generate random samples from a joint distribution over several variables, when the target distribution is either unknown or intractable. It achieves this by iteratively drawing samples from conditional distributions of the variables of interest in such a way that the target joint distribution is approximated after a period of time. This involves computing the posterior conditional probabilities of each variable conditioned on the rest.

3. While DP is a distribution over distributions, CRP is a distribution over partitions of integers. Due to De Finetti's theorem on exchangeability [3], by marginalizing out the distribution over parameters $\mathscr{G}$, DP-MM is equivalent to CRP MM [50].

4. Flatness in a regressive function can mean being less sensitive to errors in measurement/random shocks/non-stationarity of the input variables. This is encoded as maximizing the prediction error as much as possible within an ε-tube.

Appendix C

1. For a good introduction on to the topic of robotics, the reader is encouraged to consult the Handbook of Robotics [137] and its main relevant chapters ("Kinematics," "Motion Planning," "Motion Control," and "Force Control").

2. This solution cannot be applied if the robot's path goes through singular configurations (i.e., configurations where the Jacobian is singular) in practice, we recommend to narrow the space of the robot so that the Jacobian remains non-singular during manipulations.

3. λ into two (or more) groups $= 0$ would imply the pseudoinverse method.

Bibliography

[1] Amir Ali Ahmadi, Anirudha Majumdar, and Russ Tedrake. Complexity of ten decision problems in continuous time dynamical systems. In *American Control Conference*, (pp. 6376–6381). IEEE, 2013.

[2] Mostafa Ajallooeian, Jesse van den Kieboom, Albert Mukovskiy, Martin A. Giese, and Auke J. Ijspeert. A general family of morphed nonlinear phase oscillators with arbitrary limit cycle shape. *Physica D: Nonlinear Phenomena*, 263: 41–56, 2013.

[3] D.J. Aldous. Exchangeability and related topics. In *École d'Été St Flour 1983* (pp. 1–198). Springer-Verlag, 1985.

[4] W. Amanhoud, M. Khoramshahi, M. Bonnesoeur, and A. Billard. Force adaptation in contact tasks with dynamical systems. In *Proceedings of the IEEE International Conference on Robotics and Automation*, 2020.

[5] Walid Amanhoud, Mahdi Khoramshahi, and Aude Billard. A dynamical system approach to motion and force generation in contact tasks. In *Proceedings of Robotics: Science and Systems*. 2019. Available at http://roboticsproceedings.org/.

[6] Pierre Apkarian, Pascal Gahinet, and Greg Becker. Self-scheduled H_∞ control of linear parameter-varying systems: A design example. *Automatica*, 31(9): 1251–1261, 1995.

[7] B. Argall, S. Chernova, M. Veloso, and B. Browning. A survey of robot learning from demonstration. *Robotics and Autonomous Systems*, 57(5): 469–483, 2009.

[8] Anil Aswani, Humberto Gonzalez, S. Shankar Sastry, and Claire Tomlin. Provably safe and robust learning-based model predictive control. *Automatica*, 49(5):1216–1226, 2013.

[9] Bassam Bamieh and Laura Giarre. Identification of linear parameter varying models. *International Journal of Robust and Nonlinear Control*, 12(9): 841–853, 2002.

[10] Mokhtar S. Bazaraa, Hanif D. Sherali, and C. M. Shetty. *Nonlinear Programming: Theory and Algorithms*. Wiley-Interscience, 2006.

[11] Mehdi Benallegue, Adrien Escande, Sylvain Miossec, and Abderrahmane Kheddar. Fast C1 proximity queries using support mapping of sphere-torus-patches bounding volumes. In *IEEE International Conference on Robotics and Automation* (pp. 483–488). IEEE, 2009.

[12] Michele Benzi, Gene H. Golub, and Jörg Liesen. Numerical solution of saddle point problems. *Acta Numerica*, 14:1–137, 2005.

[13] Felix Berkenkamp, Riccardo Moriconi, Angela P. Schoellig, and Andreas Krause. Safe learning of regions of attraction for uncertain, nonlinear systems with Gaussian processes. In *IEEE 55th Conference on Decision and Control (CDC)*, (pp. 4661–4666). IEEE, 2016.

[14] Aude Billard. On the mechanical, cognitive, and sociable facets of human compliance and their robotic counterparts. *Robotics and Autonomous Systems*, 88: 157–164, 2017.

[15] Aude Billard and Daniel Grollman. Robot learning by demonstration. *Scholarpedia*, 8(12): 3824, 2013.

[16] Aude G. Billard, Sylvain Calinon, and Rüdiger Dillmann. Learning from humans. In *Springer Handbook of Robotics*, (pp. 1995–2014). Springer, 2016.

[17] J. Bilmes. A gentle tutorial on the EM algorithm and its application to parameter estimation for Gaussian mixture and hidden Markov models. University of California–Berkeley, 1997.

[18] M. C. Bishop. *Pattern Recognition and Machine Learning*. Springer, 2007.

[19] David M. Blei and Peter I. Frazier. Distance-dependent Chinese restaurant processes. *Journal of Machine Learning Research*, 12: (November) 2461–2488, 2011.

[20] J. F. Bonnans, J. C. Gilbert, C. Lemaréchal, and C. A. Sagastizábal. *Numerical Optimization—Theoretical and Practical Aspects*. Springer Verlag, 2006.

[21] Karim Bouyarmane, Kevin Chappellet, Joris Vaillant, and Abderrahmane Kheddar. Quadratic programming for multirobot and task-space force control. *IEEE Transactions on Robotics*, 35(1): 64–77, 2018.

[22] Christopher Y. Brown and H. Harry Asada. Inter-finger coordination and postural synergies in robot hands via mechanical implementation of principal components analysis. In *IEEE/RSJ International Conference on Intelligent Robots and Systems* (pp. 2877–2882). IEEE, 2007.

[23] Daniel S. Brown and Scott Niekum. Machine teaching for inverse reinforcement learning: Algorithms and applications. In *Proceedings of the AAAI Conference on Artificial Intelligence*, vol. 33 (pp. 7749–7758). Association for the Advancement of Artificial Intelligence, 2019.

[24] Adam Bry and Nicholas Roy. Rapidly-exploring random belief trees for motion planning under uncertainty. In *IEEE International Conference on Robotics and Automation* (pp. 723–730). IEEE, 2011.

[25] Jonas Buchli, Freek Stulp, Evangelos Theodorou, and Stefan Schaal. Learning variable impedance control. *International Journal of Robotics Research*, 30(7):820–833, 2011.

[26] S. Calinon, F. Guenter, and A. Billard. On learning, representing, and generalizing a task in a humanoid robot. *IEEE Transactions on Systems, Man, and Cybernetics, Part B: Cybernetics*, 37(2): 286–298, 2007.

[27] S. Calinon, I. Sardellitti, and D. G. Caldwell. Learning-based control strategy for safe human-robot interaction exploiting task and robot redundancies. In *IEEE/RSJ International Conference on Intelligent Robots and Systems*, October (pp. 249–254), 2010.

[28] Shu-Guang Cao, Neville W. Rees, and Gang Feng. Analysis and design for a class of complex control systems Part I: Fuzzy modelling and identification. *Automatica*, 33(6): 1017–1028, 1997.

[29] Hsiao-Dong Chiang and Chia-Chi Chu. A systematic search method for obtaining multiple local optimal solutions of nonlinear programming problems. *IEEE Transactions on Circuits and Systems I: Fundamental Theory and Applications*, 43(2): 99–109, 1996.

[30] Howie Choset. Coverage for robotics–a survey of recent results. *Annals of Mathematics and Artificial Intelligence*, 31(1–4): 113–126, 2001.

[31] Adam Coates, Pieter Abbeel, and Andrew Y. Ng. Learning for control from multiple demonstrations. In *Proceedings of the 25th International Conference on Machine Learning* (pp. 144–151), 2008.

[32] David A Cohn, Zoubin Ghahramani, and Michael I Jordan. Active learning with statistical models. *Journal of Artificial Intelligence Research*, 4: 129–145, 1996.

[33] Douglas E. Crabtree and Emilie V. Haynsworth. An identity for the Schur complement of a matrix. *Proceedings of the American Mathematical Society*, 22(2):364–366, 1969.

[34] Christian Daniel, Herke van Hoof, Jan Peters, and Gerhard Neumann. Probabilistic inference for determining options in reinforcement learning. *Machine Learning*, 104(2): 337–357, 2016.

[35] Agostino De Santis, Bruno Siciliano, Alessandro De Luca, and Antonio Bicchi. An atlas of physical human–robot interaction. *Mechanism and Machine Theory*, 43(3): 253–270, 2008.

[36] Paul Evrard, Elena Gribovskaya, Sylvain Calinon, Aude Billard, and Abderrahmane Kheddar. Teaching physical collaborative tasks: Object-lifting case study with a humanoid. In *9th IEEE-RAS International Conference on Humanoid Robots* (pp. 399–404). IEEE, 2009.

[37] Salman Faraji, Philippe Müllhaupt, and Auke Ijspeert. Imprecise dynamic walking with time-projection control. *arxiv*, November 9, 2018.

[38] Salman Faraji, Hamed Razavi, and Auke J. Ijspeert. Bipedal walking and push recovery with a stepping strategy based on time-projection control. *International Journal of Robotics Research*, 38(5): 587–611, 2019.

[39] Gang Feng. A survey on analysis and design of model-based fuzzy control systems. *IEEE Transactions on Fuzzy Systems*, 14(5): 676–697, 2006.

[40] J. Fiala, M. Kočvara, and M. Stingl. PENLAB: A MATLAB solver for nonlinear semidefinite optimization. *arXiv e-prints*, November 20, 2013.

[41] N. Figueroa and A. Billard. Learning complex manipulation tasks from heterogeneous and unstructured demonstrations. In *Proceedings of Workshop on Synergies between Learning and Interaction, IEEE/RSJ International Conference on Intelligent Robots and Systems*, 2017. Available at https://pub.uni-bielefeld.de/.

[42] Nadia Figueroa. *From High-Level to Low-Level Robot Learning of Complex Tasks: Leveraging Priors, Metrics and Dynamical Systems*. PhD thesis, École polytechnique fédérale de Lausanne, Switzerland, 2019.

[43] Nadia Figueroa and Aude Billard. A physically-consistent Bayesian non-parametric mixture model for dynamical system learning. In Aude Billard, Anca Dragan, Jan Peters, and Jun Morimoto (eds.), *Proceedings of the 2nd Conference on Robot Learning*, Vol. 87 of *Proceedings of Machine Learning Research* (pp. 927–946). PMLR, 2018.

[44] Nadia Figueroa, Salman Faraji, Mikhail Koptev, and Aude Billard. A dynamical system approach for adaptive grasping, navigation and co-manipulation with humanoid robots. In *IEEE International Conference on Robotics and Automation (ICRA)* (pp. 7676–7682). IEEE, 2020.

[45] David Fridovich-Keil, Sylvia L. Herbert, Jaime F. Fisac, Sampada Deglurkar, and Claire J Tomlin. Planning, fast and slow: A framework for adaptive real-time safe trajectory planning. In *IEEE International Conference on Robotics and Automation (ICRA)* (pp. 387–394). IEEE, 2018.

[46] Andrej Gams, Martin Do, Aleš Ude, Tamim Asfour, and Rüdiger Dillmann. On-line periodic movement and force-profile learning for adaptation to new surfaces. In *10th IEEE-RAS International Conference on Humanoid Robots* (pp. 560–565). IEEE, 2010.

[47] Andrej Gams, Tadej Petrič, Martin Do, Bojan Nemec, Jun Morimoto, Tamim Asfour, and Aleš Ude. Adaptation and coaching of periodic motion primitives through physical and visual interaction. *Robotics and Autonomous Systems*, 75: 340–351, 2016.

[48] Ming-Tao Gan, Madasu Hanmandlu, and Ai Hui Tan. From a Gaussian mixture model to additive fuzzy systems. *IEEE Transactions on Fuzzy Systems*, 13(3):303–316, 2005.

[49] Andrew Gelman, John B. Carlin, Hal S. Stern, and Donald B. Rubin. *Bayesian Data Analysis*. Chapman and Hall/CRC, 2003.

[50] Samuel J. Gershman and David M. Blei. A tutorial on Bayesian nonparametric models. *Journal of Mathematical Psychology*, 56(1):1–12, 2012.

[51] Daniel H Grollman and Odest Chadwicke Jenkins. Learning robot soccer skills from demonstration. In *IEEE 6th International Conference on Development and Learning* (pp. 276–281). IEEE, 2007.

[52] John Guckenheimer and Philip Holmes. *Nonlinear Oscillations, Dynamical Systems, and Bifurcations of Vector Fields*. vol. 42, Springer, 1983.

[53] Maoan Han and Pei Yu. Chapter 2. In *Hopf Bifurcation and Normal Form Computation* (pp. 7–58). Springer, 2012.

[54] Frank E. Harrell Jr., Kerry L. Lee, Robert M. Califf, David B. Pryor, and Robert A. Rosati. Regression modelling strategies for improved prognostic prediction. *Statistics in Medicine*, 3(2): 143–152, 1984.

[55] Neville Hogan. Impedance control: An approach to manipulation. *Journal of Dynamic Systems, Measurement, and Control*, 107: 17, 1985.

[56] Eberhard Hopf. Abzweigung einer periodischen lösung von einer stationären lösung eines differentialsystems. *Berlin Mathematics-Physics Klasse, Sachs Akademische Wissenschaft Leipzig*, 94: 1–22, 1942.

[57] L. Huber, A. Billard, and J. Slotine. Avoidance of convex and concave obstacles with convergence ensured through contraction. *IEEE Robotics and Automation Letters*, 4(2): 1462–1469, 2019. extended results available at: https://arxiv.org/abs/2105.11743

[58] Hans Jacob, S. Feder, and Jean Jacques E. Slotine. Real-time path planning using harmonic potentials in dynamic environment. In *Proceedings of the IEEE International Conference on Robotics and Automation* (pp. 874–881), 1997.

[59] Tony Jebara. Images as bags of pixels. In Proceedings of International Conference of Computer Vision, IEEE, (pp. 265–272), 2003.

[60] Xing Jin, Biao Huang, and David S Shook. Multiple model LPV approach to nonlinear process identification with EM algorithm. *Journal of Process Control*, 21(1): 182–193, 2011.

[61] Thorsten Joachims and Chun-Nam John Yu. Sparse kernel SVMs via cutting-plane training. *Machine Learning*, 76(2–3): 179–193, 2009.

[62] Steven G Johnson. *The NLopt nonlinear-optimization package*, 2015. http://github.com/stevengj/nlopt.

[63] Michael I Jordan. Computational aspects of motor control and motor learning. In *Handbook of Perception and Action* (vol. 2, pp. 71–120). Elsevier, 1996.

[64] Michael I. Jordan. Dirichlet processes, Chinese restaurant processes and all that. In *Proceedings of the 19th Annual Conference on Neural Information Processing Systems* (NIPS 2005), MIT Press, 2005.

[65] Lydia Kavraki and J. C. Latombe. Randomized preprocessing of configuration for fast path planning. In *Proceedings of the 1994 IEEE International Conference on Robotics and Automation* (pp. 2138–2145). IEEE, 1994.

[66] Lydia E. Kavraki and Steven M. LaValle. Motion Planning (pp. 109–131). In Steven M. LaValle, *Planning Algorithms*, Cambridge University Press, 2008.

[67] Haruhisa Kawasaki, Tsuneo Komatsu, and Kazunao Uchiyama. Dexterous anthropomorphic robot hand with distributed tactile sensor: Gifu hand II. *IEEE/ASME Transactions on Mechatronics*, 7(3): 296–303, 2002.

[68] Francois Keith, Nicolas Mansard, Sylvain Miossec, and Abderrahmane Kheddar. Optimization of tasks warping and scheduling for smooth sequencing of robotic actions. In *IEEE/RSJ International Conference on Intelligent Robots and Systems* (pp. 1609–1614). IEEE, 2009.

[69] F. Khadivar, I. Lauzana, and A. Billard. Learning dynamical systems with bifurcations. *Robotics and Autonomous Systems*, 136, 2020.

[70] Hassan Khalil. *Nonlinear Systems*. Prentice Hall, 2002.

[71] S.-M. Khansari-Zadeh and A. Billard. A dynamical system approach to real-time obstacle avoidance. *Autonomous Robots*, 32(4): 433–454, 2012. 10.1007/s10514-012-9287-y.

[72] S. Mohammad Khansari-Zadeh and A. Billard. Learning stable nonlinear dynamical systems with Gaussian mixture models. *IEEE Transactions on Robotics*, 27(5): 943–957, 2011.

[73] S. Mohammad Khansari-Zadeh and Aude Billard. BM: An iterative algorithm to learn stable non-linear dynamical systems with Gaussian mixture models. In *IEEE International Conference on Robotics and Automation* (pp. 2381–2388). IEEE, 2010.

[74] S. Mohammad Khansari-Zadeh and Aude Billard. *The Derivatives of the SEDS Optimization Cost Function and Constraints with Respect to the Learning Parameters*. EPFL, 2011.

[75] S. Mohammad Khansari-Zadeh and Aude Billard. Learning control Lyapunov function to ensure stability of dynamical system–based robot reaching motions. *Robotics and Autonomous Systems*, 62(6): 752–765, 2014.

[76] S. Kim, A. Shukla, and Aude Billard. Catching objects in flight. *IEEE Transactions on Robotics*, 30(5): 1049–1065, 2014.

[77] Jens Kober, J. Andrew Bagnell, and Jan Peters. Reinforcement learning in robotics: A survey. *International Journal of Robotics Research*, 32(11): 1238–1274, 2013.

[78] Torsten Koller, Felix Berkenkamp, Matteo Turchetta, and Andreas Krause. Learning-based model predictive control for safe exploration. In *IEEE Conference on Decision and Control (CDC)* (pp. 6059–6066), IEEE, 2018.

[79] Milan Korda, Didier Henrion, and Colin N. Jones. Controller design and region of attraction estimation for nonlinear dynamical systems. *IFAC Proceedings Volumes*, 47(3): 2310–2316, 2014.

[80] Aaron C. W. Kotcheff and Chris J. Taylor. Automatic construction of eigenshape models by direct optimization. *Medical Image Analysis*, 2(4): 303–314, 1998.

[81] Oliver Kroemer, Christian Daniel, Gerhard Neumann, Herke Van Hoof, and Jan Peters. Towards learning hierarchical skills for multi-phase manipulation tasks. In *IEEE International Conference on Robotics and Automation (ICRA)* (pp. 1503–1510), IEEE, 2015.

[82] K. Kronander and A. Billard. Passive interaction control with dynamical systems. *IEEE Robotics and Automation Letters*, 1(1): 106–113, 2016.

[83] K. Kronander and A. Billard. Stability considerations for variable impedance control. *IEEE Transactions on Robotics*, 32(5): 1298–1305, 2016.

[84] Klas Kronander and Aude Billard. Learning compliant manipulation through kinesthetic and tactile human-robot interaction. *IEEE Transactions on Haptics*, 7(3): 367–380, 2014.

[85] Klas Kronander, Mohammad Khansari, and Aude Billard. Incremental motion learning with locally modulated dynamical systems. *Robotics and Autonomous Systems*, 70: 52–62, 2015.

[86] Klas Kronander, Mohammad SM Khansari-Zadeh, and Aude Billard. Learning to control planar hitting motions in a minigolf-like task. In *IEEE/RSJ International Conference on Intelligent Robots and Systems (IROS)* (pp. 710–717). IEEE, 2011; S. M. Khansari-Zadeh, K. Kronander, and A. Billard. Learning to play minigolf: A dynamical system-based approach. *Advanced Robotics*, 26(17), 1967–1993, 2012.

[87] Klas Jonas Alfred Kronander. This is a PhD thesis, published by EPFL. 2015.

[88] Dana Kulić, Wataru Takano, and Yoshihiko Nakamura. Incremental learning, clustering and hierarchy formation of whole body motion patterns using adaptive hidden Markov chains. *International Journal of Robotics Research*, 27(7): 761–784, 2008.

[89] Jean-Claude Latombe. *Robot Motion Planning*, vol. 124. Springer Science+Business Media, 2012.

[90] GC Layek. *An Introduction to Dynamical Systems and Chaos*. Springer, 2015.

[91] Jaewook Lee. Dynamic gradient approaches to compute the closest unstable equilibrium point for stability region estimate and their computational limitations. *IEEE Transactions on Automatic Control*, 48(2): 321–324, 2003.

[92] Cindy Leung, Shoudong Huang, Ngai Kwok, and Gamini Dissanayake. Planning under uncertainty using model predictive control for information gathering. *Robotics and Autonomous Systems*, 54(11): 898–910, 2006.

[93] J. Lofberg. YALMIP: A toolbox for modeling and optimization in MATLAB. In *IEEE International Conference on Robotics and Automation (IEEE Cat. No.04CH37508)* (pp. 284–289). 2004.

[94] Luka Lukic, José Santos-Victor, and Aude Billard. Learning robotic eye–arm–hand coordination from human demonstration: A coupled dynamical systems approach. *Biological Cybernetics*, 108(2): 223–248, 2014.

[95] Roanna Lun and Wenbing Zhao. A survey of applications and human motion recognition with Microsoft Kinect. *International Journal of Pattern Recognition and Artificial Intelligence*, 29(5): 1555008, 2015.

[96] Kevin M. Lynch and Frank C. Park. *Modern Robotics: Mechanics, Planning, and Control*. Cambridge University Press, 2017.

[97] Ian R. Manchester, Mark M. Tobenkin, Michael Levashov, and Russ Tedrake. Regions of attraction for hybrid limit cycles of walking robots. *IFAC Proceedings Volumes*, 44(1): 5801–5806, 2011.

[98] Jacob Mattingley and Stephen Boyd. CVXGEN: A code generator for embedded convex optimization. *Optimization and Engineering*, 13(1): 1–27, 2012.

[99] José R Medina and Aude Billard. Learning stable task sequences from demonstration with linear parameter varying systems and hidden Markov models. In *Conference on Robot Learning* (pp. 175–184). Proceeding of Machine Learning Research (PMLR) volume 79, 2017.

[100] Seyed Sina Mirrazavi Salehian. *Compliant Control of Uni/ Multi- Robotic Arms with Dynamical Systems*. PhD thesis, École polytechnique fédérale de Lausanne 2018.

[101] Seyed Sina Mirrazavi Salehian, Nadia Barbara Figueroa Fernandez, and Aude Billard. Coordinated multi-arm motion planning: Reaching for moving objects in the face of uncertainty. In *Proceedings of Robotics: Science and Systems*, 2016.

[102] Seyed Sina Mirrazavi Salehian, Nadia Barbara Figueroa Fernandez, and Aude Billard. A dynamical system approach for softly catching a flying object: Theory and experiment. *IEEE Transactions on Robotics*, 32(2): 462–471, 2016.

[103] Seyed Sina Mirrazavi Salehian, Nadia Barbara Figueroa Fernandez, and Aude Billard. Dynamical system-based motion planning for multi-arm systems: Reaching for moving objects. In *International Joint Conference on Artificial Intelligence*, 2017. Available at http://roboticsproceedings.org/.

[104] Seyed Sina Mirrazavi Salehian, Nadia Barbara Figueroa Fernandez, and Aude Billard. A unified framework for coordinated multi-arm motion planning. *International Journal of Robotics Research*, 37(10): 1205–1232, 2017.

[105] Kevin P. Murphy. *Conjugate Bayesian Analysis of the Gaussian Distribution*. University of British Columbia, 2007.

[106] J. Nakanishi, R. Cory, M. Mistry, J. Peters, and S. Schaal. Comparative experiments on task space control with redundancy resolution. In *IEEE/RSJ International Conference on Intelligent Robots and Systems* (pp. 3901–3908). IEEE, 2005.

[107] Klaus Neumann, Matthias Rolf, and Jochen J. Steil. Reliable integration of continuous constraints into extreme learning machines. *International Journal of Uncertainty, Fuzziness and Knowledge-Based Systems*, 21(Suppl 2): 35–50, 2013.

[108] Klaus Neumann and Jochen J. Steil. Learning robot motions with stable dynamical systems under diffeomorphic transformations. *Robotics and Autonomous Systems*, 70: 1–15, 2015.

[109] Duy Nguyen-Tuong, Jan Peters, Matthias Seeger, and Bernhard Schölkopf. Learning inverse dynamics: A comparison. In *European Symposium on Artificial Neural Networks*, IEEE, 2008.

[110] Sylvie C. W. Ong, Shao Wei Png, David Hsu, and Wee Sun Lee. Planning under uncertainty for robotic tasks with mixed observability. *International Journal of Robotics Research*, 29(8): 1053–1068, 2010.

[111] P. Orbanz and Y. W. Teh. Bayesian nonparametric models. In *Encyclopedia of Machine Learning*. Springer, 2010.

[112] Romeo Ortega and Mark W Spong. Adaptive motion control of rigid robots: A tutorial. *Automatica*, 25(6):877–888, 1989.

[113] Nicolas Perrin and Philipp Schlehuber-Caissier. Fast diffeomorphic matching to learn globally asymptotically stable nonlinear dynamical systems. *Systems & Control Letters*, 96(Suppl C): 51–59, 2016.

[114] Nicolas Perrin, Olivier Stasse, Léo Baudouin, Florent Lamiraux, and Eiichi Yoshida. Fast humanoid robot collision-free footstep planning using swept volume approximations. *IEEE Transactions on Robotics*, 28(2): 427–439, 2011.

[115] Luka Peternel and Jan Babič. Humanoid robot posture-control learning in real-time based on human sensorimotor learning ability. In *IEEE International Conference on Robotics and Automation*, (pp. 5329–5334). IEEE, 2013.

[116] H. Poincaré. *Les méthodes nouvelles de la mécanique céleste: Méthodes de MM.* Newcomb, Glydén, Lindstedt et Bohlin. 1893. Vol. 2. Gauthier-Villars it fils, 1893.

[117] Lawrence Rabiner. A tutorial on hidden Markov models and selected applications in speech recognition. *Proceedings of the IEEE*, 77(2):257–286, 1989.

[118] Carl Edward Rasmussen and Christopher K. I. Williams. Gaussian processes for machine learning. 2006. *MIT Press*, 2006.

[119] Siddharth S Rautaray and Anupam Agrawal. Vision based hand gesture recognition for human computer interaction: A survey. *Artificial Intelligence Review*, 43(1): 1–54, 2015.

[120] Harish Ravichandar, Athanasios S. Polydoros, Sonia Chernova, and Aude Billard. Recent advances in robot learning from demonstration. *Annual Review of Control, Robotics, and Autonomous Systems*, 3, 297–330, 2020.

[121] Harish Chaandar Ravichandar and Ashwin Dani. Learning position and orientation dynamics from demonstrations via contraction analysis. *Autonomous Robots*, 43(4): 897–912, 2019.

[122] Harish Chaandar Ravichandar, Iman Salehi, and Ashwin P. Dani. Learning partially contracting dynamical systems from demonstrations. In Sergey Levine, Vincent Vanhoucke, and Ken Goldberg (eds.), *CoRL*, volume 78 of *Proceedings of Machine Learning Research* (pp. 369–378). PMLR, 2017.

[123] E. Rimon and D. E. Koditschek. Exact robot navigation using artificial potential functions. *IEEE Transactions on Robotics and Automation*, 8(5):501–518, 1992.

[124] Oren Salzman and Dan Halperin. Asymptotically near-optimal RRT for fast, high-quality motion planning. *IEEE Transactions on Robotics*, 32(3): 473–483, 2016.

[125] R. M. Sanner and J. E. Slotine. Stable adaptive control of robot manipulators using "neural" networks. *Neural Computation*, 7(4):753–790, 1995.

[126] Marco Santello, Martha Flanders, and John F Soechting. Postural hand synergies for tool use. *Journal of Neuroscience*, 18(23): 10105–10115, 1998.

[127] Stefan Schaal, Christopher G. Atkeson, and Sethu Vijayakumar. Scalable techniques from nonparametric statistics for real time robot learning. *Applied Intelligence*, 17(1): 49–60, 2002.

[128] Bernhard Scholkopf and Alexander J. Smola. *Learning with Kernels: Support Vector Machines, Regularization, Optimization, and Beyond*. MIT Press, 2001.

[129] Gregor Schöner. Timing, clocks, and dynamical systems. *Brain and Cognition*, 48(1):31–51, 2002.

[130] Gregor Schöner. Dynamical systems approaches to cognition. In *Cambridge Handbook of Computational Cognitive Modeling* (pp. 101–126), 2008.

[131] Yonadav Shavit, Nadia Figueroa, Seyed Sina Mirrazavi Salehian, and Aude Billard. Learning augmented joint-space task-oriented dynamical systems: A linear parameter varying and synergetic control approach. *IEEE Robotics and Automation Letters*, 3(3): 2718–2725, 2018.

[132] Krishna V Shenoy, Maneesh Sahani, and Mark M Churchland. Cortical control of arm movements: A dynamical systems perspective. *Annual Review of Neuroscience*, 36:337–359, 2013.

[133] Aaron P. Shon, Keith Grochow, and Rajesh P. N. Rao. Robotic imitation from human motion capture using Gaussian processes. In *Humanoids* (pp. 129–134). IEEE, 2005.

[134] Ashwini Shukla and Aude Billard. Augmented-SVM: Automatic space partitioning for combining multiple non-linear dynamics. In *Advances in Neural Information Processing Systems* (pp. 1016–1024). Curran Associates, 2012.

[135] Ashwini Shukla and Aude Billard. Coupled dynamical system based arm–hand grasping model for learning fast adaptation strategies. *Robotics and Autonomous Systems*, 60(3): 424–440, 2012.

[136] Bruno Siciliano and Oussama Khatib. *Springer Handbook of Robotics*. Springer, 2016.

[137] Jean-Jacques E Slotine, et al. *Applied Nonlinear Control*, vol. 199. Prentice Hall, 1991.

[138] Nicolas Sommer, Klas Kronander, and Aude Billard. Learning externally modulated dynamical systems. In *Proceedings of the IEEE/RSJ International Conference on Intelligent Robots and Systems*, IEEE, 2017.

[139] M. Song and H. Wang. Highly efficient incremental estimation of Gaussian mixture models for online data stream clustering. In K. L. Priddy (ed.), *Society of Photo-Optical Instrumentation Engineers (SPIE) Conference Series*, vol. 5803 (pp. 174–183), March 2005.

[140] Dagmar Sternad. Debates in dynamics: A dynamical systems perspective on action and perception, *Human Movement Science*, 19(4): 407–423, 2000.

[141] Arthur H Stroud. *Numerical Quadrature and Solution of Ordinary Differential Equations: A Textbook for a Beginning Course in Numerical Analysis*, vol. 10. Springer Science & Business Media, 2012.

[142] Jos F Sturm. Using SeDuMi 1.02, a MATLAB toolbox for optimization over symmetric cones. *Optimization Methods and Software*, 11(1–4): 625–653, 1999.

[143] Erik B. Sudderth. *Graphical Models for Visual Object Recognition and Tracking*. PhD thesis, Massachusetts Institute of Technology, 2006.

[144] Tomomichi Sugihara and Yoshihiko Nakamura. A fast online gait planning with boundary condition relaxation for humanoid robots. In *Proceedings of the 2005 IEEE International Conference on Robotics and Automation* (pp. 305–310). IEEE, 2005.

[145] Hsi G. Sung. *Gaussian Mixture Regression and Classification*. PhD thesis, Rice University, 2004.

[146] Jun Tani. Model-based learning for mobile robot navigation from the dynamical systems perspective. *IEEE Transactions on Systems, Man, and Cybernetics, Part B (Cybernetics)*, 26(3): 421–436, 1996.

[147] C. Thore. FMINSDP—a code for solving optimization problems with matrix inequality constraints, 2013. Available at https://www.researchgate.net/publication/259339847_FMINSDP_-_a_code_for_solving_optimization_problems_with_matrix_inequality_constraints.

[148] Toru Tsumugiwa, Ryuichi Yokogawa, and Kei Hara. Variable impedance control based on estimation of human arm stiffness for human-robot cooperative calligraphic task. In *Proceedings 2002 IEEE International Conference on Robotics and Automation (Cat. No. 02CH37292)* (vol. 1, pp. 644–650), IEEE, 2002.

[149] Lucia Pais Ureche and Aude Billard. Constraints extraction from asymmetrical bimanual tasks and their use in coordinated behavior. *Robotics and Autonomous Systems*, 103: 222–235, 2018.

[150] V Vapnik and A Lerner. Pattern recognition using generalized portrait method. *Automation and Remote Control*, 24, 1963.

[151] Vladimir N. Vapnik. *The Nature of Statistical Learning Theory*. Springer-Verlag, 1995.

[152] Luigi Villani and Joris De Schutter. Force control. In *Springer Handbook of Robotics* (pp. 161–185). Springer, 2008.

[153] Andreas Wächter and Lorenz T Biegler. On the implementation of an interior-point filter line-search algorithm for large-scale nonlinear programming. *Mathematical Programming*, 106(1): 25–57, 2006.

[154] Li Wang, Evangelos A Theodorou, and Magnus Egerstedt. Safe learning of quadrotor dynamics using barrier certificates. In *IEEE International Conference on Robotics and Automation (ICRA)* (pp. 2460–2465). IEEE, 2018.

[155] Wei Wang and Jean-Jacques E. Slotine. On partial contraction analysis for coupled nonlinear oscillators. *Biological Cybernetics,*, 92(1): 38–53, 2005.

[156] Rui Wu and Aude Billard. Learning from demonstration and interactive control of variable-impedance cutting skills. *Autonomous Robots*, IEEE/ASME Transactions on Mechatronics, 2021, In Press.

[157] Shao Zhifei and Er Meng Joo. A survey of inverse reinforcement learning techniques. *International Journal of Intelligent Computing and Cybernetics*, 5(3): 293–311, 2012.

Index

Note: Page numbers followed by "f" indicate figures, "t" indicate tables.

Intelligent Robotics and Autonomous Agents
Edited by Ronald C. Arkin